THE ECOLOGY AND SILVICULTURE OF MIXED-SPECIES FORESTS

FORESTRY SCIENCES

Volume 40

The titles published in this series are listed at the end of this volume.

The Ecology and Silviculture of Mixed-Species Forests

A Festschrift for David M. Smith

Edited by

MATTHEW J. KELTY
University of Massachusetts

BRUCE C. LARSON
Yale University

and

CHADWICK D. OLIVER
University of Washington

SPRINGER-SCIENCE+BUSINESS MEDIA, B.V.

Library of Congress Cataloging-in-Publication Data

The Ecology and silviculture of mixed-species forests : a festschrift
for David M. Smith / edited by Matthew J. Kelty, Bruce C. Larson,
Chadwick D. Oliver.
p. cm. -- (Forestry sciences)
ISBN 978-90-481-4135-7 ISBN 978-94-015-8052-6 (eBook)
DOI 10.1007/978-94-015-8052-6
1. Forests and forestry--Congresses. 2. Forest ecology-
-Congresses. I. Kelty, Matthew J. II. Larson, Bruce C.
III. Oliver, Chadwick Dearing, 1947- . IV. Series.
SD391.E26 1992
634.9--dc20 92-3410

ISBN 978-90-481-4135-7

Printed on acid-free paper

Contents

III. Productivity of Mixed-Species Stands

IV. Silviculture and Management of Mixed-Species Stands

V. Concluding Remarks

Contributors

P. Mark S. Ashton. School of Forestry and Environmental Studies, Yale University, New Haven, Connecticut 06511.

Peter S. Ashton. The Arnold Arboretum, Harvard University, Cambridge, Massachusetts 02138.

Gordon L. Baskerville. Faculty of Forestry, University of New Brunswick, Fredericton, New Brunswick, Canada E3B 6C2.

Peter Burschel, Hany El Kateb, and Reinhard Mosandl. Department of Silviculture and Forest Management, University of Munich, D-8000 Munich, Germany.

Nora N. Devoe. Institute of Pacific Islands Forestry, U.S. Forest Service, Honolulu, Hawaii 96813.

Matthew J. Kelty. Department of Forestry and Wildlife Management, University of Massachusetts, Amherst, Massachusetts 01003.

Bruce C. Larson. School of Forestry and Environmental Studies, Yale University, New Haven, Connecticut 06511.

Michael B. Lavigne. Forestry Canada, Newfoundland and Labrador Region, St. John's, Newfoundland, Canada A1C 5X8.

David A. Marquis. U.S. Forest Service, Northeastern Forest Experiment Station, Warren, Pennsylvania 16365.

Chadwick D. Oliver. College of Forest Resources, University of Washington, Seattle, Washington 98195.

Akira Osawa. Forestry and Forest Products Research Institute, 1 Hitsujigaoka, Toyohira, Sapporo 004, Japan.

Robert S. Seymour. Department of Forest Biology, University of Maine, Orono, Maine 04469-0125.

David M. Smith. School of Forestry and Environmental Studies, Yale University, New Haven, Connecticut 06511.

Jeffrey P. Thorpe. Applied Plant Ecology Section, Saskatchewan Research Council, Saskatoon, Saskatchewan, Canada S7N 2X8.

Frank H. Wadsworth. U.S. Forest Service, Institute of Tropical Forestry*, Rio Piedras, Puerto Rico 00928-2500. (*operated in cooperation with the University of Puerto Rico)

Acknowledgements

The editors wish to express their thanks to Dean John C. Gordon and the staff of the Yale School of Forestry and Environmental Studies for their assistance with organizing the symposium that was the basis for this book. Stephen Broker in particular was instrumental in making the symposium successful.

Thanks are also due to Martha M. Kelty for her work in the design and production of this book.

Preface

The papers in this festschrift were presented at a symposium held on August 1-2, 1990 to mark the occasion of David M. Smith's retirement from his position as Morris K. Jesup Professor of Silviculture at Yale University. The objectives of the symposium and book are twofold: to provide a forum for the presentation of ideas and research on the structure, development, productivity, and silvicultural treatment of mixed-species forests; and to honor the contributions of Prof. Smith to the fields of silviculture and forest ecology in general, and in particular to the study of complex species mixtures.

Although mixed-species stands dominate much of the world's forested area, silvicultural theory and practice have largely concentrated on single-species stands. This focus is partly a consequence of the history of the development of silviculture. Many of the ideas of modern silviculture originated in the eighteenth and nineteenth centuries in Europe, where foresters were faced with the problem of creating new forests on land that had been deforested, in many cases for centuries. Thus, many silvicultural principles were initially developed for the management of artificial forests, most often monocultures of spruce and pine, with the principal objective of meeting large demands for timber. The problems and solutions of European foresters were not unique. Converting overcultivated, overgrazed, or overburned land to forests has commonly and effectively been done with monocultures for centuries in many parts of the world. The reliance upon single-species stands still predominates in much of forestry, mostly for the benefits of simplicity of management and predictability of yield. In many parts of the world, this kind of forestry now consists not of afforesting lands, but of removing natural complex forests and replacing them with plantation monocultures.

Potential benefits of mixed-species stands have been discussed in the forestry literature for over a century, and management of mixtures has been tried and studied for as long in some places. Mixed stands have often been assumed to be uneven-aged because of differences in height and diameter among species. Where even-aged mixed stands were intentionally regenerated, they were usually established as "patches" of single species, so most trees competed with individuals of the same species. The idea that many mixed-species stands previously assumed to be uneven-aged were in fact even-aged was presented by David Smith in his 1962 edition of *The Practice of Silviculture*; he further recognized that the species were intermixed, rather than occurring in single-species patches. He described stands with layered canopy structure as "stratified mixtures" and attributed the development of this structure to the inherent differences in growth patterns among the species, rather than differences in age. These ideas were an important contribution to understanding the development of complex mixtures, and encouraged a line of research emphasizing the analysis of growth patterns and interactions of individual trees in mixed stands.

General interest in mixed forests has increased greatly in the past decade, largely as a result of exploitation of tropical forests, in the form of both timber removals and conversion to agricultural use. This has intensified the longstanding controversies that exist in forest management nearly everywhere, which arise from the fact that forests serve as a source of wood products, as well as providing environmental benefits such as protection of water, soil, and climate. Recently a new appreciation has also grown for the role that forests play as a reservoir for a large fraction of the world's plant and animal species. The old idea that sound forest management systems should serve both of these needs in a sustainable fashion has received renewed emphasis. Current political trends often seem to favor dividing these functions among different forests, by creating intensively managed plantations to serve wood products needs, while protecting natural forests from wood extraction of any sort. This approach certainly will continue to be important, and has the virtue of concentrating management on a smaller fraction of available land. However, it is unlikely that plantations will fully meet the demands for timber for a variety of reasons, including the high investment cost of plantation establishment, the difficulty of growing many important timber species in plantation, and the vast supply of young, unmanaged second-growth forests in many areas. Added to this is the fact that in many regions of the world, forests which do not provide some revenue from product extraction are converted to other uses, thus eliminating the environmental functions of the forest as well. This points to the continuing vital need to develop silvicultural techniques for complex natural or semi-natural forests to protect biodiversity and enhance environmental functions of forests, and yet provide the raw materials that help to build or maintain high standards of living.

Developing appropriate silvicultural systems for mixed forests was a major part of David Smith's interest and work long before current interest in tropical forest management brought it into more widespread attention. He recognized that scientific knowledge is needed to translate good intentions into realized goals, and that any actions in the treatment of forests need to be based on an understanding of how the various stands grew naturally and how they could be manipulated effectively. The authors of the papers in this book—colleagues and former students of Prof. Smith's—present syntheses of past work as well as current studies on mixed stands. Topics range from analyses of forest dynamics in unmanaged stands to descriptions of silvicultural systems and experiments in complex mixtures, with examples drawn from tropical, temperate, and boreal regions. Within this diversity of subject matter, there exists a common thread in the general approach of the authors. It involves the realization that the understanding of forest dynamics and the development of silvicultural systems should be continually evolving processes. These processes are advanced by a mode of analysis that does not rely upon broad ecological generalities, but concentrates on the forest itself, and appreciates the details of tree structure and physiology, soils, climate, natural disturbances, and the human use of forests that affect the development of any particular stand of trees. It emphasizes that forests are made up of individual trees, and that an understanding of forest dynamics depends upon studying interactions of those individuals with one another. It is through this kind of analysis that general patterns can be discerned concerning the growth and development of both treated and untreated stands. These recurring themes owe a great deal to the influence of David Smith.

❧

Section I of this volume contains two papers that provide an introduction to the ideas of stand dynamics. Bruce Larson presents an overview of the developmental pathways that result from the competitive interactions among species in mixed stands. The similarity of the processes across forest regions is then discussed by Chadwick Oliver, together with suggestions for improvement in the efficiency of research and management through a process of "adaptive management."

The five papers of Section II present individual case studies of stand dynamics. Akira Osawa analyzes the development of stands in northern Japan that regenerated after windthrow, and makes comparisons to patterns of stand development that followed a catastrophic hurricane in northeastern U.S., which has received considerable attention in forest ecology literature. Peter Ashton describes patterns of forest structure and species composition in dipterocarp forests in northwest Borneo, in relation to the factors of soils, topography, and disturbance that exert control over these patterns. In constrast to these tropical forests of high diversity, Jeffrey Thorpe examines the relationship of species composition, structure, and succession in the relatively species-poor boreal forests of Canada. The final two papers of this section concentrate on the regeneration phase of mixed stands. Nora Devoe presents a study of tree establishment patterns in and around gaps created in tropical wet forests in Puerto Rico. Mark Ashton focuses on the advance-regeneration mode of reproduction, comparing ecophysiological adaptations of species of *Shorea* in Sri Lanka and *Quercus* in Connecticut.

Section III presents two papers dealing with questions of productivity of mixtures. Matthew Kelty uses the approach of researchers in the field of agricultural intercropping to compare yields of monocultures and mixtures. Michael Lavigne reviews the recent advances in physiological modeling of forest stands (developed for monocultures), and considers the modifications that would be needed to apply them to mixed-species stands.

In Section IV, attention is turned to the complexity of silvicultural treatment of species mixtures. The first two papers examine the relationship between stand development processes and the history of management in two forest types in the northeastern U.S., and describe the effect these have on current silvicultural options: David Marquis discusses mixed hardwood forests in Pennsylvania, while Robert Seymour deals with spruce-fir mixtures in Maine. The third paper, by Peter Burschel, Hany El Kateb, and Reinhard Mosandl, focuses of the effects of different regeneration methods applied to mountain forests of Bavaria, with the objective of determining the methods that will maintain the full complement of tree species in the original stands. In his paper on silviculture of tropical forests, Frank Wadsworth traces its roots to temperate zone silviculture, and describes the current linkage between temperate and tropical silvicultural ideas and practices. The final paper in this section provides a quite different outlook from the previous four; Gordon Baskerville takes the same principles used in the analysis of stand dynamics and applies them to the problem of managing an entire forest, demonstrating how understanding and practice can be improved at the forest management level by this approach.

The single paper in Section V of this volume represents a break with tradition in festschrifts, in that the person receiving the well-deserved accolades is usually not asked to do any work. However, at the editors' invitation, David Smith agreed to add his thoughts on mixed stand ecology and silviculture. He has done this, and also recounted the development of his ideas on the topic, in the final paper entitled "Ideas about mixed stands."

This collection of papers is presented in recognition of Professor David M. Smith's contributions to the science and practice of silviculture. It is also done in appreciation of the kindness, generosity, and understanding that he has shown in working with us as students and colleagues. We look forward to continuing to learn from him and to enjoying his company.

Matthew J. Kelty, Bruce C. Larson, Chadwick D. Oliver

David M. Smith
Morris K. Jesup Professor Emeritus of Silviculture
Yale University

Profile of David M. Smith

David M. Smith held the Morris K. Jesup Chair in Silviculture at Yale University from 1967 until his retirement in 1990. He was responsible for silviculture instruction at Yale for even longer, having joined the faculty of the School of Forestry in 1948. Prior to that, he had received a B.S. in botany from the University of Rhode Island in 1941, and a Certificate in Meteorology from New York University in 1942, thereafter serving as a meteorologist in the U.S. Army Air Force in North Africa and Italy from 1942 to 1945. He then came to Yale to study forestry, receiving his M.F. in 1946 and his Ph.D. in 1950. His doctoral work, conducted under the guidance of Prof. Harold J. Lutz, focused on the regeneration of white pine as affected by microsite variation. He received an honorary Doctor of Science degree from Bates College in 1986. He was awarded the title of Morris K. Jesup Professor Emeritus of Silviculture by Yale in 1990.

During his years on the Yale faculty, Prof. Smith strongly influenced the practice of silviculture through his teaching of hundreds of graduate and professional students majoring in forestry and related fields of environmental studies. Much of the development of his ideas in silvicultural and ecological research has come through his work with the 36 doctoral students whose studies he directed over that time. His writings have also been influential, particularly those that have synthesized the research findings of a field of study. One paper of note is "Maintaining timber supply in a sound environment" in the 1973 report of the President's Advisory Panel on Timber and the Environment; this report was a part of the deliberations eventually leading to the drafting of the National Forest Management Act of 1976. He also influenced the forestry literature during his terms as Associate Editor of *Journal of Forestry* and *Forest Science*.

Prof. Smith has always considered it important to be an active forest manager as well as a professor of silviculture. From 1949 to 1990, he was in charge of management of the Yale Forests (approximately 4000 ha) located in the states of Connecticut, New Hampshire, and Vermont. In addition, he has been a management consultant to Baskahegan Company, owner of 40,000 ha in eastern Maine since 1967.

It has been through his textbook *The Practice of Silviculture*, with editions in 1954 (with Ralph C. Hawley), 1962, and 1986, that David Smith has become widely regarded as a leading authority in silviculture. This book serves as the standard university text and professional reference throughout North America, and is widely read in Latin America, Africa, and other parts of the world in its English edition and in several translations. Through this text, he has put the practice of silviculture on a sound scientific basis, encouraging readers to inquire, think, and develop creative answers to problems, rather than rely upon authority and tradition in making decisions concerning the treatment of forests. It has also been through the text that he has advanced his ideas on the value of mixed-species forests and the need to understand them and to develop new methods of silviculture for their management.

Stand Structure and Dynamics: Overview of Principles

I

Pathways of development in mixed-species stands

1

BRUCE C. LARSON

Introduction

Forest stands are constantly being impacted by disturbances. When the disturbances are "lethal", growing space is released which can then be occupied by the remaining trees or by new vegetation. Lethal disturbances can be either natural (e.g., fire or wind) or anthropogenic (e.g., harvesting or herbicides) events where trees are killed. The structure of the stand at the time of the disturbance as well as the type, intensity, and magnitude of the disturbance will dictate the future development of the stand. The mechanism driving development is simply change in competitive advantage of the surviving and the regenerating individuals. Just as development after a particular disturbance in single-species stands can be predicted quite well, development of mixed-species stands is far from unpredictable randomness, but is complicated by the species differences.

The fundamental difference between single-species stands and mixed-species stands is the amount of genetic diversity. The amount of genetic variation between species is greater than within species. Mixed-species stands therefore have much more variation between individuals in shade tolerance, height-growth patterns, root growth, and regeneration capabilities. Competition in stands occurs between neighboring individuals, so mixed-species stands have more possible types of competitive interactions. Therefore, there must be more deviation and variation from general development pathways because, not only is the type and frequency of disturbance stochastic, but the spatial patterns of possible species interactions also have a stochastic element.

The best framework for describing development of even-age forest stands after a catastrophic disturbance is the four stages first proposed by Oliver (1981): stand initiation, stem exclusion, understory reinitiation, and old growth. Stands become uneven-aged during the last two stages and usually also become mixed-species stands at this time if they were monospecific at initiation. During every stage mixed-species stands are obviously more complex than single-species stands. Since tree growth is the expression of gene-environment interactions, the large genetic variability of mixed-species stands results in much greater variation in growth rates among individual trees in the stand. In addition, because the "environment" of each tree is greatly modified by the neighboring individuals, there is greater environmental variation within mixed species stands. The combination of both sources of increased variation in individual tree response (genetic and environmental) does not mean that development patterns cannot be discerned, but rather, that there are more possibilities for divergence from the general pathway.

Mixed-species stands almost always stratify in height by species because height-growth patterns and the shade tolerance characteristics differ among species. When shade intolerant species are surpassed in height by more shade tolerant species, they become suppressed and die, reducing the number of species in the

M. J. Kelty (ed.), The Ecology and Silviculture of Mixed-Species Forests, 3–10.

stand. Otherwise, most species will continue to be represented and the stratification pattern either 1) remains the same for a long period of time or 2) changes if the relative height-growth rates of the species change significantly over the course of stand development. This paper explores the mechanisms that lead to stratification, and reviews situations where the height dominance shifts between species and those where it appears to remain constant.

Stratification mechanisms

Understanding stratification patterns in a well-studied forest region is greatly facilitated by knowledge of the silvical characteristics of the species. The results of competition between species are just a manifestation of the environment "perceived" by each stem and the response of the individual to this environment (the silvical characteristics).

One of the most important silvical attributes affecting competitive interaction is the characteristic height-growth pattern of each species. Early height advantages are difficult to overcome because a taller tree shades the shorter and increases its competitive advantage. This advantage may not be maintained, however, if other characteristics such as shade tolerance allow a shorter tree to maintain adequate photosynthesis for continued height growth and epinastic control (i.e., the control a terminal shoot exerts over the length and angle of lateral shoots).

There are four basic height-growth patterns in trees. The first pattern is rapid juvenile growth followed by a slowdown. In many ways this is the stereotypical "pioneer" height-growth pattern. The most common pattern is sigmoidal height growth. Trees with this pattern start slowly and then go through a "grand period of growth" with almost linear height growth. Finally the tree slows in growth and enters a mature phase of relatively slow height growth. The final two patterns of height growth are both linear growth. Some species have almost linear growth throughout their lives and the annual height increase is quite great. Other species have linear height growth, but are slow-growing no matter what the light and moisture conditions are.

Stratification patterns become apparent when the intrinsic height-growth pattern of the species is affected by competition with neighbors. A species that normally has a linear, fast height-growth pattern may have a slowdown if shaded by neighboring trees. It is important to consider the direct effect of competition with neighboring trees.

Much of this response can be explained by understanding the shade tolerance of the species involved. If the shorter tree is quite tolerant, shading by a neighbor may not have a major effect on the intrinsic height-growth pattern. Shade intolerant species may greatly slow in height growth, lose epinastic control, or even die. Part of the shade tolerance question is also the adaptability or plasticity of the species. Some species appear to have a range of tolerances which may be partially attributed to the relative amount and distribution of sun and shade foliage. It is also important to consider the shade tolerance of the taller neighbor. A shade intolerant species has a relatively small leaf area index because its own foliage will be shaded in the lower parts of the crown. On the other hand, a shade tolerant species has a high leaf area index and casts much deeper shade on the shorter neighbors

Drought tolerance is also important and affects the ability of each species to achieve its potential height-growth rate. Drought-tolerant species will be more likely to dominate on harsh sites, with drought-intolerant species dominating on more mesic sites. An additional effect of drought is crown dieback by drought-intolerant species with indeterminate shoot growth; frequent dying of the terminal shoot on one species will result in a severe height disadvantage.

Temporal (age effects) and spatial relationships between individuals in a stand can be a major factor in either the maintenance or shift in height dominance between species. Since competition in a stand occurs only between interacting neighbors, height dominance is a reflection of the silvic (genetic) characteristics of the two species and the suppression of this potential resulting from a reduction in the available growth resources by surrounding individuals.

Even a small difference in age can result in a height (and possibly root) advantage to the older individual. This physical advantage can result in a competitive advantage where the older individual is able to restrict growth resources available to a younger individual. The most obvious example is a height advantage by an older individual and the resulting shading effect on a younger neighbor. Even if the younger individual inherently has a faster juvenile growth rate, this may be thwarted by the age difference.

There is a strong interrelationship between the temporal effect on interspecific competition and the effect of regeneration mode of the species. It is much easier to understand either effect independently (e.g. younger versus older, seed regeneration versus sprouting) than the combination effect. These variations are often the result of seed rain size caused by either the distance to seed sources, or the relative seed crop sizes of each species.

The relative competitive vigor of a tree of one species will be impacted by major age differences in uneven-age stands. For example, a shade intolerant species may normally have rapid juvenile height growth and the ability to rapidly exploit available growth resources such as light or water following a disturbance. The amount of light available to this species will be reduced if the disturbance has been a minor one which has left taller residual trees of an older age cohort. A more shade tolerant species may dominate the regeneration in this case, whereas the shade intolerant species would dominate following a major disturbance with removed all or most of the older trees.

Spatial effects have two primary effects on the development of dominance patterns. The first effect is similar to the temporal effect described above; an early competitive advantage by one individual. In this case one tree grows on a superior microsite and has faster early growth. This advantage may be conveyed by a small patch of bare mineral soil, a slight advantage in water relations (either a small depression or a small rise depending on the circumstances), or an area with greater organic matter. In these examples, the type of disturbance which destroyed the previous stand and made the growing space available determines the spatial pattern of the new stand. A wind storm which results in windthrow mounds will impact all three of the early growth advantage situations described.

The other spatial pattern effect has to do with the specific characteristics of neighbors. This effect has two components: distance and species. As trees get taller, their zone of influence increases. The influence on a neighbor is partially determined by the silvical characteristics of the species, but the effect of one tree on another will also be tempered by the distance betweent trees. After generalizations are developed concerning the interrelation of two species and the resulting effect on stand development and stratification patterns, many exceptions will be apparent in each stand because of the spatial pattern. A high density of one species means that all trees are affected by this species. If stand density is reduced, the general relationship may not hold.

Growth vigor changes as trees age, as does the transformation of this vigor into response to release; these changes are species-specific. Growth and vigor is based on three primary characteristics: the amount of photosynthetic area compared to the respiratory area of the tree; the number of apical meristems (buds) which directly results in the growth potential; and the size of the root structure both in terms of amount of soil volume occupied and the amount of stored carbohydrate.

Photosynthetic capability is a direct function of the amount of foliage. This can be directly inferred from the depth and density of the crown. Other factors such as leaf morphology and leaf size will also be factors, but to a smaller degree. Trees with a larger amount of foliage (large, dense crowns) will respond much more quickly and vigorously to release. Respiration, although resulting from all living cells, is a function of the surface area of the stem and branches. Trees with a large amount of respiratory area but a limited amount of photosynthetic capability will respond slowly to release, if at all. Release of growing space may be the result of a minor disturbance to the stand as a whole or simply the death of a neighboring tree.

Growth of a tree after release also depends on number, vigor, and position of apical meristems (buds) within the crown. A large number of buds will offer more opportunities for immediate crown expansion because the amount of primary growth that can arise from any individual bud is limited. The presence of epicormic buds within the crown will also affect the potential increase in crown density after release.

The general crown morphology will also influence growth after growing space is made available. The most notable attribute is the amount of epinastic control that the terminal exerts over the length and orientation of lateral branches. Trees of some species growing in shaded understory positions tend to lose epinastic control, resulting in rounded crowns with no single central stem. Following an overstory disturbance, these trees may expand horizontally and gain control of a large amount of growing space without the vertical growth that would lead to a change in dominance or stratification. In contrast, some species tend to maintain epinastic control even while suppressed; trees of these species may respond to disturbance by growing rapidly in height and moving into the overstory ahead of newly regenerated stems or other species that were in the understory at the time of the disturbance.

Tree vigor and ability to respond to release are partially controlled by the size and development of the root system. Photosynthetic capacity is also controlled by sufficient water being available to keep the stomates open. Increased sunlight will lead to an increase in temperature. Without adequate water the stomates will close and the temperature rise will continue, resulting in increased levels of respiration and less photosynthate available for growth and expansion.

Mixed-species stands with changing dominance

There are many examples of mixed-species stands where height dominance changes from one species to another during development. This major pathway has many variations and can occur in both single-age stands and also in stands with more than one age cohort.

One of the best documented cases of this pathway is the mixed hardwood forests of southern New England (U.S.A.). These stands, whose development was first documented by Oliver (1978), consist of many species, but primarily have red maple (*Acer rubrum*), black birch (*Betula lenta*), and red oak (*Quercus rubra*). In this case oak is the same height or shorter than maple and birch during the early years of the stand, but eventually dominates the stand. In addition, very few oak die during the stem exclusion stage; the reduction in stem number results from maple and birch mortality. Kittredge (1988) has shown that, in fact, growth (or suppression) of individual oaks is primarily determined by the other oaks in the stand rather than the non-oak stem density.

Sometimes the species that is dominant in height growth will slow down in height as a subordinate species overtakes it. This is the case in mixtures of Douglas-fir (*Pseudotsuga menziesii*) and grand fir (*Abies grandis*). In this combination Douglas-fir is the taller species during the early years, but grand fir eventually catches up (Larson 1986). As grand fir crowns approach the crowns of Douglas-fir, the Douglas-firs do slow in height perhaps because of crown abrasion. Even when this species combination is found in a multi-age stand, the stratification and changing dominance is seen within each of the various age strata.

Sometimes a species has rapid initial growth and then not only slows in growth, but dies out of the stand. An example of this is pin cherry (*Prunus pensylvanica*) in northern hardwood stands in Northeastern U.S.A. Young stands have a complete overstory of cherry, with all other species being shorter (Marks 1974). Later in development, during the stem exclusion stage, all the pin cherry dies. A similar situation occurs in the western United States with cherry and willow (*Salix*) dominating stands of Douglas-fir and grand fir (Oliver and Larson 1982). In some situations enough growing space is released by the death of overstory cherry that new stems may regenerate; these new stems tend to be more shade tolerant. This is an example of a situation in which the described stages of development are somewhat ambiguous because there appears to be a brief period of "understory reinitiation" during the stage of "stem exclusion".

In the situations described above, the shorter tree species did not have an acceleration in growth, but rather the taller trees slowed down. As indicated in these examples the smaller trees in the stand are often the same age as the overstory; they may either be surpassed trees that once dominated the stands, or (in the case of younger stands) may be a species with slower initial height growth which may dominate the stand in the future. If the dominating species control a large amount of growing space and then most of the trees die over a short period of time, new regeneration may occur. However, the death of the taller trees is usually affected by the varying intensity of competition with other trees; the transfer of growing space is thus a slow process which allows residual trees to respond rather than allowing regeneration of new stems capable of moving into the overstory.

There are even documented cases where a younger tree is able to overtop older trees of a different species. In the southern U.S. example studied by O'Hara (1986) tulip-poplar (*Liriodendron tulipifera*) was able to outgrow white oaks (*Quercus alba*) of an older age class after partial cutting. This study clearly shows the ability of different trees to respond to release. Being older and bigger is not necessarily an advantage if an individual is physiologically or morphologically unable to quickly respond after an increase in available growing space.

The same pattern is also apparent in the more complicated case of red spruce (*Picea rubens*) and balsam fir (*Abies balsamea*) in Maine documented by Davis (1989). Here, advance regeneration spruce may be unable to quickly respond to partial cutting and be overtopped by fir of a younger cohort. This process may be repeated over a series of partial cuttings, with the spruce getting larger each time and eventually becoming able to retain a place in the overstory with fir below.

A change in height dominance from one species to another is a common phenomenon in mixed-species stands. This change may lead to false conclusions about the age structure of the stand. The timing and the mechanisms of both stratification and change in dominance are most important to silviculturists as they plan operations in these stands and try to match the patterns with the production of different timber products and wildlife habitats dependent upon specific forest structures.

Mixed-species stands without changing dominance

When deciding whether there is a change in species dominance during stand development, the most important question is one of time frame. Different species have different life spans and some are very long. Sometimes a transfer in dominance is related to a relatively infrequent disturbance event. It is easy for a change in dominance to take longer than the lifetime of any one researcher. To understand the pattern of dominance shift, it is important for the forester to understand when a series of stands represents a true chronosequence.

Mixed stands of red alder (*Alnus rubra*) and Douglas-fir in the Northwestern U.S.A. have been studied by Stubblefield (1978) who showed that red alder can dominate and shade Douglas-fir for a long enough period that the Douglas-fir dies. In this situation the relatively short-lived species lives long enough to dominate the site. The comparison of this type of mixed stand with the previously described example of pin cherry and northern hardwoods leads to a different perceived conclusion because of the management implications of the life span of the early dominating species, and the interval between catastrophic disturbances in the two regions.

Species shifts can also be retarded by silvical characteristics which affect direct physical interactions among species. One example of this is abrasion of the terminal leaders of a shorter species by the branches of individuals of a taller species. Even if a shorter species has a potential height-growth rate that exceeds that of the taller species given the size, light and available water, the shorter species may not be able to realize this potential and overtop the taller species because of this physical limitation. Sometimes a recurring event primarily affects one species in a mixed-species stand. Crowns of tulip-poplar are easily broken by wind in northeastern United States, affecting their overall ability to outgrow other species.

Probably the best studied example of physical imposition of dominance is in mixed stands of Douglas- fir and western hemlock (*Tsuga heterophylla*). Wierman and Oliver (1979) showed that wind abrasion of hemlock terminals by Douglas-fir prevented any transition in dominance. The ability of Douglas-fir to maintain dominance is also enhanced by its long, almost linear, height-growth period. This same effect was also documented by Kelty (1986) with red oak keeping eastern hemlock (*Tsuga canadensis*) suppressed. In both these cases the relatively long life of the dominating species is also required for the continued dominance.

Sometimes the height growth curves of different species are such that no transfer of dominance ever occurs. Some species grow slower throughout their life. The most obvious examples of this are tree species which never become true overstory species. Dogwood (*Cornus florida*) in the eastern United States and Sitka alder (*Alnus sinuata*) in western North America are two examples of this pattern.

Several of these factors may work in concert. In northern California there is little shift of dominance in mixed stands of redwood (*Sequoia sempervirens*) and western hemlock. In this example the redwood has a very long lifespan and has an almost linear height growth pattern for much of this time. In addition, the thick bark makes this species very resistant to low ground fires which are fatal to the shorter, thinner-barked hemlock. Suppression of low natural fires may eventually lead to catastrophic fires which will destroy both species but hemlock may eventually overtop redwood if minor disturbances kill only scattered overstory redwood trees.

A better example of different age structures of the two species occurs in parts of Germany, where stratified stands of overstory oak (*Quercus petraea*) above a lower canopy of beech (*Fagus sylvatica*) occur. These stands are sometimes managed on a rotation for oak that exceeds 200 years, with the beech being 40-100 years younger than the oaks. Most of these stands are intensively managed and any beech that appeared to be in a position of overtop the oak has been removed during thinning, so it is impossible to know if this lack of change in dominance would persist over several centuries if no management intervention occurred.

Another factor which can affect the possible lack of transfer in dominance, at least between certain individuals in the stand, is the spatial pattern of the stand. Oliver et al. (1989) showed in mixed stands of sycamore(*Platanus occidentalis*) and southern red oak (*Quercus falcata*) that spacing between trees is an important factor in the shift of dominance. In these stands the sycamore quickly overtopped oak that were in close proximity, but were unable to outgrow the oaks which were much further away. Although direct competition resulted in a transfer of dominance, an analysis of the tallest trees in the stand would show that many oaks were as tall as the sycamores. These patterns reflect a combination of factors: different shapes of the height-growth curves of the two species, with slower initial growth in the oaks; a difference in shade tolerance; and a limit to the horizontal expansion of the sycamore.

Height-growth patterns, physical control, longevity, age differences, and differential mortality from minor disturbances can all contribute to stands which will not show a transfer of dominance between species. In most cases, it is impossible to determine if a transfer would occur later in time because in these examples, the frequency of a catastrophic disturbance (which would initiate an entirely new stand) is shorter than the time period which might be necessary for a transfer to take place.

Conclusions

Most of the mixed-species stands that have been studied show a stratified height structure. Predictions of future stand structure demand a knowledge of whether the height dominance pattern will shift or remain the same. The development patterns in these stands is complicated, but predictable. These predictions can be made with a combination of autecological knowledge and an understanding of the competitive interactions between the species. As more situations are studied, the general patterns become more apparent and the competitive interactions that are responsible for deviations from the general patterns become clearer. Forest managers can then use this information not only to predict future stand structures, but to plan silvicultural operations to produce desired stand structures.

Literature

Davis, W. C. 1989. Role of released advance growth in the development of spruce-fir stands in eastern Maine. Unpublished doctoral dissertation, Yale University, New Haven, CT. 150 pp.

Kelty, M. J. 1986. Development patterns in two hemlock-hardwood stands in southern New England. Canadian Journal of Forest Research 16:885-891.

Kittredge, D. B. 1988. The influence of species composition on the growth of individual red oaks in mixed species stands in southern New England. Canadian Journal of Forest Research 18:1550:1555.

Larson, B. C. 1986. Development and growth of even-age stands of Douglas-fir and grand fir. Canadian Journal of Forest Research 16:367-372.

Larson, B. C. and C. D. Oliver. 1982. Forest dynamics and fuelwood supply of the Stehekin Valley, Washington, in Ecological Research in National Parks of the Pacific Northwest Oregon State University Forest Research Laboratory Publication, Corvallis, Oregon, pp 127-134.

Marks, P. L. 1974. The role of pin cherry (*Prunus pensylvanica*) in the maintenance of stability in northern hardwood ecosystems. Ecological Monongraphs 44:73-88.

O'Hara, K. L. 1986. Development patterns of residual oaks and oak and yellow-poplar regeneration after release in upland hardwood stands. Southern Journal of Applied Forestry 10:244-248.

Oliver, C. D. 1978. Development of northern red oak in mixed-species stands in central New England, Yale University School of Forestry and Environmental Studies Bulletin No. 91, 63pp.

Oliver, C. D. 1981. Forest development in North America following major disturbances. Forest Ecology and Management 3:153-168.

Oliver, C. D., W. K. Clatterbuck and E. C. Burkhardt. 1990. Spacing and stratification patterns of cherrybark oak and American sycamore in mixed, even-aged stands in the southeastern United States. Forest Ecology and Management 31:67-79.

Stubblefield, G. W. 1978. Reconstruction of a red alder/Douglas-fir/western hemlock/western red cedar mixed stand and its biological and silvicultural implications. Unpublished Master of Science thesis, College of Forest Resources, University of Washington, Seattle, Washington. 141 pp.

Wierman C. A. and C. D. Oliver. 1979. Crown stratification by species in even-aged mixed stands of Douglas-fir/western hemlock. Canadian Journal of Forest Research 9:1-9.

Similarities of stand structures and stand development processes throughout the world—some evidence and applications to silviculture through adaptive management

2

CHADWICK D. OLIVER

Introduction

Much progress has been made during the past four decades in understanding how forest stands develop. The understanding increases both our basic ecological knowledge and our ability to do silvicultural manipulations for increasingly varied objectives.

Forest stands occur in various structures. Stand structure refers to "the physical and temporal distribution of trees in a stand" (Oliver and Larson 1990). Stand structures can have patterns in the number, rate of change, or spatial distribution of total trees, tree parts, species, or age classes.

This paper will show that stands exist in many structural patterns, but similar patterns are found in many parts of the world. It will show that similarities are because similar stand development processes occur throughout the world. It will then describe how awareness of the similarities of patterns and processes can improve silvicultural practices and ecological knowledge.

Similar patterns of stand structures

There are six evolutionarily distinct groups of plants--"floristic realms"--in the world (Walter 1973). These floristic realms occupy geographically distinct areas. One realm, the Holarctic, occupies the northern hemisphere's temperate and boreal forests. The Neotropic realm occupies Central and South America. The Paleotropic realm occupies Africa and southern Asia, while the Australia realm occupies Australia; the Capensis, extreme southwestern Africa; and the Antarctic, coastal southwest South America and the subantarctic islands.

Each realm generally contains angiosperm genera different from the others. The conifer family Cupressaceae is found in all continents. The family Podocarpaceae and the genus *Araucaria* occur naturally only in the southern hemisphere, while the Pinaceae and nearly all Taxodiaceae occur naturally only in the northern hemisphere (Walter 1973). The differences are because the realms evolved in relative isolation (Gentry 1982, Tallis 1991).

Some genera are common in certain environments within each realm. *Pinus* species dominate droughty sites in western and eastern North America, Europe, and western, central, and eastern Asia. *Picea* and *Abies* are generally found in cooler sites throughout the northern hemisphere. Similar dominance of different genera occur in Neotropic (Bruenig and Y-w Huang 1989, Gentry 1989, Junk 1989), Paleotropic (Ashton 1989, Geesink and Kornet 1989) and other floristic realms.

Where similar soils and climates exist in geographically isolated places within a realm, stand structures can be very similar since related species often have similar

M. J. Kelty (ed.), The Ecology and Silviculture of Mixed-Species Forests, 11–26.

physiological and morphological processes. For example, the Holarctic realm contains arid *Pinus/Populus/Salix/Artemisia* stands in parts of the Rocky Mountains and eastern Cascade Range which closely resemble the *Pinus/Populus/Salix/Artemisia* stands of Inner Mongolia and similar stands of central Asia Minor. The mixed oak/maple (*Quercus/Acer*) stands of the eastern United States contain many structural features like those of mixed hardwood stands of central and northern Japan, the central elevations of the Himalayas of Nepal, central Europe, and north-central Turkey. Similar climates and soils create analogous stand structures in the *Alnus/Rubus/Picea/Abies/Rhododendron* forest communities of northeastern Turkey and the *Alnus/Pseudotsuga/Abies/Rubus* forest communities of the Pacific Northwestern United States. The boreal forests are similar throughout the northern hemisphere (Bonan and Shugart 1989). Geographically separated Dipterocarpaceae species also form analogous structures in the Paleotropic realm (Ashton 1988).

Similar stand structures are also found where genera are not evolutionarily related. For example, stands in warm, humid environments often contain many tree species and analogous structures whether in the Holarctic forests of the southeastern United States (Putnam et al. 1960) or the Neotropic or Paleotropic jungles of South America, Africa, or Asia (Gentry and Emmons 1987, Gentry 1988, 1989). Forests in cooler or drier climates generally have fewer species and slower growth rates. Similar community structures are also found in genetically unrelated shrub communities (Milewski 1983).

There are common stand structure patterns and variations in patterns which can be found repeatedly throughout the world, as will be discussed below.

Long-term species composition patterns

Forest species compositions have changed naturally and dramatically over the past few thousand years in such diverse places as the Amazon basin (Gentry 1989, Irion 1989, Uhl et al. 1990), the eastern and Pacific Northwestern United States (Davis 1981, Cwynar 1987, Ruddiman 1987), and Asia Minor (Aytug et al. 1975).

The earth's climate has changed dramatically, even in the last 15,000 years. The land area free of glaciers, above sea level, and otherwise suitable for forest growth has also changed; and plant species are constantly migrating in response to climate and soil changes (Tallis 1991).

Plant species do not migrate in concert (Van Devender and Spaulding 1979), so what has been viewed as a stable forest community in the past is usually just a recent amalgamation of species (Prance 1982). Consequently, species composition and dominance in most forest areas have changed dramatically over the past few thousand years--a short time compared to the average life spans of trees. (See Ashton, P.S.[1992] for exceptionally stable community in Borneo.)

Age distribution patterns

Tree ages within a stand are commonly found in groups (or cohorts) slightly younger than the ages of various disturbances. This aggregated pattern of tree ages following disturbances has been found in even-aged (single cohort) and uneven-aged (multiple cohort), mixed and pure species forests of hardwoods, conifers, or both, in eastern and western North America from Mexico to Alaska (Oliver and Stephens 1977, Oliver 1978, 1981, Stubblefield and Oliver 1978, Frelich and Lorimer 1985, Cobb 1988, Deal et al. 1991, Segura-Warnholtz and Snook 1992), as well as in Japan (Nakashizuka and Numata 1982) and the Himalayas (Oliver and Sherpa 1990), Australia (Ashton 1976), Puerto Rico (Crow 1980), and the Andes (Veblen and Ashton 1978, Veblen et al. 1980). Analyses of stand structures in southeastern Asia (Whitmore 1989) and the Amazon (Uhl et al. 1988, Lang and Knight 1983) suggest that trees in these regions also primarily regenerate following disturbances.

Development patterns following disturbances

Following disturbances which destroy all trees in a stand or which create gaps, the new tree cohort changes in similar ways in many temperate and tropical forests (Isaac 1940, Jones 1945, Raup 1946, Watt 1947, Whitmore 1975; for additional references, see pp. 142-143 , Oliver and Larson 1990).

- Immediately after the disturbance, many trees begin growing from seeds, sprouts, and other regeneration forms. The time when the stand (or newly beginning cohort, if the stand still contains older cohorts) has this newly initiating structure has been referred to as the "aggradation phase" (Bormann and Likens 1981), the "stand initiation stage" (Oliver 1981), or the "establishment phase" (Peet and Christensen 1987) in Holarctic forests. Similar structures following gap disturbances in tropical rain forests have been referred to as the "gap phase" in southeastern Asia (Whitmore 1975, 1989) and South America (Hartshorn 1980).

- The stand's structure changes as the trees reoccupy the site and compete with each other. The newly formed structure has been referred to as the "transition phase" (Bormann and Likens 1981), "stem exclusion stage" (Oliver 1981), or "thinning stage" (Peet and Christensen 1987) in Holarctic forests and the "building phase", "small sapling phase", or "pole phase" in southeastern Asian (Whitmore 1975, 1989) or South American (Hartshorn 1980) tropical rain forests.

- As trees grow taller and variations in tree heights and diameters increase, the new structures have been separated into two phases--the "understory reinitiation" and "old growth" stages by Oliver (1981) and the "transition" and "steady state" phases by Peet and Christensen (1987). Bormann and Likens (1981) referred to both structures as the "steady state phase" in temperate forests, and Whitmore (1975, 1989) and Hartshorn (1980) considered similar structures the "mature phase" in tropical rain forests.

Patterns of species dominance

The stand structure on an area often changes through repeated cycles, with the same species dominating a cycle as long as a specific disturbance type occurs periodically. The stand can shift dramatically to different species if the disturbance type changes (Whitmore 1989, Oliver and Larson 1990). The new disturbance type can also recur repeatedly, creating cycles of the new structure. These cycles can be anticipated, but not accurately predicted. For example, forests dominated by white pines (*Pinus strobus* L.) continued for at least two cycles after hurricanes followed by fires in central New England. When weather patterns were not conducive to a fire after a hurricane, the forest began a long cycle dominated by hardwoods (Henry and Swan 1974). Analogous shifts in cycles have been observed in the Cascade

Range of North America, where a stand can be dominated by either *Abies* or *Tsuga*, depending on events surrounding the disturbances (Oliver et al. 1985). Forests in the Himalayas of eastern Nepal can similarly shift between domination by *Abies* and domination by *Rhododendron*, depending on events occurring during the few years following a stand-replacing disturbance (Oliver and Sherpa 1990). Studies of changing species domination following different types of disturbances in rain forests in southeastern Asia (Bruenig and Y-w Huang 1989) and South America (Swaine and Whitmore 1988, Whitmore 1989, Uhl et al. 1990) suggest a similar pattern is occurring there.

Stratification patterns of mixed species stands

Mixed species stands often contain trees in horizontal strata in tropical, temperate, and boreal forests (Richards 1952, Halle et al. 1978, Oliver and Larson 1990). Arguments over the existence of strata in some forests are largely concerned with suitable explanations of the stratification process (Whitmore 1975, Halle et al. 1978). At times, various strata are predominated by different species, while at other times the same species are common to several strata. Both conditions occur in single cohort and multiple cohort stands (Oliver and Larson 1990).

The stratification pattern occurs following a single, stand-replacing disturbance, where all trees are in a single cohort and one or several species dominate others and relegate them to lower strata (Oliver 1978, Clatterbuck and Hodges 1988, Cobb 1988). The stratification pattern also occurs where several, minor disturbances allow several cohorts to be present (Oliver and Stephens 1977). Both single cohort and multiple cohort patterns have been found in Holarctic forests with diverse species, climates, and disturbance histories (Oliver and Larson 1990). The single-cohort stratification pattern is more widespread than has often been assumed (Oliver 1980, 1981).

Even where stands contain many species, few species usually predominate, as has been found in deciduous forests in eastern North America (Putnam et al. 1960) and Asia Minor as well as in tropical humid forests of the Amazon (Hubbell and Foster 1986, Balslev and Renner 1989, Hartshorn 1989b) and southeastern Asia (Ashton 1989).

Tree size/number/age patterns

Single and mixed species stands follow a general pattern of tree spacing, stand age, and tree volume (Drew and Flewelling 1977, 1979, White 1985, McFadden and Oliver 1988, O'Hara and Oliver 1988). A similar relationship has been found in natural and artificial herbaceous communities (Kira et al. 1953, Yoda et al. 1963, White and Harper 1970). There is an upper limit to the number of trees of a given mean size which can live in an area. If some trees grow larger, others generally die. The slopes of the upper limit line have been surprisingly consistent (with some disagreement of the amount of variation [Weller 1987, Osawa and Sugita 1989]) for species or species groups from Asia, Europe, and North America.

Patterns associated with crowding

Crowding in single-cohort stands creates similar stand structure patterns. Often some trees become excessively crowded, and the others differentiate above them (Oliver and Larson 1990). At other times, trees become more uniformly crowded and the whole stand becomes physically unstable, is killed by insects, or dramatically slows in growth. Tree falls, tree buckling, insect outbreaks, stagnation, and/or fires are common in crowded stands throughout North America (see Oliver and Larson 1990) and in Europe, Japan (Ishikawa et al. 1986), the Andean Highlands (*Pinus radiata* plantations, Galloway 1991), and the Asian and South American tropics (Halle et al. 1978, Liegel et al. 1985, Bruenig and Y-w Huang 1989).

Differences in patterns

Differences in structural patterns also occur. "Alpine" tropical communities are not like high elevation temperate forests, but contain "rosette trees", tussock grasses, and small-leaf shrubs (Smith and Young 1987). Different species of the same genus can behave differently and so contribute to different stand structures. Noble fir (*Abies procera* Rehd.) is relatively shade intolerant compared to most *Abies* species. Unlike American beech (*Fagus grandifolia* Ehrh.) or Oriental beech (*Fagus orientalis* Lipsky), European beech (*Fagus sylvatica* L.) does not produce root sprouts. There is probably as much variation in different oak (*Quercus*) species' abilities to survive as advance regeneration in the eastern United States as there is variation in *Quercus* species throughout the world. Behaviors of animals and herbaceous plant species vary even more.

There seems to be no pattern to the differences described above; however, other differences occur more predictably. Growth and decomposition rates of dead organic material vary greatly, but in patterns. The more polar, upper elevations and interior parts of continents have the lowest decomposition rates, while tropical and eastern parts of continents below about 40 degrees latitude have quite rapid decomposition rates. The number of tree species generally increases as one moves to the equator and to more humid climates. There are more tree ferns and tree monocots as one moves to the tropics. Bogs are found on uplands at high latitudes or elevations (Klinger et al. 1990), but primarily in stagnant water at low latitudes (Whitmore 1975). Trees with flat crowns are common to tropical latitudes, conical or rounded crowns are common to mid-latitudes, and columnar crowns are common to polar latitudes (Kuuluvainen and Pukkala 1989).

Causes of the similar patterns of stand structures

The similar patterns in diverse areas indicate similar processes occur. The study of processes causing observed phenomena can be described as a hierarchy (Salisbury and Ross 1969, Oliver 1982), with more fundamental processes being the cause of more integrated (less fundamental) phenomena. Consequently, mathematical principles are the basis for the less fundamental discipline of physics. Physics is the basis for chemistry and biochemistry. Chemistry and physics are the bases for the physical environment--climate, soils, and disturbances. Biochemistry forms one of the bases for subcellular physiology and morphology which, in turn, underlie whole tree physiologal and morphological phenomena. Physiological and morphological patterns of trees interact with the physical environment to create stand development patterns. These stand development patterns become the processes which create stand structures.

Influence of the physical environment

A stand's physical environment consists of its climatic and soil conditions as well as the disturbances affecting it. Understanding the patterns and causes of climates, soils, and disturbances should help clarify their influences on stand development processes and structural patterns.

Climatic patterns. Climates are repeated over the earth and can be predicted from physical effects of sunlight (heat), positions of continents, and influences of the earth's rotation on behaviors of the gaseous atmosphere and the liquid oceans (MacArthur 1972). As examples, deserts are generally on the west sides of continents at about 30 degrees latitude; Mediterranean climates are poleward from deserts; temperate rain forests are poleward from these; warm, humid forests are found on the southeastern sides of continents in the northern hemisphere and the northeastern sides in the southern hemisphere. The same influences which cause monsoon rains in summer in Asia cause late summer rains in the southwestern and southeastern United States.

Soil patterns. Soil patterns can also be understood and predicted from factors leading to their formation—parent material, climate, disturbance history, topography, and time. Parent materials are produced by geologic events and show repeated patterns over large areas. Ash soils are common downwind from areas of high volcanic activity—such as in parts of the Rocky Mountains, the Andes, and Asia Minor. Broad river floodplains are common in flat terrain—such as in the southeastern United States, eastern South America, and southeastern Asia. Loess deposits are common near presently or previously braided rivers. Glaciated areas are common in many mountain areas and polar regions. Glaciated valleys contain coarse sands and gravels and few finely textured soils, in contrast to non-glaciated valleys. Old beach deposits cover much of the earth and can form coarse sandy soils of low nutrient value and low cation exchange capacity which last for millions of years.

Disturbance patterns. Forests throughout the world are impacted by various disturbances (White 1979, Oliver 1981, Sousa 1984, Oliver and Larson 1990, Uhl et al. 1990). Fires occur naturally in boreal forests, in dry and wet temperate forests, and in dry and wet tropical forests. Many tropical areas previously assumed invulnerable to fires are proving to have periodic "dry times" caused by "El Niño." During these and other dry times, the southeastern Asian and Central and South American rain forests burn (Davis 1984, Sanford et al. 1985, Lopez-Portillo et al. 1990). Windthrows of trees occur in all regions—from Alaska (Lutz 1940, Deal et al. 1991) to the Himalayas (Oliver and Sherpa 1990) to the Amazon (Halle et al. 1978, Hartshorn 1978, 1980, Veblen and Ashton 1978) to the humid tropics of southeastern Asia (Whitmore 1978, Ashton 1989). Landslides, siltation, flooding, and other disturbances similarly occur throughout the world (Garwood et al. 1979, Glynn 1988, Bruenig and Y-w Huang 1989, Gentry 1989, Irion 1989, Junk 1989, Salo and Rasanen 1989). Disturbances by humans (Edwards 1986) and animals similarly occur throughout the world.

The disturbances are not random, nor are they completely predictable. Certain disturbance types occur more frequently and cover larger areas in some regions than in others. In areas such as western Borneo, disturbances may occur infrequently and on a small scale (Ashton, P.S., 1992); elsewhere, they can be large and/or frequent (Smith 1946, Cooper 1960, Meier 1964, Agee 1981, Yamaguchi 1986, Davis 1984, Zhou 1987, Christensen et al. 1989, Oliver and Larson 1990, Uhl et al. 1990). Disturbances in most areas occur at shorter time intervals than the potential life spans of the trees.

Disturbance frequency and type can be caused by climatic and geologic processes. Windstorms of different intensities are common in all regions. Fire frequency and severity are the result of droughts and ignition sources (lightning storms for most natural fires), which are both influenced by climatic patterns. Erosion and flooding result from climatic and geologic processes; and volcanic activities are produced by geologic processes.

Behavior of the component species

Similar genetic and environmental processes partly explain similarities of stand development processes and resultant structures in many places; however, stands often have analogous patterns among genetically unrelated species and in places with dramatically different soils, climates, and disturbance histories.

Physiological characteristics. Most green plants behave similarly at the subcellular physiology and morphology level and more fundamental ones (e.g., biochemical and molecular). At the above-cellular level of physiology, all terrestrial woody plants behave similarly (Larcher 1983). They all require light of similar wavelengths and a range of intensities for growth. They all require moisture between field capacity and several atmospheres of tension. They all require temperatures for growth between slightly below freezing (0°C, 32°F) and 55°C (130°F). They all require oxygen and the same mineral nutrients (Salisbury and Ross 1969).

The main differences in physiological processes between tree species at the above-cellular level are their abilities to endure extremes of nutrients, moisture, oxygen, sunlight, or temperature. Different species can endure different extremes of conditions but generally grow when conditions are moderate (Levitt 1980).

Morphological characteristics. All plants have the same organs--roots, stems, leaves, and reproductive organs. Woody plants differ in their distributions of plant parts and their allocation of growth to the different parts.

Forms of regeneration. Species of diverse genetic and geographic locations regenerate by one or several sexual and asexual forms (Oliver and Larson 1990). Many Holarctic *Abies* (Blum 1973), *Quercus* (Merz and Boyce 1956), and other species; many Paleotropic Dipterocarpaceae (Ashton,P.M.S., 1992); and many Neotropic species (Uhl et al. 1988, Uhl and Guimaraes 1989) have the advanced regeneration form. Some gymnosperm and many angiosperm tree species from most floristic realms regenerate by sprouting from the root collar or roots. Many *Quercus* and Dipterocarpaceae (Ashton, P.M.S.,1992) regenerate from heavy seeds, while Holarctic *Populus species* and Neotropic Kapok (*Ceiba pentandra*) regenerate by light seeded mechanisms (Junk 1989). Variations of these and other regeneration forms—such as cladoptosis (Galloway and Worrall 1979), layering, bulbs, and tubers (Halle et al. 1978)—exist some trees.

Forms of photosynthate allocation. Trees throughout the world also have similar priorities of photosynthate allocation—with highest priority given to respiration; second priority given to fine root and needle growth; third, to flower and fruit growth; fourth, to height growth; and last, to diameter growth and mechanisms of resistance to insects and diseases (Oliver and Larson 1990). When crowded, a tree slows in diameter growth and becomes susceptible to buckling, loses resistance to insects and diseases, then slows in height growth, and sometimes dies from lack of photosynthate, from being blown over, or from insects or diseases. Species vary in the degree of allocation to the different priorities and in the time before the tree becomes susceptible to buckling or to insect and disease attacks.

Tree architecture forms. Trees of different species allocate growth differently to produce different patterns of stem elongation; branching; height growth; root depth, elongation, and branching; xylem, phloem, and bark growth; leaf sizes and shapes; branch thicknesses; and fluting and buttressing forms. These variations create differences in the geometry of limbs, stems, foliage, and roots. In spite of the thousands of tree species, Halle and Oldeman found they could encompass all species with about 30 generalized tree architecture models, regardless of their geographic, climatic, or evolutionary origin (Halle et al. 1978).

Tree stem and crown forms. Tree stems and crowns also have generally been classified into a few forms (Zimmerman and Brown 1971). Stems can be excurrent or decurrent and can have strong or weak epinastic control; branching patterns can be opposite, whorled, neither, or nonexistent (in the case of some palms; Dransfield 1978); and crowns can be flat-topped, conical or rounded, or columnar.

Forms in response to "high shade." Trees growing in diffuse sunlight ("high shade") far beneath an upper canopy develop a much different, noticeably flat-topped crown shape in many parts of the world (Busgen and Munch 1929, Trimble 1968, Kohyama 1980, 1983, Oliver and Larson 1990). This flat top is either because the terminal becomes noticeably short relative to the lateral branches or because the terminal grows laterally and forms a branch. This flat-topped behavior in either stratified, single-cohort stands or multiple cohort stands gives forests similar appearances even when the species are unrelated. Release from high shade causes a distinct branching pattern, forking, or crook at the height of the stem when released (Oliver and Larson 1990) and may explain the "kinks" in tree trunks in southeastern Asian forests (Whitmore 1975).

Forms of height growth. Different species within an area have different relative height growth patterns, although climate, soil, and regeneration forms influence height growth. Some species commonly have rapid early height growth while others grow slowly at first.

Summary of species behaviors. Nearly all trees in the world contain one or several of relatively few forms for each physiological and morphological characteristic, although there are variations within each form. Species of different geographic ranges and genetic histories often follow similar forms. On the other hand, some forms are only found in certain areas. For example, the non-branching architectures common to palms are generally not found in boreal or northern temperate forests.

With the few physiological and morphological forms for each characteristic, it is not surprising to have the same forms appear in many different forest types, respond similarly to the environment and to other species, create similar stand structures, and so indicate similar stand development processes are occurring.

Interactions between the tree and the environment

Tree species evolve slowly because of the long times between generations (Gentry 1989, Oliver and Larson 1990). Tree species have evolved various physiological and morphological adaptations to the physical environment–climate, latitude, soils, and disturbances; however, they generally have not co-evolved specific adaptations with other tree species. Tree species are found together because each species is physiologically and morphologically versatile and resilient enough to survive in niches where the others can not live or compete as successfully–NOT because the species have developed interdependencies through coevolution (Oliver and Larson 1990).

Some animals (especially insects) and herbaceous plant species have probably evolved to co-exist with certain tree species–or groups of species (Gentry 1989). The shorter genetic turn-around time of these animals and plants allow them to evolve through many generations while the tree species develop through very few.

Throughout the world, the primary interaction between trees is competition for necessities such as sunlight, moisture, oxygen, and certain mineral nutrients. Unlike animal communities which exist in a food chain or food web, trees are generally not interdependent for energy, vitamins, nutrients, or other essential factors. Where trees grow close together, they compete for the various necessities. Inherited forms of the different physiological and morphological characteristics allow certain individuals to survive and gain a competitive advantage at a particular time and place. The soil, climate, type of disturbance, and other factors help determine which forms have the competitive advantage at the particular time and place.

Causes of similar stand structure patterns

Each forest area contains tree species with differences in morphological forms and physiological abilities to endure extreme conditions. These differences cause patterns in the stand structures as different species have competitive advantages under local environments (McKnight et al. 1981, Junk 1989). The same, relatively few physiological and morphological forms are repeated in many related and unrelated species in the world. Consequently, two species which are different in genetic background, geographic location, and environment but have a similar morphological form or physiological characteristic can be found in similar stand structure patterns. The similarities give both species a competitive advantage relative to their neighbors in the same circumstances. For example, many Holarctic *Quercus* in the temperate Northern Hemisphere and Paleotropic Dipterocarpaceae in tropical Asia have heavy seeds and advanced regeneration and in many ways respond similarly to disturbances (Ashton, P.M.S.,1992). Yellow birch (*Betula alleghaniensis* Britton) in northern New England (Hill 1977) and Sitka spruce (*Picea sitchensis* [Bong.] Carr.)in coastal Alaska (Deal et al. 1991) both regenerate from windblown seeds on mineral soils, and so are found frequently on windthrow mounds.

A species can be similar to a geographically isolated, unrelated species in one physiological or morphological characteristic and different in another. Distant stand structures will be analogous in features where similar characteristics of species in the two stands allow both stands to develop the same way. The same stands will be different in structural features that are strongly influenced by physiological or morphological characteristics which are different between species in the two stands. For example, northern red oak (*Quercus rubra* L.) and western larch (*Larix occidentalis* Nutt.) similarly occupy the upper stratum in mixed stands on many sites when growing with red maple (*Acer rubrum* L.)/black birch (*Betula lenta* L.) in New England (Oliver 1978) and with Douglas-fir (*Pseudotsuga menziesii* [Mirb.] Franco)/grand fir (*Abies grandis* [Dougl.] Lindl.) in eastern Washington (Cobb 1988), respectively. Both northern red oaks and western larches maintain a steady height growth form compared to their competitors. Alternatively, they have advantages after entirely different disturbance types, since northern red oaks have the advantage when regenerating from advanced regeneration and western larches regenerate most advantageously from light seeds on mineral soil (Oliver and Larson 1990). Similarly, both tulip-poplar (*Liriodendron tulipifera* L.) and gray birch (*Betula populifolia* Marsh.) have rapid early height growth after a disturbance and can occupy the upper strata in mixed stands soon after a disturbance. Unlike gray birch, tulip-poplar can live a long time and remain dominant in the upper strata for over 200 years.

Applications to silviculture through adaptive management

Awareness that similar stand structure patterns and development processes occur in diverse forests can be used to advance ecological knowledge and silvicultural effectiveness. Research in unstudied forests can be done by comparing stand structure patterns and resident species' physiological and morphological characteristics in the newly studied forest to those elsewhere in the world. Processes in the new area can be hypothesized from analogous structural patterns and species characteristics in stands whose development processes have been well studied. For example, the absence of annual rings makes reconstructing growth rates and ages difficult in tropical forests. The development processes in tropical stands can be hypothesized to be similar to stands with similar structural features in better studied, temperate forests. (The "null hypothesis" would be: there is no reason to believe the stand development processes are different.) More specifically, stand structures and development processes in mixed species stands in the Amazon floodplains and the Paleotropic realm can be compared with moist, temperate, mixed species stands on floodplains in the southeastern United States, where more is known or can be learned rapidly.

The hypotheses can be tested using usual research techniques of permanent plots, reconstruction studies, chronosequences, and other means. Where time and money for these techniques are limited, silvicultural manipulations can be used to test the hypotheses. Changes in structures over time following a silvicultural manipulation can be predicted according to the stand development processes hypothesized to be occurring. Components of the stand can be monitored, and the accuracy of predictions will determine the validity of the hypothesized processes. Comparisons of actual to predicted structures not only will test hypotheses, but will allow processes to be understood better and new processes hypothesized.

This approach has recently been used in fisheries management (Walters 1986) and has been proposed for forest management (Baskerville 1985, Oliver et al. 1992) as "adaptive management." The adaptive management approach is similar to other iterative management approaches such as "quality control" (Ishikawa 1982) and "time series analysis" (Bowerman and O'Connell 1987). The basis of the approach is to make a projection (hypothesis) of how the system will develop under a given set of conditions; to implement the conditions; to monitor, analyze, and learn from the results; and to adjust the conditions and knowledge base.

Many stand development processes (Oliver and Larson 1990) are known and can be compared to newly studied areas using adaptive management. The development processes form the bases for silvicultural keys, guides, and practices, which can be treated as hypotheses. Individual conditions and insights into possible processes in new areas will help determine exactly which hypotheses are appropriate. Elements of adaptive management are being incorporated into operational/research strip clearcuts, spacing regimes, and other silvicultural techniques in tropical forests (Jordan 1982, Larson and Zaman 1985, Ashton et al. 1989, Hartshorn 1989a, Anderson 1990, Galloway 1991).

As an example, silvicultural spacing guides for managing teak (*Tectona grandis* L.) in Bangladesh (Larson and Zaman 1985) and blue mahoe (*Hibiscus elatus* SW.) in Puerto Rico (Ashton et al. 1989) have been prepared by considering these species' similarities in stand development processes to northern red oak (*Quercus rubra* L.) in North America. The teak and blue mahoe were assumed to develop similarly to northern red oak; therefore, the oak guides were calibrated to teak and blue mahoe from crown and spacing data. The guides can be tested when plantations are established at a target spacing according to the guide. The general hypothesis is: there is no reason to believe that development processes are different between northern red oak stands and teak (or blue mahoe) stands. Consequently, the spacing guide would be expected to predict the future diameters correctly. Measurable future conditions (such as diameters of teak or blue mahoe at different spacings) are hypothesized at short future time intervals to determine quickly if the hypothesis is being proved. The regime is then implemented and monitored. Results of the monitoring are compared with the hypothesized growth pattern, and the validity of the hypothesis is tested. If the hypothesis is proven wrong, the data can readily be used both to develop or refine the hypothesis and to adjust the stand based on new hypotheses. Since silvicultural activities are done to different stands over many years within a region, later silvicultural regimes can be refined as information is gained from earlier ones. Similar adaptive management studies can be done to determine such things as species regeneration responses to different natural or artificial disturbances, species stratification patterns, appropriate times of thinnings (Galloway 1991), and other stand development patterns.

It will be very important NOT simply to extend unadjusted copies of known silvicultural manipulations to new species and regions. Adjustments to silvicultural systems or operations need to be made when extending them to new regions based on differences in the underlying climatic, soil, physiologic, morphologic, and stand development processes. Most of the time, silvicultural systems and operations in new forest areas will be amalgamations of several known systems taken from other areas.

Undoubtedly, variations to presently known stand structures and development processes will be discovered—as will entirely new structures and processes. New structures and processes will emerge quite rapidly and inexpensively from adaptive management when the hypothesized results of silvicultural operations are not attained. There is, however, stronger reason to believe that the stand development processes and structures in less studied forests are similar to forests already well studied than there is reason to believe the processes and structures are entirely different.

Literature cited

Agee, J.K. 1981. Fire effects on Pacific Northwest forests: Flora, fuels, and fauna. IN Proceedings of the Northwest Forest Fire Council Conference, November 23-24, 1981. Northwest Forest Fire Council, Portland, Oregon. pp.54-66.

Anderson, A.B. (editor). 1990. Alternatives to Deforestation: Steps Toward Sustainable Use of the Amazon Rain Forest. Columbia University Press, New York. 281 pp.

Ashton, D.H. 1976. The development of even-aged stands of *Eucalyptus regnans* F. Muell. in central Victoria. Australian Journal of Botany 24: 397-414.

Ashton, P.M.S. 1992. Establishment and early growth of advance regeneration of canopy trees in moist mixed-species forest. Paper in this volume.

Ashton, P.M.S., J.S. Lowe, and B.C. Larson. 1989. Thinning and spacing guidelines for blue mahoe (*Hibiscus elatus* SW.) Journal of Tropical Forest Science 2(1):37-47.

Ashton, P.S. 1988. Dipterocarp biology as a window to the understanding of tropical forest structure. Annual Review of Ecology and Systematics 19:347-370.

Ashton, P.S. 1989. Species richness in tropical forests. IN L.B. Holm-Nielson, I.C. Nielson, and H. Balslev, editors. Tropical Forests. Academic Press, New York. pp. 239-251.

Ashton, P.S. 1992. The structure and dynamics of tropical rain forest in relation to tree species richness. Paper in this volume.

Aytug, B., N. Merev, and G. Edis. 1975. Surmenet—agacbasi dolayari ladin ormanin tarihi ve gelecigi. Proje No. TOAG 113; TOAG Seri No. 39, TBTAK Yayinlari No. 252, Turkiye Bilimsel ve Teknik Arastirma Kurumu, Ankara, Turkey. 64 pp.

Balslev, H., and S.S. Renner. 1989. Diversity of east Ecuadorean lowland forests. IN L.B. Holm-Nielson, I.C. Nielsen, and H. Balslev, editors. Tropical Forests. Academic Press, New York. pp. 287-295.

Baskerville, G. 1985. Adaptive management: Wood availability and habitat availability. Forestry Chronicle 61 (2):171-175.

Blum, B.M. 1973. Some observations on age relationships in spruce-fir regeneration. USDA Forest Service Research Note NE-169.

Bonan, G.B., and H.H. Shugart. 1989. Environmental factors and ecological processes in boreal forests. Annual Review of Ecology and Systematics 20:1-28.

Bormann, F.H., and G.E. Likens. 1981. Pattern and Process in a Forested Ecosystem. Springer-Verlag, New York. 253 pp.

Bowerman, B.L., and R.T. O'Connell. 1987. Time Series Forecasting. Second Edition. Duxbury Press, Boston, Massachusetts. 540 pp.

Bruenig, E.F., and Y-w Huang. 1989. Patterns of tree species diversity and canopy structure and dynamics in humid, tropical, evergreen forests on Borneo and China. IN L.B. Holm-Nielson, I.C. Nielsen, and H.B alslev, editors. Tropical Forests. Academic Press, New York. pp. 75-88.

Busgen, M., and E. Munch. 1929. The Structure and Life of Forest Trees. (Translated by T. Thompson), Third Edition, St. Giles'Works, Norwich, Great Britain. 436 pp.

Christensen, N.L., J.K. Agee, P.F. Brussard, J. Hughes, D.H. Knight, G.W. Minshall, J.M. Peek, S.J. Pyne, F.J. Swanson, J.W. Thomas, S. Wells, S.E. Williams, and H.A. Wright. 1989. Interpreting the Yellowstone fires of 1988. BioScience 39:678-685.

Clatterbuck, W.K., and J.D. Hodges. 1988. Development of cherrybark oak and sweetgum in mixed, even-aged bottomland stands in central Mississippi, U.S.A. Canadian Journal of Forest Research 18:12-18.

Cobb, D.F. 1988. Development of mixed western larch, lodgepole pine, Douglas-fir, grand fir stands in eastern Washington. Unpublished Masters thesis. College of Forest Resources, University of Washington, Seattle. 99 pp.

Cooper, C.F. 1960. Changes in vegetation, structure, and growth of southwestern pine forests since white settlement. Ecological Monographs 30:129-164.
Crow, T.R. 1980. A rainforest chronicle: A 30-year record of change in structure and composition at El Verde, Puerto Rico. Biotropica 12 (1):42-55.
Cwynar, L.C. 1987. Fire and the forest history of the North Cascade Range. Ecology 68 (4):791-802.
Davis, J.B. 1984. Burning another empire. Fire Management Notes 45 (4):12-17.
Davis, M.B. 1981. Quaternary history and the stability of forest communities. IN D.C. West, H.H. Shugart, and D.B. Botkin, editors. Forest Succession: Concepts and Applications. Springer-Verlag, New York. pp. 132-153.
Deal, R.L., C.D. Oliver, and B.T. Bormann. 1991. Reconstruction of mixed hemlock/spruce stands on low elevation, upland sites in southeastern Alaska. Canadian Journal of Forest Research 21: 643-654.
Dransfield, John. 1978. Growth forms of rain forest palms. IN P.B. Tomlinson and M.H. Zimmerman, editors. Tropical Trees as Living Systems. Cambridge University Press, Cambridge. pp. 247-268.
Drew, T.J., and J.W.Flewelling. 1977. Some recent Japanese theories of yield-density relationships and their application to Monterey pine plantations. Forest Science 23:517-534.
Drew, T.J., and J.W.Flewelling. 1979. Stand density management: An alternative approach and its application to Douglas-fir plantations. Forest Science 25: 518-532.
Edwards, C.R. 1986. The human impact on the forest in Quintana Roo, Mexico. Journal of Forest History, July, 1986:120-127.
Frelich, L.E., and C.G. Lorimer. 1985. Current and predicted long-term effects of deer browsing in hemlock forests in Michigan, U.S.A. Biological Conservation 34:99-120.
Galloway, G. 1991. Beyond afforestation: A case study in Highland Ecuador. Unpublished Ph.D.dissertation, College of Forest Resources, University of Washington, Seattle, Washington. 245 pp.
Galloway, G., and J. Worrall. 1979. Cladoptosis: A reproductive strategy in black cottonwood. Canadian Journal of Forest Research 9:122-125.
Garwood, N.C., D.P. Janos, and Nicholas Brokaw. 1979. Earthquake-caused landslides: A major disturbance to tropical forests. Science 205:997-999.
Geesink, R., and D.J. Kornet. 1989. Speciation and Malesian Luguminosae. IN L.B. Holm-Nielson, I.C. Nielsen, and H. Balslev, editors. Tropical Forests. Academic Press, New York. pp. 135-151.
Gentry, A.H. 1982. Neotropical floristic diversity: Phytogeographical connections between central and South America. Pleistocene climatic fluctuations, or an accident of the Andean orogeny. Annals of the Missouri Botanical Garden 69: 557-593.
Gentry, A.H. 1988. Changes in plant community diversity and floristic composition on environmental and geographical gradients. Annals of the Missouri Botanical Garden 75 (1):1-34.
Gentry, A.H. 1989. Speciation in tropical forests. IN L.B. Holm-Nielson, I.C. Nielsen, and H. Balslev, editors. Tropical Forests. Academic Press, New York. pp. 113-134.
Gentry, A.H. 1987. Geographical variation in fertility, phenology, and composition of the understory of neotropical forests. Biotropica 19(3):216-227.
Glynn, P.W. 1988. El Niño-southern oscillation 1982-1983: Nearshore population, community, and ecosystem responses. Annual Review of Ecology and Systematics 19:309-345.
Halle, F., R.A.A. Oldeman, and P.B. Tomlinson. 1978. Tropical Trees and Forests: An Architectural Analysis. Springer-Verlag, New York. 441 pp.
Hartshorn, G.S. 1978. Tree falls and tropical forest dynamics. IN P.B. Tomlinson and M.H. Zimmerman, editors. Tropical Trees as Living Systems. Cambridge University Press, Cambridge. pp. 617-638.
Hartshorn, G.S. 1980. Neotropical forest dynamics. Biotropica 12 (Supplement): 23-30.
Hartshorn, G.S. 1989a. Application of gap theory to tropical forest management: Natural regeneration on strip clearcuts in the Peruvian Amazon. Ecology 70 (3):567-569.
Hartshorn, G.S. 1989b. Gap-phase dynamics and tropical tree species richness. IN L.B. Holm-Nielson, I.C. Nielsen, and H. Balslev, editors. Tropical Forests. Academic Press, New York. pp. 65-73.
Henry, J.D., and J.M.A. Swan. 1974. Reconstructing forest history from live and dead plant material—An approach to the study of forest succession in southwest New Hampshire. Ecology 55:772-783.
Hill, D.B. 1977. Relative development of yellow birch and associated species in immature northern hardwood stands in the White Mountains. Unpublished Ph.D.dissertation, School of Forestry and Environmental Studies, Yale University, New Haven, Connecticut. 135 pp.
Hubbell, S.P., and R.B. Foster. 1986. Commonness and rarity in a neotropical forest: Implications for tropical tree conservation. IN M.E. Soule, editor. Conservation Biology: The Science of Scarcity and Diversity. Sinauer Associates, Sunderland, Massachusetts. pp. 205-231.

Irion, G. 1989. Quaternary geological history of the Amazon lowlands. IN L.B. Holm-Nielson, I.C. Nielsen, and H. Balslev, editors. Tropical Forests. Academic Press, New York. pp. 23-34.

Isaac, L.A. 1940. Vegetation succession following logging in the Douglas-fir region with special reference to fire. Journal of Forestry 38:716-721.

Ishikawa, K. 1982. Guide to Quality Contol. Asian Productivity Organization. Quality Resources, White Plains, New York. 225 pp.

Ishikawa, M., R. Nitta, T. Katsuta, and T. Fujimori. 1986. Snowbreakage-mechanisms and methods of avoiding it. Reviews of Studies in Forestry, No. 83. Forest Science Organization, Tokyo. 101 pp. (In Japanese)

Jones, E.W. 1945. The structure and reproduction of the virgin forest of the North Temperate zone. New Phytologist 44:130-148.

Jordan, C.F. 1982. Amazon rain forests. American Scientist 70:394-401.

Junk, W.J. 1989. Flood tolerance and tree distribution in central Amazonian floodplains. IN L.B. Holm-Nielson, I.C. Nielsen, and H. Balslev, editors. Tropical Forests. Academic Press, New York. pp. 47-64.

Kira, T., H. Ogawa, and N. Sakazaki. 1953. Intraspecific competition among higher plants: I. Competition-density-yield interrelationship in regularly dispersed populations. Journal of the Institute of Polytechnics, Osaka City University 4:1-16.

Klinger, L.F., S.A. Elias, V.M. Behan-Pelletier, and N.E. Williams. 1990. The bog climax hypothesis: Fossil arthropod and stratigraphic evidence in peat sections from southeast Alaska, USA. Holarctic Ecology 13:72-80.

Kohyama, T. 1980. Growth patterns of *Abies mariensii* saplings under conditions of open-growth and suppression. Botanical Magazine of Tokyo 93: 13-24.

Kohyama, T. 1983. Seedling stage of two subalpine *Abies* species in distinction from sapling stage: A matter-economic analysis. Botanical Magazine of Tokyo 96:49-65.

Kuuluvainen, T., and T. Pukkala. 1989. Simulation of within-tree and between-tree shading of direct radiation in a forest canopy: Effect of crown shape and sun elevation. Ecological Modelling 49:89-100.

Lang, G.E., and D.H. Knight. 1983. Tree growth, mortality, recruitment, and canopy gap formation during a 10-year period in a tropical moist forest. Ecology 64 (5):1075-1080.

Larcher, W. 1983. Physiological Plant Ecology. Springer-Verlag, New York. 303 pp.

Larson, B.C., and M.N. Zaman. 1985. Thinning guidelines for teak (*Tectona grandis* L.). The Malaysian Forester 48 (4):288-297.

Levitt, J. 1980. Responses of Plants to Environmental Stresses: Volume II. Water, Radiation, Salt, and Other Stresses. Academic Press, New York. 607 pp.

Liegel, L.H., W.E. Balmer, and G.W. Ryan. 1985. Honduras pine spacing trial results in Puerto Rico. Southern Journal of Applied Forestry 9:69-75.

Lopez-Portilla, J., M.R. Keyes, A. Gonzalez, E. Cabrera C., and O. Sanchez. 1990. Los incendios de Quintana Roo: Catastrofe ecologica o evento periodico? Ciencia Y Desarrollo 16 (91):43-57.

Lutz, H.J. 1940. Disturbance of forest soil resulting from the uprooting of trees. Yale University School of Forestry Bulletin 45. 37 pp.

MacArthur, R.H. 1972. Climates on a rotating earth. IN Geographical Ecology: Patterns in the Distribution of Species. Harper and Row, New York. pp. 5-19.

McFadden, G., and C.D. Oliver. 1988. Three-dimensional forest growth model relating tree size, tree number, and stand age: Relation to previous growth models and to self-thinning. Forest Science 34:662-676.

McKnight, J.S., D.D. Hook, O.G. Langdon, and R.L. Johnson. 1981. Flood tolerance and related characteristics of trees of the bottomland forests of the southern United States. IN J.R. Clark and J. Benforado, editors. Wetlands of Bottomland Hardwood Forests. Elsevier Scientific Publishing Co., New York: 29-69.

Meier, M.F. 1964. Ice and glaciers. IN V.T. Chow, editor. Handbook of Applied Hydrology. McGraw-Hill, New York. pp. 16.1-16.32.

Merz, R.W., and S.G. Boyce. 1956. Age of oak "seedlings." Journal of Forestry 54:774-775.

Milewski, A.V. 1983. A comparison of ecosystems in Mediterranean Australia and southern Africa: Nutrient-poor sites at the Barrens and the Caledon Coast. Annual Review of Ecology and Systematics 14:57-76.

Nakashizuka, T., and M. Numata. 1982. Regeneration process of climax beech forests: I. Structure of a beech forest with the undergrowth of Sasa. Japanese Journal of Ecology 32:473-482.

O'Hara, K.L., and C.D. Oliver. 1988. Three-dimensional representation of Douglas-fir volume growth: Comparison of growth and yield models with stand data. Forest Science 34:724-743.

Oliver, C.D. 1978. Development of northern red oak in mixed species stands in central New England. Yale University School of Forestry and Environmental Studies Bulletin No. 91. 63 pp.
Oliver, C.D. 1980. Even-aged development of mixed-species stands. Journal of Forestry 78:201-203.
Oliver, C.D. 1981. Forest development in North America following major disturbances. Forest Ecology and Management 3:153-168.
Oliver, C.D. 1982. Stand development—its uses and methods of study. IN J.E. Means, editor. Forest Development and Stand Development Research in the Northwest. Forest Research Laboratory, Oregon State University, Corvallis. pp. 100-112.
Oliver, C.D., and B.C. Larson. 1990. Forest Stand Dynamics. McGraw-Hill, New York. 467 pp.
Oliver, C.D., and L.N. Sherpa. 1990. The effects of browsing and other disturbances on the forest and shrub vegetation of the Hongu, Inkhu, and Dudh Koshi Valleys. The Makalu-Barun Conservation Project Working Paper Publication Series, Report 9. Department of National Parks and Wildlife Conservation, HMG of Nepal, Kathmandu; and Woodlands Mountain Institute Mount Everest Ecosystem Conservation Program, Kathmandu, Nepal and Franklin, West Virginia, U.S.A. 40 pp.
Oliver, C.D., and E.P. Stephens. 1977. Reconstruction of a mixed species forest in central New England. Ecology 58:562-572.
Oliver, C.D., A.B. Adams, and R.J. Zasoski. 1985. Disturbance patterns and forest development in a recently deglaciated valley in the northwestern Cascade Range of Washington, U.S.A. Canadian Journal of Forest Research 15:221-232.
Oliver, C.D., D.R. Berg, D.R. Larsen, and K.L. O'Hara. 1992. Integrating management tools, ecological knowledge, and silviculture. IN R. Naiman and J. Sedell, editors. New Perspectives for Watershed Management. Springer-Verlag, New York. (In press)
Osawa, A., and S. Sugita. 1989. The self-thinning rule: Another interpretation of Weller's results. Ecology 70:279-283.
Peet, R.K., and N.L. Christensen. 1987. Competition and tree death. BioScience 37: 586-595.
Prance, G.T. 1982. A review of the phytogeographic evidences for Pleistocene climate changes in the Neotropics. Annals of the Missouri Botanical Garden 69:594-624.
Putnam, J.A., G.M. Furnival, and J.S. McKnight. 1960. Management and inventory of Southern hardwoods. U.S.Department of Agriculture, Agricultural Handbook No. 181. 102 pp.
Raup, H.M. 1946. Phytogeographic studies of the Athabaska-Great Slave Lake Region II. Journal of the Arnold Arboretum 27:1-85.
Richards, P.W. 1952. The Tropical Rain Forest: An Ecological Study. Cambridge University Press, Cambridge. 450 pp.
Ruddiman, W.F. 1987. North America and Adjacent Oceans During the Last Deglaciation. The Geology of North America, Vol. K-3. The Geology Society of America, Inc. Boulder, Colorado. 501 pp.
Salisbury, F.B., and C. Ross. 1969. Plant Physiology. Wadsworth Publ.Co., Belmont, California. 747 pp.
Salo, J., and M. Rasanen. 1989. Hierarchy of landscape patterns in western Amazon. IN L.B. Holm-Nielson, I.C. Nielsen, and H. Balslev, editors. Tropical Forests. Academic Press, New York. pp. 35-45.
Sanford, R.L., Jr., J. Saldarriaga, K. Clark, C. Uhl, and R. Herrara. 1985. Amazon rain-forest fires. Science 227:53-55.
Segura-Warnholtz, G., and L. Snook. 1992. Population dynamics and disturbance patterns of a pinon pine forest in Central East Mexico. Forest Ecology and Management (In press).
Smith, A.P., and T.P. Young. 1987. Tropical alpine plant ecology. Annual Review of Ecology and Systematics 18:137-158.
Smith, D.M. 1946. Storm damage in New England forests. Master of Forestry thesis. Yale University School of Forestry and Environmental Studies, New Haven, Connecticut. 173 pp.
Sousa, W.P. 1984. The role of disturbance in natural communities. Annual Review of Ecology and Systematics 15:353-391.
Stubblefield, G.W., and C.D. Oliver. 1978. Silvicultural implications of the reconstruction of mixed alder/conifer stands. IN W.A. Atkinson, D. Briggs, and D.S. DeBell, editors. Utilization and Management of Red Alder, USDA Forest Service General Technical Report PNW-70, pp. 307-320.
Swaine, M.D., and T.C. Whitmore. 1988. On the definition of ecological species groups in tropical rain forests. Vegetatio 75:81-86.
Tallis, J.H. 1991. Plant Community History: Long-term Changes in Plant Distribution and Diversity. Chapman and Hall, New York. 398 pp.
Trimble, G.R. 1968. Form recovery by understory sugar maple under uneven-aged management. USDA Forest Service Research Note NE-89. 8 pp.

Uhl, C., and I.C. Guimaraes Vieira. 1989. Ecological impacts of selective logging in the Brazilian Amazon: a case study from the Paragominas Region of the state of Para. Biotropica 21 (2):98-106.

Uhl, C., K. Clark, and P. Maquirino. 1988. Vegetation dynamics in Amazonian treefall gaps. Ecology 69 (3):751-763.

Uhl, C., D. Nepstad, R. Buschbacher, K. Clark, B. Kauffman, and S. Subler. 1990. Studies of ecosystem response to natural and anthropogenic disturbances provide guidelines for designing sustainable land-use systems in Amazonia. IN A.B. Anderson, editor. Alternatives to Deforestation: Steps Toward Sustainable Use of the Amazon Rain Forest. Columbia University Press, New York. pp. 24-42.

Van Devender, T.R., and W.G. Spaulding. 1979. Development of vegetation and climate in the southwestern United States. Science 204:701-710.

Veblen, T.T., and D.H. Ashton. 1978. Catastrophic influences on the vegetation of the Valdivian Andes, Chile. Vegetatio 36 (3):149-167.

Veblen, T.T., F.M. Schlegel, and B. Escobar R., 1980. Structure and dynamics of old-growth nothofagus forests in the Valdivian Andes, Chile. Journal of Ecology 68:1-31.

Walter, H. 1973. Vegetation of the Earth and Ecological Systems of the Geo-biosphere. Second Edition. (translated in 1979 by Joy Wieser), Springer-Verlag, New York. 274 pp.

Walters, C. 1986. Adaptive Management of Renewable Resources. Macmillan Publishing Company. 374 pp.

Watt, A.S. 1947. Pattern and process in the plant community. Journal of Ecology 35:1-22.

Weller, D.E. 1987. A reevaluation of the -3/2 power rule of plant self-thinning. Ecological Monographs 57:23-43.

White, P.S. 1979. Pattern, process, and natural disturbance in vegetation. Botanical Review 45:229-299.

White, J. 1985. The thinning rule and its application to mixtures of plant populations. IN J. White, editor. Studies on Plant Demography: A Festchrift for John Harper. Academic Press, New York. pp. 291-309.

White, J., and J.L. Harper. 1970. Correlated changes in plant size and number in plant populations. Journal of Ecology 58:467-485.

Whitmore, T.C. 1975. Tropical Rain Forests of the Far East. Clarendon Press, Oxford. 282 pp.

Whitmore, T.C. 1978. Gaps in the forest canopy. IN P.B. Tomlinson and M.H. Zimmerman, editors. Tropical Trees as Living Systems. Cambridge University Press, Cambridge. pp. 639-655.

Whitmore, T.C. 1989. Canopy gaps and the two major groups of forest trees. Ecology 70 (3): 536-538.

Yamaguchi, D.K. 1986. The development of old-growth Douglas-fir forests northeast of Mount St. Helens, Washington, following an A.D. 1480 eruption. Unpublished Ph.D.dissertation. University of Washington, Seattle. 100 pp.

Yoda, K., T. Kira, H. Ogawa, and K. Hozumi. 1963. Intraspecific competition among higher plants: II. Self-thinning in overcrowded pure stands under cultivated and natural conditions. Journal of Biology of Osaka City University 14:107-129.

Zhou, Y. 1987. More troops help control forest fire. China Daily Newspaper, March 25, 1987.

Zimmerman, M.H., and C.L. Brown. 1971. Trees: Structure and Function. Springer-Verlag, New York. 336 pp.

Stand Structure and Dynamics: Case Studies

II

Development of a mixed-conifer forest in Hokkaido, northern Japan, following a catastrophic windstorm: A "parallel" model of plant succession

3

AKIRA OSAWA

Introduction

Structural development of mixed-species forests tend to take similar patterns in temperate forests (Oliver 1981, Oliver and Larson 1990). The analysis of long-term data on population dynamics and stand development of a mixed-conifer forest in Hokkaido, northern Japan, showed many similarities of canopy stratification to those found in mixed hardwood forests of central New England (Oliver 1978). It also suggested a pattern of plant succession that has not been fully appreciated in general ecological literature. Close examination of these observations, therefore, may help expand our understanding of the theories of plant succession that have been much discussed, but are still short of reaching overwhelming acceptance (Pickett et al. 1987).

Objectives of this paper are: 1) to describe detailed patterns of stand development of a mixed-conifer forest in Hokkaido; 2) to compare the observed patterns with stand development reported elsewhere; and 3) to examine the results in relation to the ecological theories of succession.

Plant succession has intrigued ecologists since the early days of this scientific discipline. Succession in this paper refers to "the changes observed in an ecological community following a perturbation that opens up a relatively large space" (Connell and Slatyer 1977). A comprehensive theory that Clements (1916) put forward was so satisfying that it virtually monopolized the field for the next several decades (Connell and Slatyer 1977). Some objections were raised soon after its proposal (Gleason 1917, 1926, 1927); however, it was only much later that serious doubts began to appear systematically (Whittaker 1953, Egler 1954, Drury & Nisbet 1973).

The organismic model of succession (Clements 1916) suggested that plant associations are distinct units that mimic living organisms, and proposed that they, too, are born, grow, and then die. Gleason (1917, 1926) argued that there are no such distinct units in vegetation. An "individualistic concept" of plant association was thus proposed (Gleason 1917) that rejected any analogy to organisms. Lack of distinctness of plant associations was substantiated quantitatively much later (Whittaker 1967).

Another objection to Clementsian theory of succession was related to its mechanism. Clements' (1916) model has been identified with "relay floristics" in which a species that occupies a site earlier alters physical as well as biological conditions for the later species to become established and grow. In contrast, Egler (1954) proposed another mechanism of succession, "initial floristic composition." It was argued that most species that appear as dominants during the course of succession are already present at the beginning of vegetation development, or shortly thereafter. A preparatory influence of earlier plants on later ones is denied in the hypothesis of initial floristic composition.

M. J. Kelty (ed.), The Ecology and Silviculture of Mixed-Species Forests, 29–52.

It was a critical review of the subject by Drury and Nisbet (1973) that reset the discussion of plant succession. They examined available data on succession and concluded that: 1) the theory was inferred mostly from observations of spatial sequences on adjacent sites, and the direct evidence was only for the early stages of succession; 2) the facilitative effect of species already on the site was not substantiated; their effect was frequently to inhibit the species that may follow; and, 3) successional changes in structure and function result primarily from correlations between size, longevity, and growth characteristics in plants. Connell and Slatyer (1977) argued that the mechanisms producing successional sequences had not been elucidated partly because of an absence of clearly stated testable hypotheses of mechanisms; therefore, they formulated such hypotheses that can be tested with manipulative field experiments.

In the following section, detailed patterns of stand development of the mixed-conifer forest in Hokkaido are described; which will be followed by discussions of and comparisons to theories of plant succession, particularly to those of Connell and Slatyer.

Post-typhoon development of a mixed-conifer stand

Typhoons are tropical cyclones that affect the northwestern quarter of the Pacific from Indo-China through the Phillipines; the maritime regions of China, Korea, Japan; and to the Russian province of Sakhalin. They are virtually identical to hurricanes of the Atlantic from a meteorological perspective. These tropical cyclones are agents that cause extensive disturbances to forest communities (Adams et al. 1941, Smith 1946, Morodome 1955, Tamate 1956, 1959, Henry and Swan 1974, White 1979, Oliver 1981, Foster 1988a). Both in the northern hardwood as well as in the mixed conifer-hardwood type forests of North America and of Eastern Asia, devastating damage to forest vegetation has been recorded in history. The hurricane of 1938 destroyed much old timber in New England, U.S.A. (Smith 1946, Foster 1988b), amounting to 84 million m^3 (3 billion board feet; Spurr 1956) of stem volume. The typhoon of 1954 (also known as typhoon "Tohyamaru" or "Marie") destroyed extensive forest areas, including the most valuable stands in Hokkaido, the northern island of Japan (Morodome 1955), causing 27 million m^3 of timber loss (Matsukawa et al. 1959) in an area about the size of the state of Maine, U.S.A.

Figure 1 shows the paths of the 1938 hurricane in New England and the 1954 typhoon in Hokkaido, as well as the extents of forest damage (redrawn from Foster 1988b, Matsukawa et al. 1959, and Tamate 1959). Both areas are at approximately the same latitude (42°N ~ 45°N) and tend to be affected by tropical cyclones in late September: the 1938 hurricane hit on 21 September; the 1954 typhoon hit on 26 September. Highest damage occurred just to the east of the path of the eyes of the storms in both cases, because of the combined effect of the counter-clockwise movement of the air about the eye and the storm's northward movements.

Forests in both places are dominated by hardwood genera such as *Quercus*, *Acer* and *Betula* as well as by conifers such as *Abies* and *Picea* in high elevations and northern areas (Osawa 1989). *Pinus* comprises only a minor component in Hokkaido. In contrast, there were extensive *Pinus strobus* (L.) stands at the time of the 1938 hurricane in New England; however, a large percentage of the *Pinus* stands were the result of previous land use, having become established as pure stands on abandoned agricultural fields (Oliver 1978, Spurr and Barnes 1980, Foster 1988a).

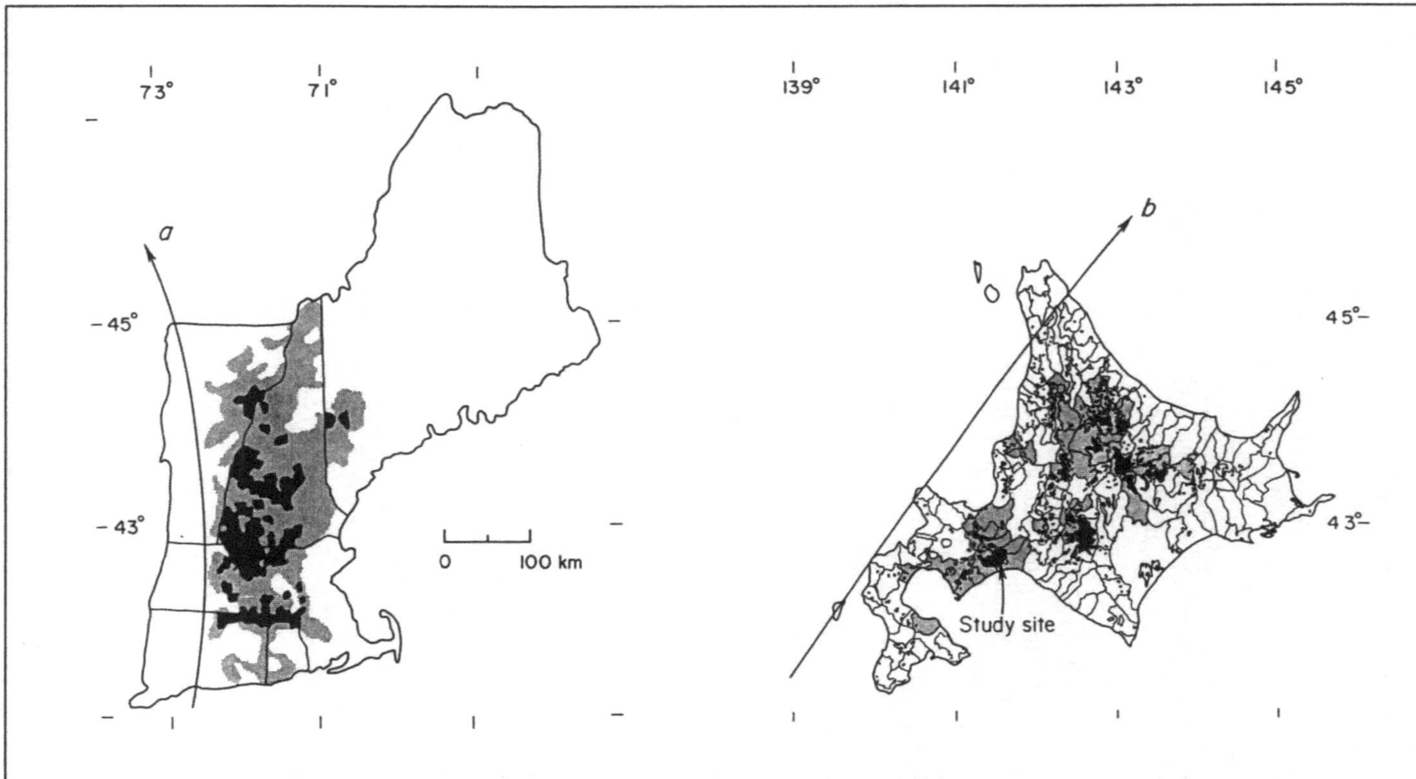

Fig. 1. Courses of (a) 1938 hurricane in New England and (b) 1954 typhoon in Hokkaido with the associated damage to forests. Dark and light shades represent heavy and moderate forest damage (subjectively defined), respectively. New England portion of the figure is based on Foster (1988b); the Hokkaido figure is drawn from data in Matsukawa et al. (1959) and Tamate (1959).

Unlike the rest of the Japanese archipelago, large-scale agriculture had started in Hokkaido only in the 1870's, after systematic migration of the Japanese to this northern territory. As a result, extensive areas of virgin or natural forests were damaged by the 1954 typhoon.

Both storms created similar effects on forests as well as on human society. In addition to the great loss of the best timber (Foster and Boose in press, Matsukawa et al. 1959), there were problems of salvage in both cases including shortages of loggers, stabilities of timber prices, temporary storage of salvaged timber, fire hazards (Clapp 1938, Matsukawa et al. 1959), and anticipated bark beetle outbreaks (Clapp 1938, Inoue 1955, Yamaguchi 1959, Matsukawa et al. 1959).

Study site

A study of stand development following the 1954 typhoon was conducted in the Tomakomai Forest District of the National Forest of Japan (43°N, 142°E), 40 km southeast of Sapporo. Annual precipitation in the area is about 1,880 mm; mean monthly air temperature is 6.0°C. Summers are characterized by heavy fogs because of the cold current along the eastern shore of Hokkaido, in much the same way as summer fogs occurring along the coast of Maine, U.S.A. and Canadian Maritimes. A total of 9,420 ha (23,280 acres) of both natural stands and plantations on the east slope of Mt. Tarumae (1,042 m elevation) were nearly completely blown over by the typhoon in the Forest District (Matsukawa et al. 1959). This district recorded the typhoon's maximum wind velocity of 31.8 m/sec (71.1 mi/hr) (Tamate 1959). Fifty-six percent of the district's forests were blown over, and 44.6% of the district's timber volume was lost (Matsukawa et al. 1959). Damage was exceptionally intense in the area because its southward aspect was exposed to the Pacific Ocean and to the strong south to southwesterly winds at the height of the typhoon. Acceleration of the wind along the conical topography of the volcano is also believed to have intensified the damage (Tamate 1959).

The area was originally covered with old growth stands of *Picea glehnii*, *Picea jezoensis*, *Abies sachalinensis* and some associated hardwoods (mainly *Quercus mongolica* var. *grosseserrata* (Blume) Rehder et. Wilson, *Betula ermanii* Chamisso and *Sorbus commixta* Hedlund). Large canopy trees were more than 30 m tall, 60 cm DBH (diameter at 1.3 m height) (Inoue 1959), and perhaps more than 300 years old. Stem volume of the old *Picea-Abies* forests was typically about 190 m^3/ha (Sato et al. 1985).

The predominance of conifers was partly because of the peculiar geology of the area. Mt. Tarumae is one of the more active volcanoes in the region, with four major eruptions during the past 323 years (Sasaki 1982); and thick tephra covers the east slope of the mountain. The eruption of 1667 was more significant than later ones and produced a new layer of pumicious rocks approximately two meters or more thick around the current study area (Ujiie 1984). The forest probably suffered devastating damage from this eruption, although there is little information to document its effect on forests. Three subsequent eruptions (in 1739, 1804-17, and 1874) were relatively minor and together have added 10 to 50 centimeters of new tephra. The combined thickness of the tephra layers varies locally; however, it is more than three meters deep in the vicinity of the study site. Trees, therefore, are growing on a thin layer (approximately 10 to 15 cm) of organic soil developed over the excessively well drained, nutrient poor, immature soil of tephra origin (Yamada 1958, Kubo and Sanada 1975, Ujiie 1984). *Picea glehni* often dominates poor sites (Tatewaki and Igarashi 1971) in a pattern similar to *P. mariana* in North America.

As a result, the pre-typhoon vegetation of the study site was an old growth stand dominated by *Picea glehnii* (Fr. Schm.) (Sato et al. 1985).

Methods

A permanent plot of 0.16 ha size (40 m X 40 m) was established in Compartment 463 at approximately 300 m elevation in the summer of 1957, nearly three years after the typhoon. The uprooted and damaged trees had been salvaged by the time of plot establishment, and only broken and unmerchantable stems and branches were left on the ground (Sato et al. 1985). The stand development patterns reported here may not represent patterns that occur naturally; rather, they represent patterns of forest regrowth that can occur under the human influence of salvage operations. This was the case for both the Hokkaido and the New England situations (Spurr 1956, Foster 1988a).

All stems over 1.3 m tall in the permanent plot were recorded by species, breast height diameter, and total tree height. The 0.16 ha plot was then divided into a grid with six meter square subplots by placing wooden stakes at the grid points (replaced by PVC poles in 1970). Two four-meter-wide strips were also established perpendicular to each other and crossing at the center of the plot. Using this grid system, all stem locations (> 1.3 m tall) and their crown projections were mapped. Major features such as the locations of residual logs left after salvaging were also recorded on the maps.

The central strips were further divided into 4 m^2 quadrats (2 m x 2 m). All stems over 10 cm tall were mapped in the central strips with records of species, total height, crown projection, and breast height stem diameter if appropriate.

A remeasurement was taken in 1970 in the central strips using the same method as in 1957. In 1970, numbered labels were placed on the individuals to follow their growth and survival. Subsequent measurements of the central strips were made in 1973, 1978, 1983, and in 1990. The following discussion of the patterns of stand development utilized the time-series data from a 64 m^2 section (4 m X 16 m) in the southeastern end of the central strips. This section comprised one arm and the cross-section of the cross-shaped central strips. The data admittedly came from a very small stand; however, stems have been relatively small during 36 years of stand development after the catastrophic windthrow (mean height of survivors was only 3.77 m in 1990).

The results probably represent one possible outcome of forest succession. The data were used exclusively for the discussion of possible processes of succession in which use of a small plot may be warranted; however, data from larger plots scattered at various locations are necessary for assessing general patterns of stand development in the area following the typhoon-caused catastrophic windthrow and statistically testing hypothesized models.

Stem maps

Stem maps for the 64 m^2 plot in consecutive measurement times are shown in Figure 2 by tree size (height) and species. A distribution map of residual stems and branches from salvage operations as well as other debris that existed in 1957 is also shown. There was a combined total of 280 listed stems of all species in 1957. The number increased to 320 in 1970 because small (< 10 cm in height) plants were not counted in 1957, but grew up to be measured by 1970. Stem height data from 1970 suggested that small tree seedlings (< 10 cm tall) were nearly absent then, although their presence at that time cannot be excluded entirely.

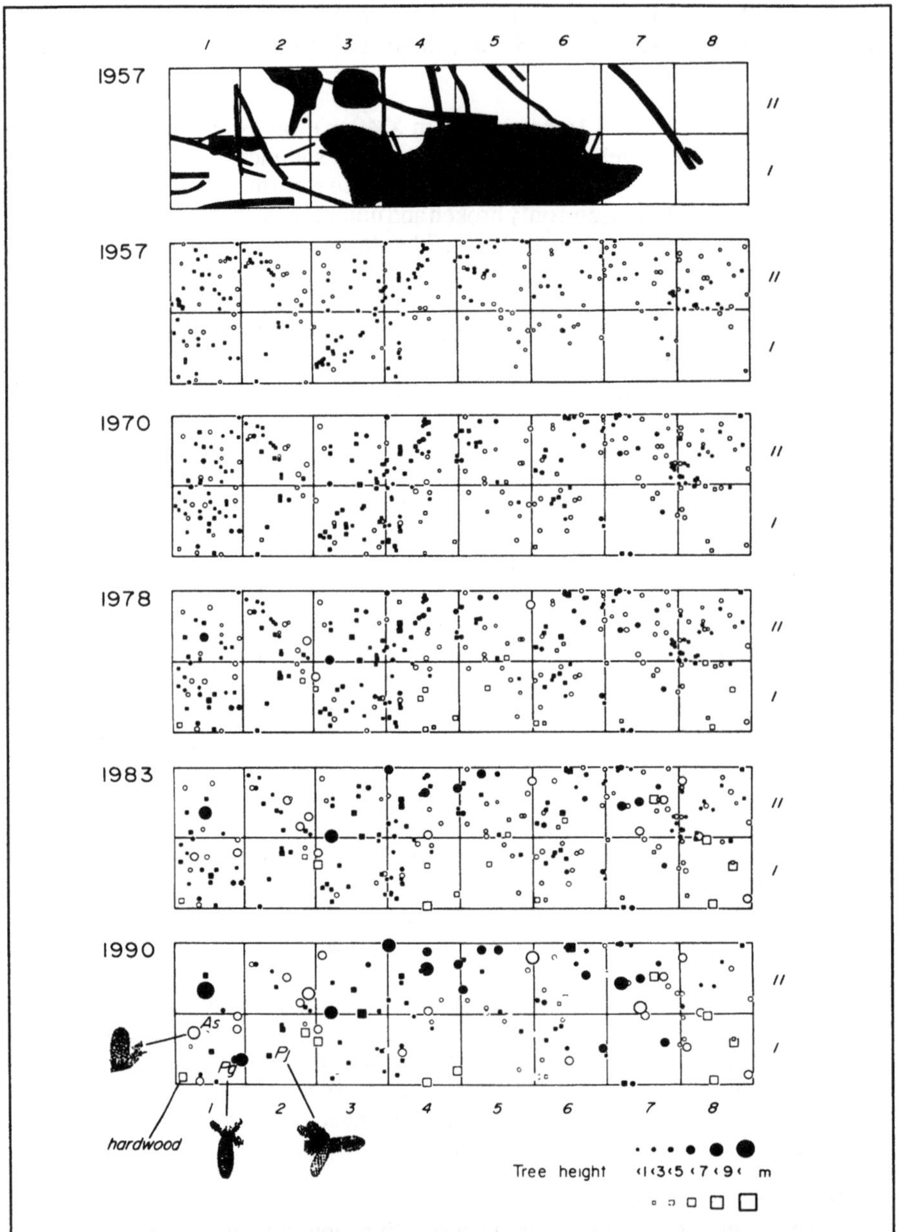

Fig. 2. Stem maps of the study plot after measurements in 1957, 1970, 1978, 1983, and 1990. Top drawing indicates distribution of stems (thick bars), branches and other debris (netted pattern), and soil disturbance (wave pattern) observed in 1957. Stem number increased somewhat in 1970 because the smallest individuals (< 10 cm tall) were not measured in 1957. Letters (I, II) and numerals (1, 2,..., 8) at the margin define 2 m x 2 m subplots.

Abies sachalinensis and *Picea jezoensis* predominated the early stages of stand development. Forty percent of the total stem number was *A. sachalinensis* in 1957, and 41% was *P. jezoensis*. *Picea glehnii* comprised only 18% of the total stem number. *Picea glehnii* gained in relative abundance by 1990 as a result of higher mortality in *P. jezoensis* and *A. sachalinensis*. Twenty-six percent of live stems were *P. glehnii* in 1990. The ratios were 37% and 26% for *A. sachalinensis* and *P. jezoensis*, respectively, in the same year. Comparison of these figures to those of 1957 indicates that the percentage changes of relative abundance in those conifer species were +8.3%, -14.6%, and -3.1% for *P. glehnii*, *P. jezoensis*, and *A. sachalinensis*, respectively. Also, some hardwood trees became established at locations where trees had been relatively sparse.

Survivorship curves

Survivorship of individuals depended on both species and initial height. Figure 3 indicates survivorship curves for the cohorts (definitions Oliver and Larson 1990) of three conifer species present in 1957. When stems of all sizes are combined, *P. glehnii* survived the best through 1990 (mean survival rate $l_x = 0.469$; 95% confidence intervals [0.494, 0.444]). Survivorship of *A. sachalinensis* ($l_x = 0.379$ [0.380, 0.378]) was significantly less than that of *P. glehnii*, but much greater than for *P. jezoensis* ($l_x = 0.218$ [0.243, 0.193]).

Trends of stem survivorship are similar for trees that were relatively large (> 40 cm tall) in 1957 (Fig. 3b). Absolute values of the survival rates were greater in originally large stems (Fig. 3b) than the combined data for all stems (Fig. 3a) in any species. In particular, very few *Picea glehnii* individuals died during the 36 years of stand development if they were initially greater than 40 cm tall. Survivorship patterns for initially small (< 40 cm tall) stems are also similar to those of all stem sizes combined (Fig. 3c); however, the 95% C.I. of the survival rates indicate that the rates were not different between *A. sachalinensis* and *P. jezoensis*, both of which were lower than survival rates of *P. glehnii*.

Only a small portion of the stems of any species died before about 1978, after which major mortality developed. It is of particular importance to assess survivorship of large individuals of each species during the period when intense competition and mortality occur, since this will control the shifts in dominance over a relatively long period of time. Survival rates of stems (> 2 m tall) between 1978 (when extensive mortality started) and 1990 are compared between the species. *Picea glehnii* (l_x = 1.0) survived significantly better than both *A. sachalinensis* ($l_x = 0.76$; $P < 0.001$, t-test) and *P. jezoensis* ($l_x = 0.84$; $0.001 < P < 0.005$, t-test). This difference in survivorship is one reason why the stand is shifting toward dominance by *P. glehnii* (Fig. 2).

Basal area growth

Development patterns of basal area (sum of stem cross-sectional area measured at 1.3 m above ground) since the typhoon are shown in Fig. 4 for the three conifer species and hardwoods. *Abies sachalinensis* has comprised the greatest basal area and has been growing steadily; however, its growth has slowed since 1983. In contrast, the basal area of *Picea glehnii* has been increasing rapidly, especially since 1973. As of 1990, the basal areas of *A. sachalinensis* and *P. glehnii* are 1190 cm^2 and 1103 cm^2, respectively, for the 64 m^2 plot. These trends indicate that *P. glehnii* may surpass *A. sachalinensis* within the next several years.

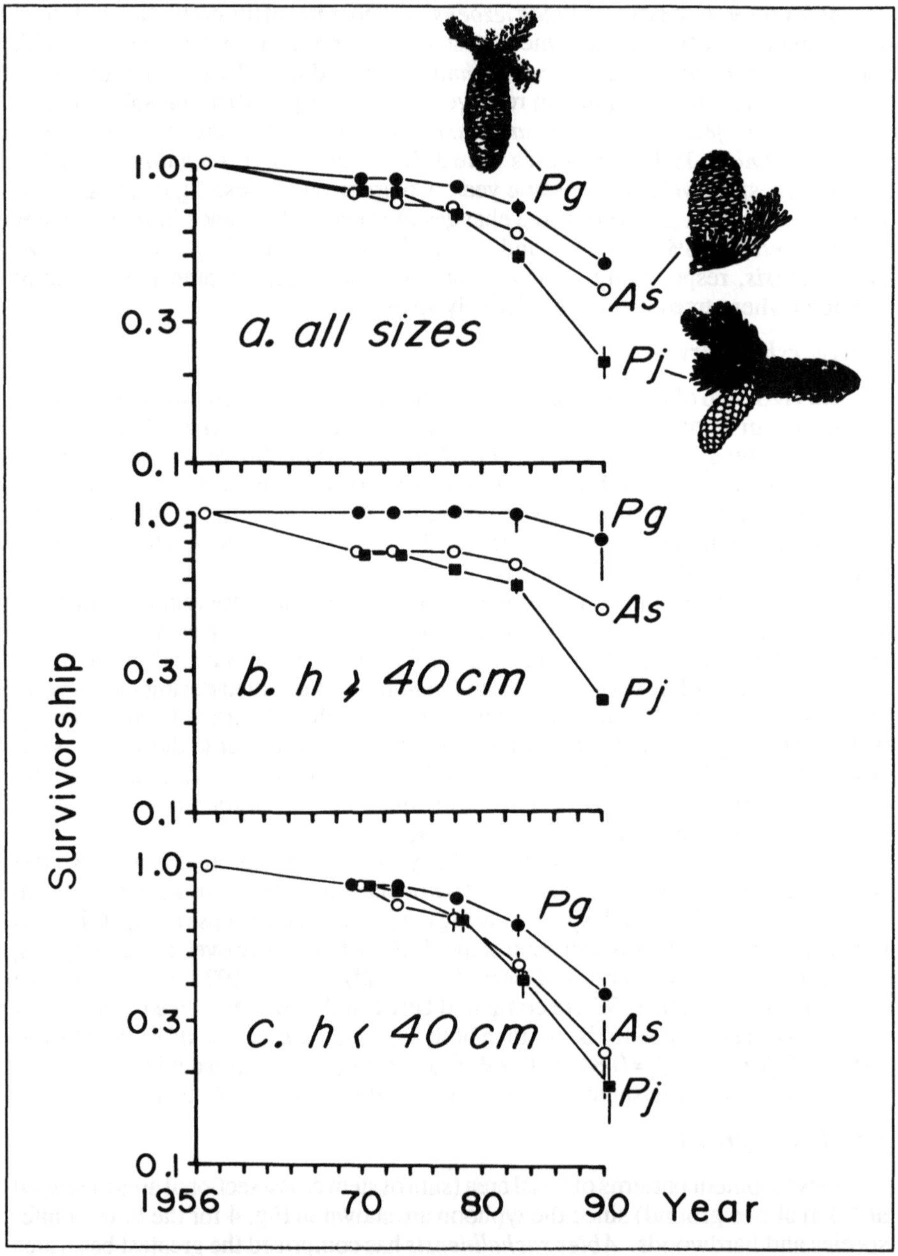

Fig. 3. Survivorship curves of the cohorts that were present in 1957; *Picea glehnii*, *Abies sachalinensis*, and *P. jezoensis*. Curves show a) individuals of all sizes combined, b) those taller than or equal to 40 cm in 1957, and c) those shorter than 40 cm in 1957. Vertical bars indicate 95% confidence intervals; some bars are too short to be visible on the figures.

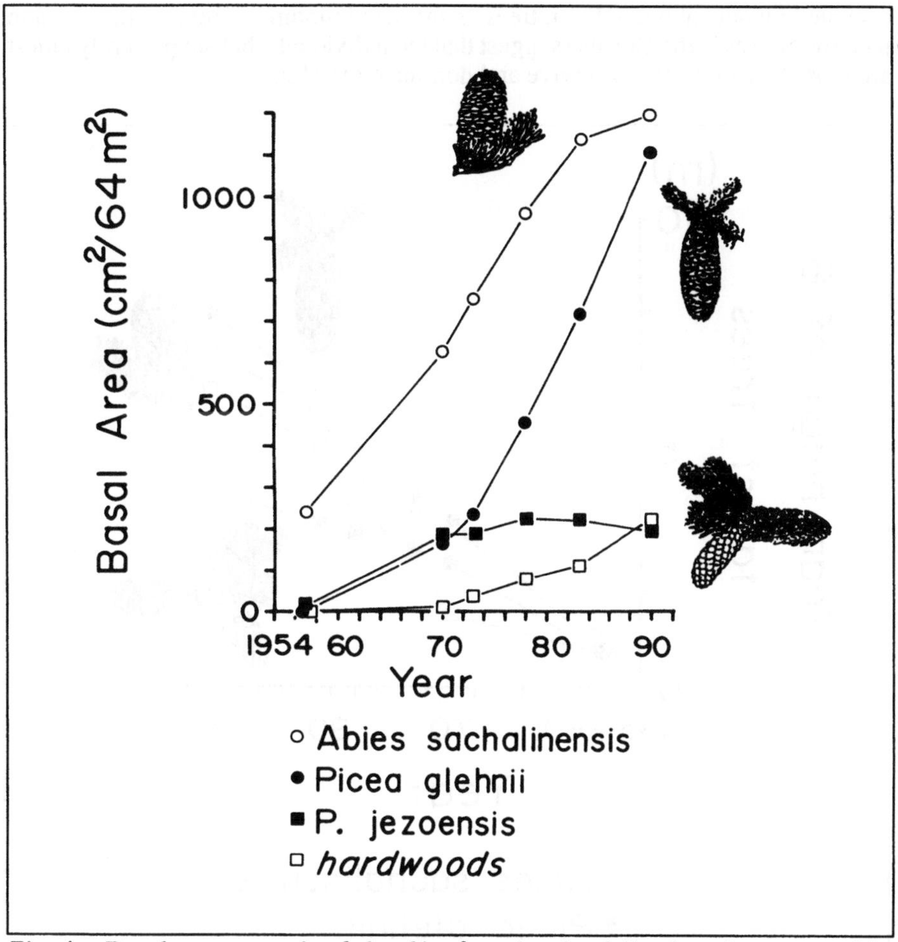

Fig. 4. Basal area growth of the 64 m² study plot following the catastrophic windthrow of 1954.

The basal area of *P. jezoensis* developed in a similar pattern to that of *P. glehnii* until the early 1970s; however, *P. jezoensis* has been in a level or declining trend since that time. In terms of basal area, hardwood species were nearly absent until about 1970, but have been gradually increasing in importance,partly because of delayed recruitment and relatively fast growth of new hardwood individuals in relatively open spots in the plot (Fig. 2).

Height growth

The increasing importance of *Picea glehnii* over time is also evident in the patterns of height growth, expressed by the mean height of the six tallest individuals within each species (Fig. 5). Those six trees were identified in 1990 data, and mean heights at previous years were obtained from older height data of those individuals. Special attention to the largest individuals is important because the 64 m² plot is likely to be occupied by only about a dozen trees as the stand matures. Recent survivorship patterns of trees in the plot also suggest that the individuals that are presently tallest are those that are likely to survive and dominate the plot.

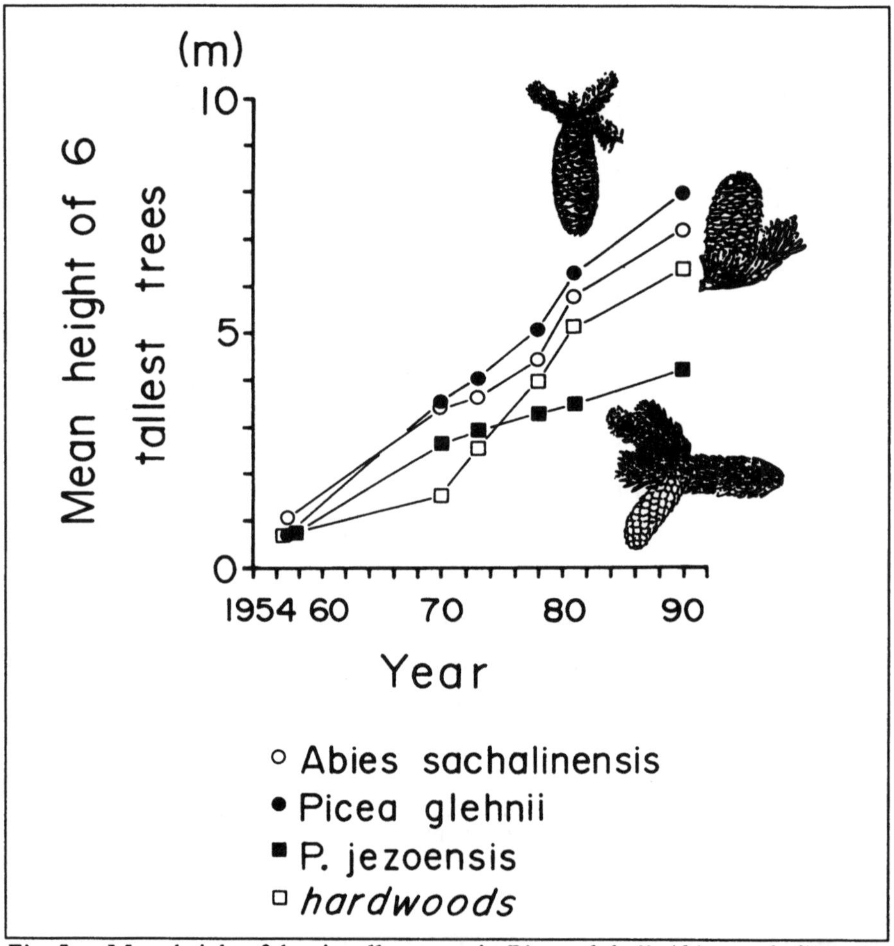

Fig. 5. Mean height of the six tallest trees in *Picea glehnii*, *Abies sachalinensis*, *Picea jezoensis*, and hardwood species.

In 1957, *A. sachalinensis* trees were the tallest, with a mean height of 106 cm. Other species were all approximately 70 cm tall with mean heights ranging between 68 cm and 73 cm. *P. glehnii* became taller than *A. sachalinensis* by 1970, and that trend has remained until 1990. In 1990, mean height of the tallest *P. glehnii* trees was 789 cm, whereas that of *A. sachalinensis* was 714 cm. *P. jezoensis* lagged behind other conifers in height at 419 cm; it was only 53% of *P. glehnii*. Hardwood species were much shorter even than *P. jezoensis* until 1973; since then, however, they have started to gain height as fast as *A. sachalinensis*. The mean height of the tallest six hardwood individuals reached 630 cm in 1990. Delay in hardwood's height growth is likely to have been caused by the delayed recruitment of these trees (Fig. 3).

Height distribution

Figure 6 shows the height distribution patterns and their change over time for the three conifer species. The rapid height growth of the tallest *P. glehnii*, although few in number, is notable. In contrast, the height growth of the tallest *P. jezoensis* lagged behind both *A. sachalinensis* and *P. glehnii*. By 1983, *P. glehnii* developed a bimodality in its height distribution, which is likely to have been created by the difference between *P. glehnii* individuals that had managed to attain the upper canopy level (approximately 2 m in height) by the beginning of the period of tree mortality that commenced about 1978, and those that had not. All stems of *P. glehnii* that were taller than 2 m in 1978 survived until 1990. On the other hand, the survival rate of shorter *P. glehnii* (< 2 m in 1978) was only 0.603 ± 0.017 (mean ± SD). The difference in survivorship between the two is highly significant ($P < 0.001$, t-test).

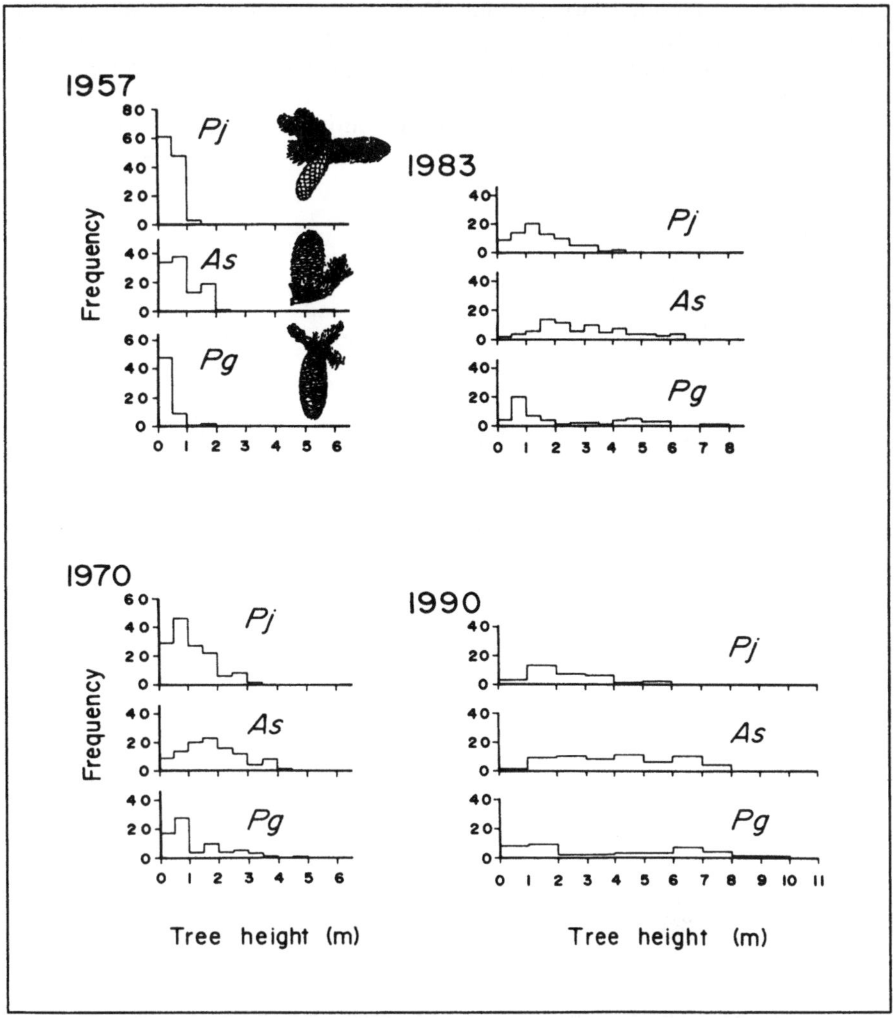

Fig. 6. Height distribution patterns of individual stems of *Picea jezoensis*, *Abies sachalinensis*, and *P. glehnii*, and their change over time.

Canopy dynamics

Development patterns of the vertical canopy structure in subsections of the 64 m^2 plot are depicted in Fig. 7. Three subplots 1, 4, and 7 are shown as the vertical profiles (viewed from the southeastern side (Fig. 2) at consecutive observations: 1957, 1970, 1978, 1983, and 1990. Different symbols with bars represent various species and heights of individuals.

In 1990, the upper canopy was dominated by *P. glehnii* and *A. sachalinensis*; however, there were substantial numbers of dominant *P. jezoensis* in 1957. The tallest individuals were either *A. sachalinensis* or *P. jezoensis* in 1957, and *P. glehnii* was inconspicuous in these subplots.

The individuals of *A. sachalinensis* that form the upper canopy layer in 1990 were already the tallest trees in 1957. In contrast, two general ways are apparent for *P. glehnii* stems to reach the top stratum. One is the continuing occupation of the tallest position in the immediate neighborhood throughout stand development similar to *A. sachalinensis*. Stems labeled as A, B, and C in Fig. 7 developed this way. There were also individuals of *P. glehnii* that were shorter than some other neighboring trees in 1957, but became the local dominants by 1990. Trees D, E, F,

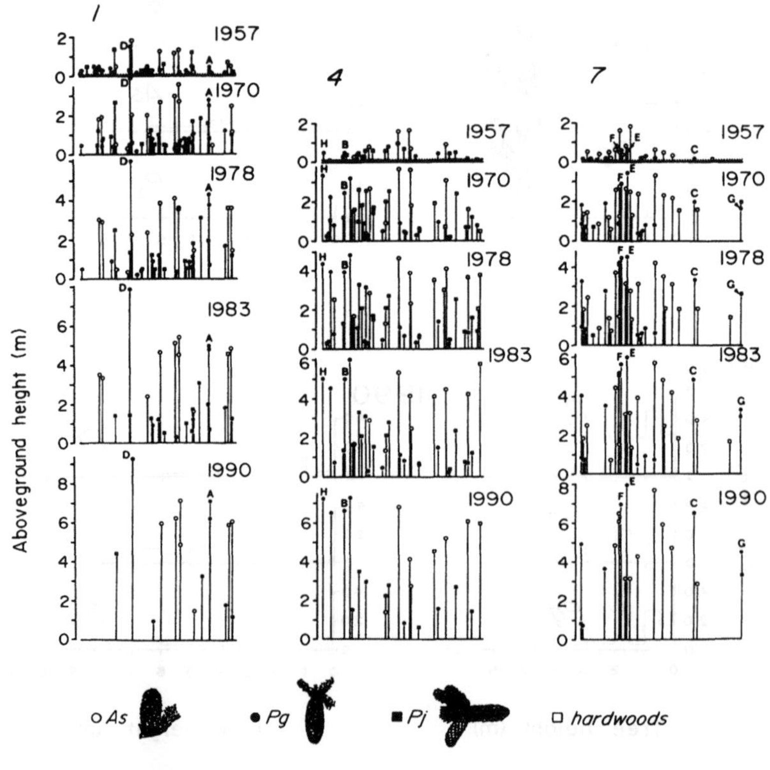

Fig. 7. Development patterns of vertical canopy structure in subplots 1, 4, and 7 of the 64 m^2 study area.

and G (Fig. 7) grew in this manner. Examination of changes in canopy profiles indicate that most individuals that became members of the upper canopy layer in 1990 were already the tallest in their vicinity in 1970. Most stems that comprised the upper layer in 1970 survived until 1990. On the other hand, many stems that were not at the tallest position in 1970 did not survive. Apparently, sorting of individuals into more dominant stems and suppressed ones had occurred during the initial 16 years of stand development after the catastrophic windthrow. The canopy profiles in 1970 and 1978 depict a stratification of individuals into two (or more) canopy layers (Fig. 7). Changes in canopy structure in subsequent years mainly consist of mortality of suppressed individuals.

Three models of the mechanisms of plant succession

Connell and Slatyer (1977) presented three models of the mechanisms of succession; their major points are summarized here. The first model was distinguished from the other two by the pattern in which new individuals are recruited after major disturbances. Model 1 assumed that only certain "early successional" species can become established at a newly opened site. Late successional species cannot be recruited unless the early species change the local environment and prepare the site for those that follow. This mechanism was advocated by Clements (1916) and has come to be identified as "relay floristics." Connel and Slatyer (1977) termed model 1 the "facilitation" model.

In contrast, model 2 and model 3 propose that any species that can survive there as adults, including those that appear later in a successional sequence, can become established when new growing space (Oliver and Larson 1990) is created for colonization, regardless of the presence of "early successional" species. Both early and late successional species are, therefore, present soon after the disturbance. Egler (1954) was the first to have proposed this process and designated it as "initial floristic composition."

The distinction between model 2 and model 3 was based on the ways early species are replaced by later ones. Model 2 assumed that the presence of early species does not affect recruitment and growth of later species. The later species are thought to withstand the cover of early species and to survive and grow to maturity. Replacement of the early species by the later species would occur by simply overgrowing the overtopping cover of the early species, as a result of contrasting life-history characteristics of the various species. This model was designated as the "tolerance" model (Connell and Slatyer 1977).

On the other hand, model 3 assumed that early species inhibit recruitment and subsequent growth of later species. The later species can grow to assume the dominant position only when the early ones are damaged or killed by physical or biological disturbances. This model was called the "inhibition" model (Connell and Slatyer 1977).

The tolerance model (#2) requires that the later species are more tolerant to shade (or to other environmental factors) than the early species; however, the inhibition model (#3) does not require this greater tolerance. According to the inhibition model, the species that follow could be the same early successional species or those that tend to dominate the site at later times after a disturbance. The observed sequence of species is then interpreted as the expected (in the statistical sense) sequence of species over time produced by differences in longevity. Even though any combination of species replacements could occur, the communities would tend to move toward those dominated by species of longer life-spans.

Evidence for these three models of succession was reviewed by Connell and Slatyer (1977). Examples of the facilitation model were found (although one needs to be cautious in making analogies among different plant and/or animal communities) in heterotrophic successions on animal carcasses, logs, dung, and litter (Savely 1939, Payne 1965), and in primary succession (Crocker and Major 1955, Lawrence et al. 1967, Cowles 1899, Olson 1958) after glacial retreat and on sand dunes. Phenomena consistent with the inhibition model of succession were observed in terrestrial plant communities (Keever 1950, Parenti and Rice 1969, Niering and Egler 1955, Niering and Goodwin 1974, Webb et al. 1972, Booth 1941) and in marine communities (O'Neill and Wilcox 1971, Sutherland 1974). In contrast, Connell and Slatyer (1977) did not find any evidence for the tolerance model of succession in the literature.

Field experiments were conducted in the context of Connell and Slater's (1977) hypotheses in intertidal marine communities (Lubchenco and Menge 1978, Sousa 1979, Dayton et al. 1984) and in terrestrial communities (Keever 1979, Hils and Vankat 1982, Armesto and Pickett 1985); however, experiments conducted inforests are almost absent. This lack of rigorous tests in forests may be partly because of the long life-span of tree species (Connell and Slatyer 1977). It is also possible that Connell and Slatyer's (1977) hypotheses have overlooked other possible mechanisms (Pickett et al. 1987).

A parallel model

There seems to be a fourth model of succession that describes the observed developmental processes of at least two distinctly different forest communities for which detailed growth data are available: one in hardwood stands of central New England, and the other in conifer forests of northern Japan.

Model 4 assumes, as in models 2 and 3, the "initial floristic composition" for species' recruitment. The new model possesses some characteristics of both model 2 and model 3, but differs in the mechanism of replacement of dominant species. Models 1, 2, and 3 all assume that early successional species are replaced by the other, later ones growing up from below; the subsequent dominants are thought to have occupied the subordinate position in the community, the lower canopy strata in the case of forests, prior to emergence to the top canopy layer. There is, however, sufficient evidence that still another model exists. It is possible that both "early" and "late" successional species grow together, sharing the same crown position (or similar size in the case of sessile animals) for extended periods, without either physically dominating or interfering substantially. The "early" species are labeled as such by the observers, simply because they are far more abundant than the "late" species. Individuals of late successional species are growing as well as, or at least not very much worse than, the early species during this period. Replacement occurs simply because the "late" species continue to grow in size, whereas the "early" species decline in growth rate at a certain stage of community development. In forests, this process results in canopy stratification in which "late" successional species overtop the "early" ones; therefore, the species in lower canopy strata are those submerged to that position from the top layer (Oliver 1978, 1981). In contrast, model 2 (the tolerance model) would interpret that the species in the lower strata are those that would eventually emerge to the top layer.

The tolerance model (# 2) required that the "late" successional species are more tolerant of shade (or of other factors), but the suggested "parallel" model (# 4), as in model 3, does not assume any difference in tolerance between early and late species.

Under model 4, shade intolerant species can become dominant after a more tolerant species has dominated. Submerged species from the top canopy layer can survive and form a lower stratum if they are shade tolerant; if not, they are killed over time.

Figure 8 depicts the difference between model 2 and model 4. Stage A represents a condition shortly after a major disturbance where early as well as late successional species are present (initial floristic composition). This condition is equivalent to the "stand initiation stage" of Oliver (1981) and is common to both models. As the community develops, individuals in the stand interfere with one another. Stage B indicates the beginning of this interacting phase corresponding to the "stem exclusion" stage (Oliver 1981). Once established after recruitment, most individuals are expected to survive to the beginning of stage B, but after that point, competition will both exclude further establishment of regeneration and, in due time, cause self-thinning among the already established trees.

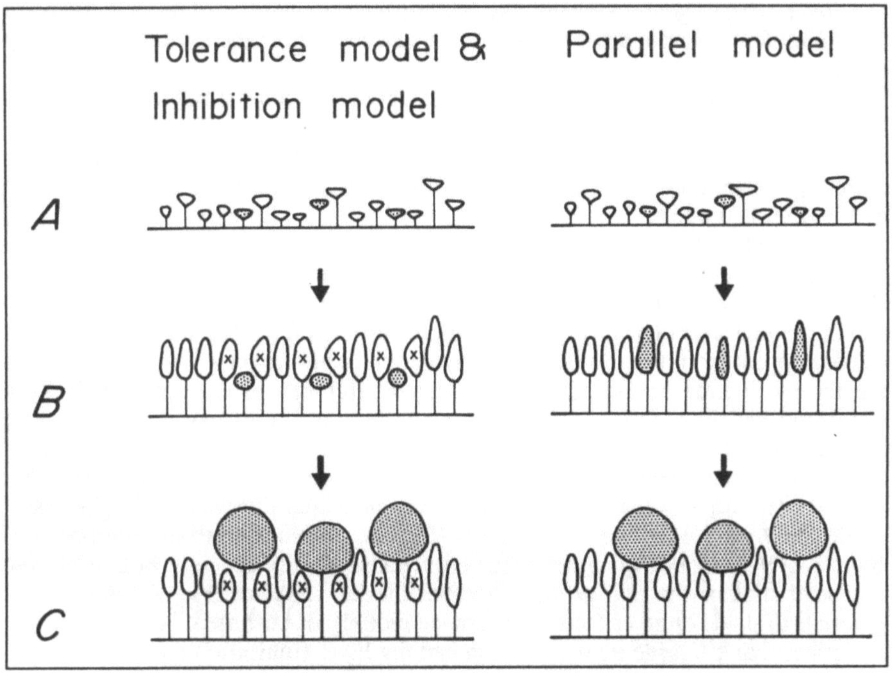

Fig. 8. Schematic representation of tolerance, inhibition, and parallel models of plant succession, indicating stand development over time: A is the initial condition immediately after a major disturbance (e.g. catastrophic windthrow), B is the beginning of intensified stem mortality, and C is the stage when the early successional species (plain crown) begin to be overtaken by another species (shaded crown). In the tolerance model, another species (shaded) will gradually outgrow the initially overtopping species (marked by "x"). In the inhibition model, another species will emerge over the initially overtopping individuals only after they are killed or damaged. The parallel model does not assume a difference in stature between early and late successional species.

Model 4 and model 2 differ in the position of "late" species during stage B. In model 4, late successional species occupy the same crown stratum as the dominant early species. Over time, the late species outgrow the early species in height (stage C). In contrast, model 2 assumes that the late species are relegated to the lower canopy strata during stage B, and an ability to endure shade is necessary for the late species to survive and gradually outgrow the early species (stage C).

Under model 4, the late successional species tend to maintain high survival rates until they attain dominance because of the lack of (or very weak) interference by the early species. Survival rates of late species are also expected to be quite high under model 2 because of their greater tolerance.

Patterns of stand development under model 3 (inhibition model) are similar to those indicated by the schematic representation of model 2 (Figure 8). The early successional species would suppress the late species during stage B. In this model, however, transition to stage C could occur only after the early species are damaged or killed by some agents. Unlike models 2 and 4, survival rates of late species prior to their attaining dominance will be quite small because of the inhibitive effect by the overtopping early successional species.

Model 4 may be referred to as a "parallel" model of succession since late species are at the same canopy position as the early successional species – "in parallel" – during early stand development. It must be emphasized that the process hypothesized in the parallel model is critically different from those in both the tolerance and inhibition models.

In mixed hardwood forests of central New England, small numbers of *Quercus* trees often occupy the continuous top canopy layer above the subcanopy layers of *Betula* and *Acer* species (Fig. 9). These stands are typically 30 years old or older and originated after clearcutting or similar disturbances. Instead of being all-aged as they appear from the stratified canopy structure and reverse J-shaped stem diameter distribution, the stands are even-aged. *Betula* and *Acer* are much more numerous, and thus "dominant", until a stand age of about 30 years. *Quercus* trees are relatively inconspicuous during this period; however, Oliver (1978) revealed that *Quercus* trees were growing as fast as other associated, more numerous species and began to surpass others in height after 30 years. *Quercus* was not "suppressed" while *Betula* and *Acer* were "dominant" (Fig. 9). Succession from dominance of *Betula-Acer* to that of *Quercus* occurred because of difference in height growth characteristics of the two groups (parallel model), not because of tolerance of *Quercus* to shade or other factors (tolerance model), or because of small scale disturbances that kill large trees in the top canopy layer (inhibition model).

Parallel development of the Abies-Picea stand

Data concerning height growth, basal area growth, and survivorship in the mixed *Abies-Picea* stand in Hokkaido (Figs. 4, 5, 7) also give evidence for the parallel model of stand development. Specifically, *Picea glehnii* moves from a relatively inconspicuous position in terms of abundance within the dominant canopy layer to one of increasing importance. This does not appear to be a result of the mechanisms described by either the tolerance or inhibition models.

The tolerance model would specify that small *P. glehnii* would be able to withstand the suppression of the overstory *Abies sachalinensis*, and gradually grow past them to assume dominance. This does not appear to be happening. Survivorship of *P. glehnii* was affected significantly by the presence of the overtopping canopy,

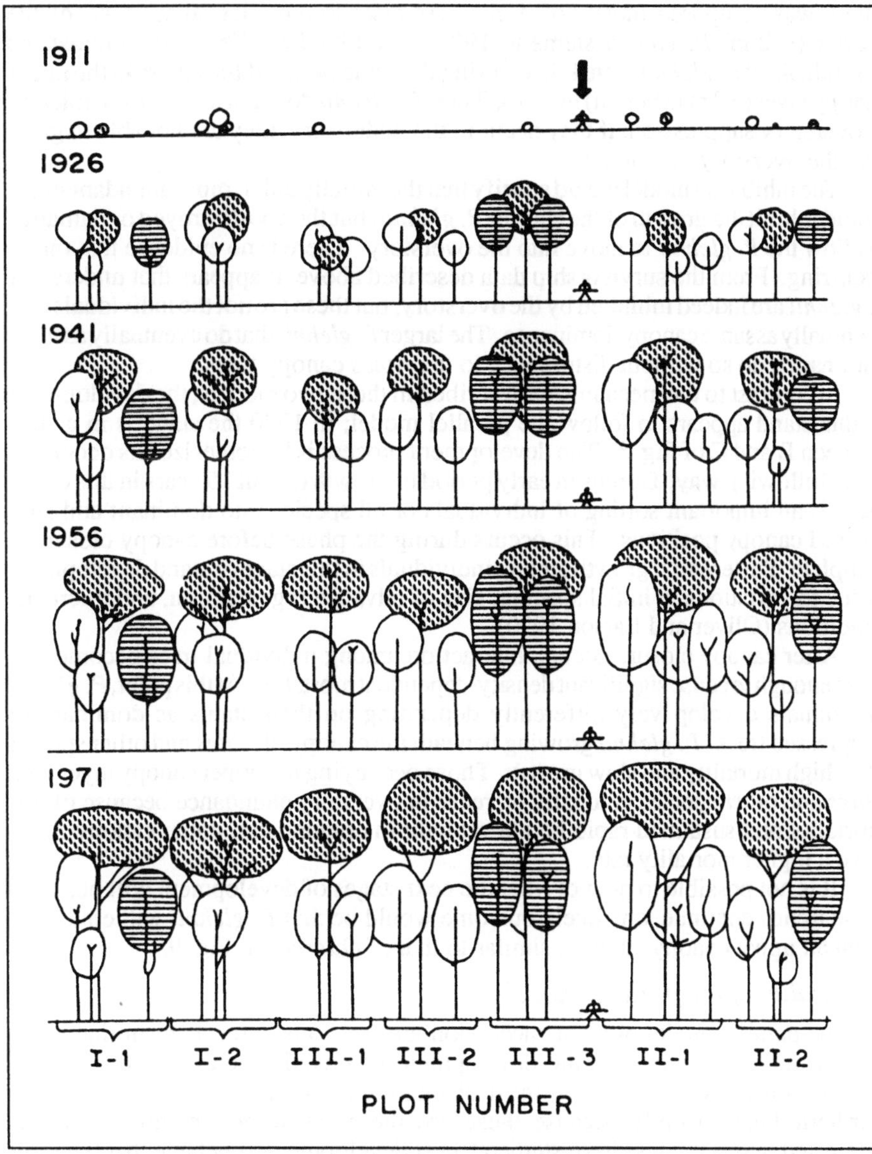

Fig. 9. Growth of even-aged composite stand consisting of seven B-stratum red oaks (dotted crowns) and tallest red maple (plain crowns) and/or black birch (horizontally striped crowns) and other red maples and black birches with stems beneath vertical projection of crown of B-stratum red oak. Person (see arrow) is six feet (1.8 m) tall and holds a stick ten feet (3.1 m) long). Reprinted from Oliver (1978) ©Yale University, School of Forestry and Environmental Studies.

which was composed mostly of *Abies*. As discussed earlier, about 40% of the shorter (< 2 m) *P. glehnii* stems in 1970 were killed by 1990. In contrast, no individuals of *P. glehnii* were killed during the same period if they were in the upper canopy layer (> 2 m in height) in 1970. Thus, *P. glehnii* does not appear to be tolerant of overstory suppression; if they occur in the understory, they are not able to grow past the overstory individuals.

The inhibition model would specify that the initially taller, more abundant *Abies* would inhibit the growth of the shorter *P. glehnii*, but then be destroyed or damaged to allow the *P. glehnii* to move into the overstory. There is no evidence that this is occurring. From the survivorship data described above, it appears that understory *P. glehnii* are indeed inhibited by the overstory, but these are not the individuals that eventually assume canopy dominance. The larger *P. glehnii* that do eventually assume dominance do so without disturbance to the *Abies* canopy trees.

In contrast to the mechanisms described in those two models, the development of this stand appears to follow the parallel model; in 1990 the stand is in a state between B and C of Fig. 8. The development process is hypothesized as occurring in the following way. During an early period of growth (about 15 years in this case), there is an important sorting of individuals of all species into dominant and suppressed canopy positions. This occurs during the phase before canopy closure is complete. Differential growth among individuals in this stage of stand development is due to variations in initial spacing, size of advance regeneration, and microsite conditions (Oliver and Larson 1990).

After canopy closure occurs, interaction among individual trees becomes intense enough to cause significant density-dependent mortality. At this point, *P. glehnii* individuals develop very differently depending on their status as dominant or suppressed trees. *P. glehnii* growing beneath the canopy of *Abies* and other species have high mortality and slow growth. Those occupying the upper canopy layer with *Abies* and other species gradually increase in overstory abundance because of low mortality and sustained rapid height growth; in comparison, upper canopy *Abies* have a higher mortality rate.

It is not possible to predict what the next stages of development will be, but if these trends continue, one likely outcome would be that *P. glehnii* will emerge to form an upper canopy stratum, similar to that of *Quercus* in Fig. 9.

Generality of the parallel model

The parallel model of plant succession appears to be a mechanism that is observed rather commonly in the development of mixed-species forests. Several examples are discussed below; most of them are from either North America or northern Japan, simply because these are the areas where scientists who are interested in this process have studied forest development. I believe that the same can be found elsewhere.

Pinus strobus can grow in a similar way as *Quercus rubra* (Fig. 9; Oliver 1978) when it is mixed with hardwood species in the northeastern United States (Smith 1986). *Pinus strobus* trees begin to emerge above other species at a stand age of about 70 years because the hardwoods' height growth starts to decline (Fig. 10). The stand will eventually develop strong canopy stratification, with *P. strobus* occurring as emergents above the hardwood canopy layer. A similar pattern has been observed in mixed forests of *Pseudotsuga menziesii* (Mirb.) Franco and *Tsuga heterophylla* (Raf.) Sarg. in the Pacific Northwest of the United States (Wierman and Oliver 1979), where *Pseudotsuga* could overtop *Tsuga* in even-aged stands.

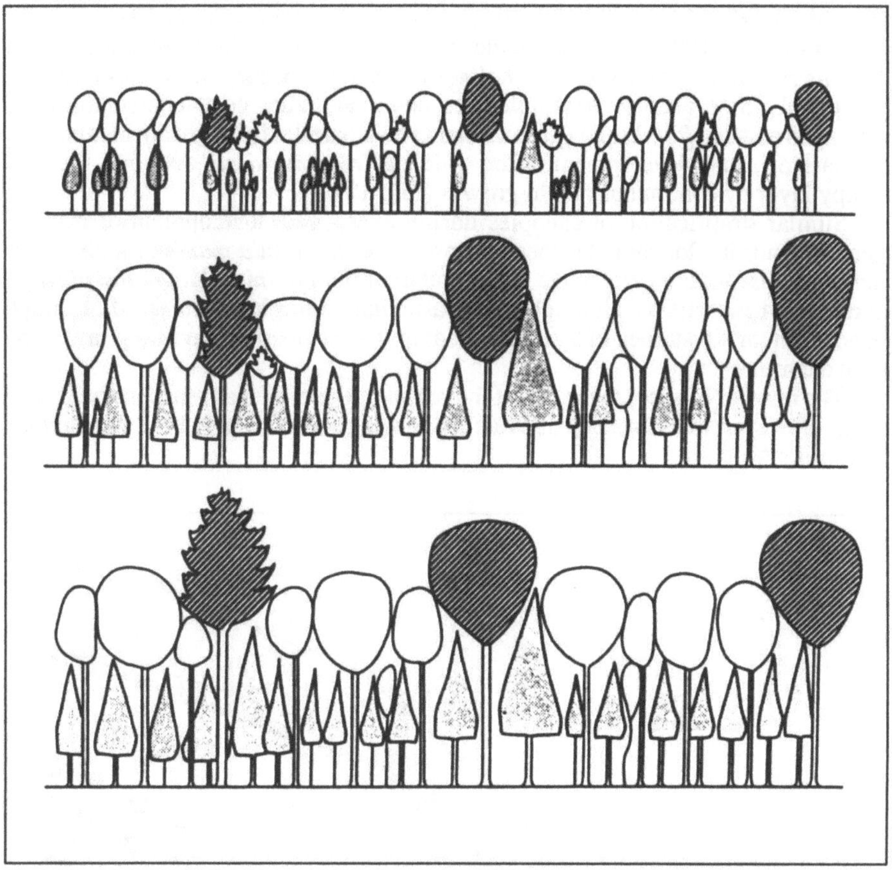

Fig. 10. Stages in the natural development of an untreated stratified mixture in an even-aged stand of the eastern hemlock-hardwood-white pine type. Upper sketch shows the stand at 40 years with the hemlock (gray crowns) in the lower stratum beneath an undifferentiated upper stratum. By the seventieth year (middle sketch) the emergents (hatched crowns) have ascended above the rest of the main canopy, except for the white pine which has only started to emerge. The lower sketch shows the stand as it would look after 120 years with the ultimate degree of stratification developed. Reprinted from Smith, D.M., *The Practice of Silviculture*. Copyright ©1986 by John Wiley & Sons. Used with permission.

Mixed-species stands of *Quercus falcata* var. *pagodifolia* Ell. and *Liquidambar styraciflua* L. (Clatterbuck and Hodges 1988) in Mississippi, U.S.A. and those of *Picea sitchensis* (Bong.) Carr. and *Tsuga heterophylla* in Alaska, U.S.A. (Deal et al. 1991) have also been shown to exhibit development patterns of canopy stratification that resemble the patterns of *Quercus-Betula-Acer* forests of New England.

Possible examples of the parallel development of dominant tree species in Hokkaido have also been observed (in addition to the *Picea-Abies* stand discussed at length in the present paper). Kushima (1989) studied a mixed hardwood-conifer forest in central Hokkaido that developed after a severe forest fire of 1911. Stands were stratified in 1984 into three distinct canopy layers with *Betula maximowiczii*, *Quercus mongolica*, and *Abies sachalinensis* at top, middle, and bottom layers, respectively. Stem analyses of sample trees, however, indicated that the stands were even-aged and that *Betula* and *Quercus* had grown parallel in height during the first 20 years of stand development, after which *B. maximowiczii* overtook the top canopy layer by continuous height growth (Fig. 11).

Similar stratification of canopies during even-aged development of mixed-species stands in Hokkaido has been reported for the *Betula maximowiczii-Abies sachalinensis* mixture (Watanabe 1985) and the *Tilia japonica-Magnolia obovata* mixture (Ishizuka 1981), although these particular examples were based on small plots. Systematic studies of even-aged canopy stratification are likely to reveal more examples.

The number of cases that suggest the parallel model of plant succession is still small; however, this is sufficient evidence to suggest that it may be a mechanism of significance.

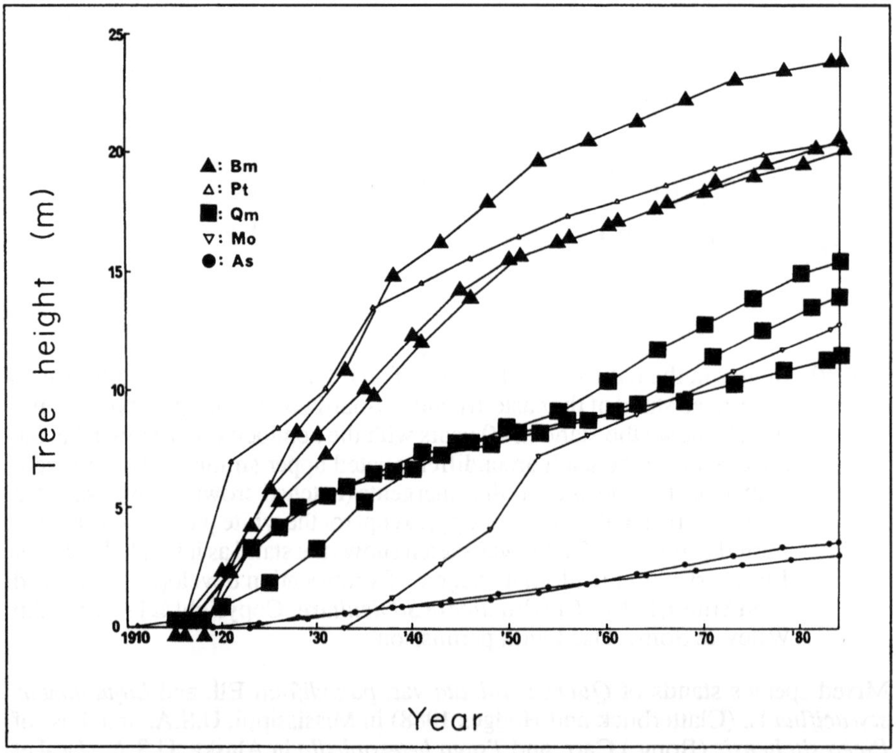

Fig. 11. Post-fire development of tree heights in three stands of mixed hardwood-conifer forest in central Hokkaido (from Kushima 1989).

Conclusions

Development of canopy stratification and stem-size variation in even-aged stands of mixed-species forests has been explained by simultaneous establishment of various species following major disturbances, and subsequent variation in their height growth patterns (Oliver 1978, 1981, Wierman and Oliver 1979, Kelty 1984, Smith 1986, Clatterbuck and Hodges 1988). When the processes of canopy stratification are re-examined in relation to various models of succession, however, it becomes apparent that several different mechanisms could produce even-aged canopy stratification, as described in the second half of this paper. In addition to the models previously described by Connell and Slatyer (1977), a mechanism that had not been previously recognized is likely to be creating patterns of succession in *Quercus-Betula-Acer* stands of New England, *Picea-Abies* forest of Hokkaido, and possibly many others. This new mechanism is formally hypothesized as the parallel model of plant succession in this paper.

We have tendencies to fit observations into pre-existing models when being faced with the necessity to understand complex phenomena. For a time, all forest succession patterns were forced into the Clementsian model stressing a facilitative mechanism, until Egler (1954), Connell and Slatyer (1977) and others developed the ideas contained in the tolerance and inhibition models. However, this can subsequently lead to the problem of forcing all observations of succession into one of these three models. I propose the parallel model as a fourth possibility, which, of course, should not preclude the possibility of there being additional ones not yet described.

One problem that may sometimes complicate the accurate description of successional patterns is the fact that terminology is sometimes poorly defined and has a tendency to evolve by itself. The term "dominant" is particularly problematic, because it is used to mean two different things. When we think of succession, it is "the changes observed in an ecological community following a perturbation that opens up a relatively large space" (Connell and Slatyer 1977). That is, it is the changes in the "dominant" species that comprise a community. If one species is considered "dominant" at one stage of succession, the others are, by definition, "subordinate." This use of these terms generally refers to the relative abundance of each species at each stage of succession. However, in the development of forest stands, these terms are used to describe the competitive relationship (usually judged by heights) between two individuals or species, as in the terms "dominant canopy stratum" and "subordinate canopy stratum." In some successional studies, these two uses of the term have been erroneously equated; discussions of tolerance to, and inhibition by, the "dominant" canopy layer followed naturally, culminating in the proposal of tolerance and inhibition models of succession.

These may well be the operating mechanisms of succession at times, but they may not at other times. In the parallel model, the "subordinate" species does not have to occupy the "subordinate canopy stratum." It can grow with the "dominant" species in the same canopy layer in parallel. It is subordinate to the dominant species in number, but not in stature. Assuming such species are growing in the "subordinate stratum" is likely to have been caused by lack of data, and to some extent, observer's bias in the use of the term "dominant." Where data exist on relative heights and crown structure over time (from either permanent plots or stem reconstruction studies), one can see that such an assumption does not always hold.

I conclude that the parallel model of succession describes a real possibility, and needs to be tested with other models that explain mechanisms of plant succession.

Acknowledgments

I am indebted to a number of people who made it possible for me to write this paper. Much of the present data came from continued efforts by the scientists who worked at Government Forest Experiment Station and at Forestry and Forest Products Research Institute in Sapporo, Japan. They established and maintained the permanent plot for over 30 years, and include M. Nakano, H. Toyooka, K. Hayashi, A. Sato, M. Ishizuka, and S. Sugawara. I also thank Y. Kanazawa, S. Ohba, and H. Kushima for assistance in the sampling of 1990. Contribution by A. Sato and M. Ishizuka are especially acknowledged, because of their samplings of 1978 and 1983, and detailed summary of data gathered previously. The present paper was based only on data from a small portion of the permanent plot, from which the stand development patterns were inferred. More detailed patterns and variations of stand development of the current study area will be reported elsewhere. A part of the present paper was presented also at the Annual Lectures of Hokkaido Research Center, Forestry and Forest Products Research Institute in Sapporo in March 1991.

The manuscript was improved a great deal by critical reading by M.J. Kelty, C.D. Oliver and C.D. Canham. Special thanks are due to D. M. Smith, C. D. Oliver and R.S. Seymour for useful discussions and ideas on stand development.

Literature cited

Adams, W.R., R.I. Ashman, H.I. Baldwin, E.S. Bryant, A.C. Cline, G.W.I. Creighton, H.J. Lutz, L.C. Swain, and M. Westveld. 1941. Report of committee on silviculture, New England Section of American Foresters: a silvicultural policy for the hurricane area. J. For. 39:376-378.

Armesto, J.J., and S.T.A. Pickett. 1985. Experiments on disturbance in oldfield plant commenities: Impact on species richness and abundance. Ecology 66:230-240.

Booth, W.E. 1941. Revegetation of abandoned fields in Kansas and Oklahoma. Amer. J. Bot. 28:415-422.

Clapp, R.T. 1938. The effects of the hurricane upon New England forests. J. For. 36:1177-1181.

Clatterbuck, W.K., and J.D. Hodges. 1988. Development of cherry-bark oak and sweet gum in mixed, even-aged bottomland stands in central Mississippi, U.S.A. Can. J. For. Res. 18:12-18.

Clements, F.E. 1916. Plant succession. An analysis of the development of vegetation. Carnegie Inst. Washington, Publ. 242. Washington, D.C.

Connell, J.H., and R.O. Slatyer. 1977. Mechanisms of succession in natural communities and their role in community stability and organization. Amer. Natur. 111:1119-1144.

Cowles, H.C. 1899. The ecological relations of the vegetation on the sand dunes of Lake Michigan. Bot. Gaz. 27:97-117, 167-202, 281-308, 361-391.

Crocker, R.L., and J. Major. 1955. Soil development in relation to vegetation and surface age at Glacier Bay, Alaska. J. Ecol. 43:427-448.

Deal, R.L., C.D. Oliver, and B.T. Bormann. 1991. Reconstruction of mixed hemlock/spruce stands in coastal southeastern Alaska. Can. J. For. Res. (in press).

Drury, W.H., and I.C.T. Nisbet. 1973. Succession. Journal of the Arnold Arboretum 54:331-368.

Egler, F.E. 1954. Vegetation science concept. I. Initial floristic composition. A factor in old-field vegetation development. Vegetatio 4:412-417.

Foster, D.R. 1988a. Species and stand response to catastrophic wind in central New England, U.S.A. J. Ecol. 76:135-151.

Foster, D.R. 1988b. Disturbance history, community organization and vegetation dynamics of the old-growth Pisgah Forest, Southwestern New Hampshire, U.S.A. J. Ecol. 76:105-134.

Foster, D.R., and E.R. Boose. Patterns of forest damage resulted from catastrophic wind in central New England, U.S.A. J. Ecol. (in press).

Gleason, H.A. 1917. The structure and development of the plant association. Bull. Torrey Club. 44:463-481.

Gleason, H.A. 1926. The individualistic concept of the plant association. Bull. Torrey Club. 53:7-26.

Gleason, H.A. 1927. Further views on the succession-concept. Ecology 8:299-326.

Henry, J.D., and J.M.A. Swan. 1974. Reconstructing forest history from live and dead plant material - an approach to the study of forest succession in southwest New Hampshire. Ecology 55:772-783.

Hils, M.H., and J.L. Vankat. 1982. Species removals from a first year oldfield plant community. Ecology 63:705-711.

Inoue, M. 1955. Prevention of bark beetle damage on wind-damaged trees and use of aircrafts. Hoppo Ringyo 7:9-12.*

Inoue, Y. 1959. Forest management (1). In: A report of the scientific investigations of the forests wind-damaged in 1954, Hokkaido, Japan. K. Matsukawa (ed.) Japan Forest Technical Association.*

Ishizuka, M. 1981. Development of mixed conifer-hardwood forests–analysis at Jozankei. In: Analysis of stand structure and regeneration in natural forests (2). Hokkaido Regional Forestry Office. 216p.*

Keever, C. 1950. Causes of succession on old fields of the Piedmont, North Carolina. Ecol. Monogr. 20:229-250.

Keever, C. 1979. Mechanisms of succession on old fields of Lancaster County, Pennsylvania. Bull. Torrey Bot. Club. 106:229-250.

Kelty, M.J. 1986. Development patterns in two hemlock-hardwood stands in southern New England. Can. J. For. Res. 16:885-891.

Kubo, T., and E. Sanada. 1975. Some physical properties of the volcanogeneous regosol originated from the coarse volcanic ejecta of Mt. Tarumae. Shinrin Ritchi 17(1):1-5.*

Kushima, H. 1989. Canopy stratification and its development of post-fire stands at Tokyo University Hokkaido Experimental Forest. In: Proccedings of the 36th annual meeting of Japanese Ecological Society, Kushiro.*

Lawrence, D.B., R.E. Schoenike, A. Quispel, and G. Bond. 1967. The role of *Dryas drummoidi* in vegetation development following ice recession at Glacier Bay, Alaska, with special reference to its nitrogen fixation by root nodules. J. Ecol. 55:793-813.

Lubchenco, J., and B.A. Menge. 1978. Community development and persistence in a low rocky intertidal zone. Ecol. Monogr. 59:67-94.

Matsukawa, K., S. Karasawa, and S. Matsubara. 1959. General. In: A report of the scientific investigations of the forests wind-damaged in 1954, Hokkaido, Japan. K. Katsukawea (ed.) Japan Forest Technical Association.*

Morodome, T. 1955. Typhoon No. 15 and forest damage. Hoppo Ringyo 7:29-31.*

Niering, W.A., and F.E. Egler. 1955. A shrub community of *Viburnum lentago*, stable for twenty-five years. Ecology 36:356-360.

Niering, W.A., and B.H. Goodwin. 1974. Creation of relatively stable shrublands with herbicides: arresting "succession" on right-of-way and pastureland. Ecology 55:784-795.

Oliver, C.D. 1978. The development of northern red oak in mixed stands in central New England. Yale University, School of Forestry and Environmental Studies, Bulletin No. 91.

Oliver, C.D. 1981. Forest development in North America following major disturbances. For. Ecol. Manage. 3:153-168.

Oliver, C.D., and B.C. Larson. 1990. Forest Stand Development, 1st ed. McGraw-Hill, New York. 467p.

Olson, J.S. 1958. Rates of succession and soil changes on southern Lake Michigan sand dunes. Bot. Gaz. 119:125-170.

O'Neill, T.B., and G.L. Wilcox. 1971. The formation of "primary films" on materials submerged in the sea at Port Hueneme, California. Pacific Sci. 25:1-12.

Osawa, A. 1989. Forests of North America – Comparison of New England and Hokkaido forests - Hoppo Ringyo 41:2-9.*

Parenti, R.I., and E.L. Rice. 1969. Inhibitional effects of *Digitaria sanguinalis* and possible role in old-field succession. Bull. Torrey Bot. Club 96:70-78.

Payne, J. 1965. A summer carrion study of the baby pig *Sus scrofa* Linnaeus. Ecology 46:592-602.

Pickett, S.T.A., S.L. Collins,and J.J. Armesto. 1987. A hierarchical consideration of causes and mechanisms of succession. Vegetatio 69: 109-114.

Sasaki, T. 1982. (ed.) Tephra of Hokkaido. Committee for naming Hokkaido tephra.*

Sato, A., M. Ishizuka, S. Sugawara, H. Toyooka, and K. Hayashi. 1985. Growth dynamics of advanced seedlings after damage to forest canopy by Typhoon No. 15 in 1954 – Analytical case study on Mt. Tarumae in Hokkaido. In: Proceedings of the 96th annual meeting of the Japanese Forestry Society, Sapporo.*

Savely, H.E. 1939. Ecological relations of certain animals in dead pine and oak logs. Ecol. Monogr. 9:321-385.

* in Japanese

Smith, D.M. 1946. Storm damage in New England forests. M.F. Thesis, Yale University.
Smith, D.M. 1986. The practice of silviculture, 8th ed. John Wiley and Sons, New York. 527p.
Sousa, W.P. 1979. Experimental investigations of disturbance and ecological succession in a rocky intertidal algal community. Ecol. Monogr. 49:227-254.
Spurr, S.H. 1956. Natural restocking of forests following the 1938 hurricane in central New England. Ecol. 37:443-451.
Spurr, S.H., and B.V. Barnes. 1980. Forest Ecology, 3rd ed. John Wiley and Sons, New York. 687p.
Sutherland, J.P. 1974. Multiple stable points in natural communities. Amer. Natur. 108:859-873.
Tamate, S. 1956. Catastrophic windthrow and its prevention. Hoppo Ringyo 8:180-183.*
Tamate, S. 1959. Meteorology. In: A report of the scientific investigations of the forests wind-damaged in 1954, Hokkaido, Japan. K. Matsukawa (ed.) Japan Forest Technical Association.*
Tatewaki, M., and T. Igarashi. 1971. Forest vegetation in the Teshio and the Nakagawa district experiment forests of Hokkaido University, Prov. Teshio, n. Hokkaido, Japan. Research Bull. Coll. Exp. For., Hokkaido University 28:1-192.*
Ujiie, M. 1984. Studies on the soils derived from volcanic deposits in Tomakomai district. Research Bull. Coll. Exp. For. Hokkaido University 41:149-190.
Watanabe, S. 1985. Dendrosociological studies of the natural forest in Hokkaido, Japan. Hokkaido Regional Forestry Office. 196p.*
Webb, L.J., J.G. Tracey, and W.T. Williams. 1972. Regeneration and pattern in the subtropical rain forest. J. Ecol. 60:675-695.
Whittaker, R.H. 1953. A consideration of climax theory: the climax as a population and pattern. Ecol Monogr. 23:41-78.
Whittaker, R.H. 1967. Vegetation of the Great Smoky Mountains. Ecol. Monogr. 26:1-80.
White, P.S. 1979. Pattern, process, and natural disturbance in vegetation. Bot. Rev. 45:229-299.
Wierman, C.A., and C.D. Oliver. 1979. Crown stratification by species in even-aged mixed stands of Douglas-fir–western hemlock. Can. J. For. Res. 9:1-9.
Yamada, S. 1958. Soil productivity of tephra origin. Hoppo Ringyo 10:197-199.*
Yamaguchi, H. 1959. Development of bark beetle damages following catastrophic windthrow. Hoppo Ringyo 11:95-99.*

The structure and dynamics of tropical rain forest in relation to tree species richness

4

PETER S. ASHTON

Foresters, in the quest for management and harvesting prescriptions, have traditionally sought to bring order out of the chaotic diversity of tropical evergreen forest by treating silvicultural guilds or economic groups rather than species. Floristic ecologists and systematists, on the other hand, revel in the diversity itself. Now, thanks to Alf Leslie, I.T.T.O. and the current revolution in resource economics, we all seem to be coming together at some middle ground in which some of the species diversity, at least, is of universal interest (Leslie 1987, Ashton and Panayotou 1988). I refer to the new hope offered to our beleaguered rain forests by the growing realization among policy-makers that they may, after all, represent more than standing timber, and the current ingenious efforts to set economic values on their hydrological, edaphic, wildlife, touristic, educational and even mythological functions. I offer the observations that follow in that spirit!

The mechanistic approach in research requires that causal relationships can only be established by examining one variable at a time. As a first approximation to understanding the relationships between the many aspects of physical environment and biological variables in a forest, the silvicultural or floristic ecologist can at best attempt to draw conclusions by deduction: By comparative studies at different sites in which as many environmental variables as possible are held constant. Few more appropriate regions for such an approach are available than in the rain forests of northwest Borneo. There, biogeographical and some paleontological evidence suggests that the regional climate has been uniform since late Miocene times (Morley and Flenley 1987). Rainfall there on average exceeds evapotranspiration every month of the year. In the lee of the mountains, the region escapes the periodic ravages of El Niño droughts; and it is south of the typhoon zone. But the generally shallow and infertile soils of the region, which mantle a young topography caused by continuing tectonic uplift, are extraordinary in their variability (Baillie et al. 1987). The forest varies with soils in structure, species composition and, we have recently discovered, dynamics in ways which evoke comparison with broadleaf forests of other regions.

Species richness in mixed dipterocarp forests

That northwest Borneo supports a great diversity of lowland forests, including the majestic forests of the vast oligotrophic peat swamps and the curious even-canopied Heath forests on podsols (Whitmore, 1984) has been known since the great naturalist Eduardo Beccari (1904) explored Sarawak in the 1850's. I will confine my attention here to the forests on the prevailing yellow/red soils of the uplands below 700 m. There, forests are generally known as lowland dipterocarp forests in Malaya on account of the canopy dominance of that family. I have termed them Mixed Dipterocarp forests to distinguish them from the swamp forests in which the dipterocarp *Shorea albida* forms pure stands, and other forests dominated

M. J. Kelty (ed.), The Ecology and Silviculture of Mixed-Species Forests, 53–64.

by single dipterocarp species on limiting soils, seasonal climates, or at the edge of the geographical range of the family (Ashton 1964).

Mixed Dipterocarp forests, though diverse, share in common are emergent overstorey of varying density comprised of at least 80% Dipterocarpaceae (Ashton 1964), whose stature varies between 40-75 m. They are the forests richest in tree species, with between 150-350 species in samples of 1000 trees, of which 10-60 may be dipterocarps.

Visiting these lofty forests for the first time, their majesty evoked timelessness in me, their structural complexity chaos. The fact that they are dynamic and that structure and dynamics might vary with site never then occurred to me. As a researcher within the Brunei, and later the Sarawak Forest Service, I concentrated on getting to know the species, and it was through them that I came to realize that dramatic variation in composition exists: Tree species composition sharply differentiates forests on sticky clay soils, slippery under foot and usually on broken topography, from those on yellow sands to sandy clays with densely root-matted surface raw humus and on gentle topography (Ashton 1964). This floristic variation occurs at geographical scale and is correlated with geology. It is far more dramatic than the subtle floristic changes that occur from ridge to valley, except where the sedimentary rock itself covaries with topography, which is sometimes the case. Aerial photographs clearly show though that forest canopy structure varies primarily with topography, and with soil physical properties and geology only second, something that I could not observe from the ground. Ridges support the trees with the biggest crowns where adjacent slopes are steep and their soils truncated by erosion; but they support trees with uniformly small crowns where it is their soils which are shallowest. Floristic composition in these forests therefore varies largely independently, and at a different scale from structure.

Some hundreds of 0.6 and 0.4 ha plots, grouped into fifteen sites each on uniform geology, were laid out in forests throughout Sarawak and Brunei as a means to describe variation in tree species composition in detail (Ashton 1964, 1973). Because the sites were located on the full range of volcanic and sedimentary rocks and topography there, soils varied in depth from less than 20 cm mixed with decomposing substrate, to in excess of 1 m, from sands to clays of varying friability. They included two broad classes: humult ultisols, generally derived from sandstones and acid volcanic rocks, leached and with low mineral soil nutrient concentrations between 30-400 ppm total phosphorus and 100-1200 ppm total magnesium, with a distinct surface horizon of root matted raw humus with pH 3.7-4.2; and less leached udult ultisols, inceptisols and oxisols, richer in clay and with up to 1200 ppm total phosphorus and 3500 ppm magnesium; in which the mineral soil is visible directly below the litter, with pH 4.0-4.8 near the surface. Though mean annual rainfall varies between 2-4 m, and in no month does mean evaloptranspiration exceed mean precipitation, periodic dry spells occur which may induce water stress in shallow and freely draining soils, particularly on ridges (Baillie 1976). Quantitative analysis revealed that there are two species series, on humult and on udult soils as I had observed in the field, irrespective of whether the substrate is sedimentary or volcanic. Species composition varies relatively more, and mainly in relation to mineral soil nutrient levels on the lower nutrient humults; and relatively less, and in relation to topography and by inference soil water economy, on udults (Ashton, in preparation). Leading nutrient correlates in our analysis were total magnesium and phosphorus concentrations in the mineral soil (Baillie et al,

1987). Species richness, here measured as the number of species per one thousand individuals in order to separate the effect of variable stand density on the species/area curve, has a peaked distribution in relation to nutrient concentrations (Ashton 1977) (Fig. 3a). This is reminiscent of Tilman's (1982) predictive model of species richness based on competition for multiple resources, in which a peak occurs towards the lower end of the resource range. Interestingly, our peak is apparently at or very close to the point where humus accumulation and decomposition are in equilibrium, with richness positively correlated with nutrient concentration on humult and negatively on udult soils. David Smith would be the first to point out though, that such a distinctive pattern is unlikely to have a single cause, and that one must first get to know the forest in detail and over a range of habitats, before designing experiments which may ultimately rigorously resolve its causes. We are at the observational stage, and I dedicate what follows as a mark of respect to Dr. Smith's insistence on careful and extensive field observation, in space and – perhaps even more important – over time.

Species richness in Borneo rain forest also shows a peak in relation to the within-sample variance of nutrient concentration, suggesting that species richness on low nutrient soils is due, at least in part, to fairly small-scale beta diversity. There is high negative correlation between species richness on udults, and population densities of the most abundant species, but there is no such correlation on humults. There is also negative correlation between our measure of species richness on udults, and the relative basal area of subcanopy species with large leaf sizes; and positive correlation between species richness and subcanopy density throughout the range of soils. Together, these observations support Tilman's prediction that species richness will climb to the threshold beyond which no resources are limiting, after which it will become suppressed by the dominance of one of a few species which competitively exclude others in completition for the one ubiquitous resource, which is light. But there is much variation. Udults on ridge sites consistently support higher species richness and lower dominance than on gently lower slopes. Periodic water stress on ridges and puggy clay soils may be expected to lower canopy density and, aided by the greater oblique sunlight, canopy penetration on ridges will increase the amount and spatial heterogeneity of light on the forest floor. This brings us back to forest structure and gaps.

Forest structure

The tallest known dipterocarp, at 76 m, happens to be a *Dryobalanops lanceolata* in one of our permanent plots, on a gentle basalt slope (Fig. 1c illustrates this forest type). There, the forest is the most magnificent I have ever seen, with stands of mighty dipterocarps, of species which more typically stand above the main forest canopy as emergents, here growing side by side, their crowns touching, providing light continuous shade for a large leaved but scattered subcanopy far beneath whose deep shade little else grows. There is a sense of cavernous space, and even the calls of the Argus pheasants seem to echo further through these lofty halls. In places, these forests bear an emergent canopy as continuous as that of the mighty alan (*S. albida*) forests at the margin of the peat swamps, where the canopy can average up to 70m over considerable areas. The shade cast by the diffuse leafage of alan is less though and, far below, there is a dense and heterogenous thicket of mixed species in the subcanopy and understorey.

I was surprised to find stands only slightly less tall, at 68 m, on the deep fine sand humult utisols of the neogene coastal hills. There, the emergents are scattered or in clusters, and mostly on the ridges, while in between their crowns is a heterogenous main canopy, at c. 40 m but very variable. Beneath again, there is a patchy but generally dense understorey in which pole-size advance regeneration of emergents is relatively frequent. Understorey light intensity is notably greater than on the more nutrient rich udult ultisols and oxisols. Forests on some more puggy udult ultisols are similar to those on these sandy humults in canopy structure, but seldom exceed 50m while their understorey is more shady. Therefore, I conclude again that forest stature is largely independent of soil nutrient status, and is apparently primarily related to the cost, under average conditions of soil moisture, of extracting and lifting soil water up to these heights. In this respect it is interesting that leaf size does not vary among the tallest trees in these forests in a manner which correlates with either nutrients or topography, but in the subcanopy it correlates with both. It is not clear whether nutrients or water stress are the primary cause of the prevailing small leaf sizes on podsols and, to a lesser extent, humult ultisols. That peat swamp and Heath forests on podsols share mostly the same xeromorphous tree flora, notwithstanding contrasting soil water regimes, suggests that nutrients are influential. Variation in leaf sizes within species, though, is correlated with soil depth in Heath forest and variability of water table in peat swamps.

What factors maintain the spacing of emergents on intermediate sites (fig. 1)? The great droughts in eastern Borneo during 1982-83 resulted in selective culling of emergents, with up to 30% dying on ridges (Leighton and Wirawan 1986). The same has been observed in ridge forests in peninsular Malaysia (Tang and Chang 1979) and among the tallest *Shorea robusta* trees in seasonal dipterocarp forests in India (Seth et al. 1960). It is possible that such periodic culling of emergents on the most droughty sites also occurs in northwest Borneo, although the climate there is more equable than in these examples. On the other hand, the frequent concentration of emergents on ridges seems to be due to more frequent windthrows and soil movement including landslips, and the generally darker, less favorable conditions for regeneration on steep adjacent slopes. Whatever the cause, I suspect that scattering and clumping of emergents bespeaks past catastrophe.

Canopy gap formation

The nature of canopy gaps also differs among mixed dipterocarp forests (Ashton and Hall, in press). In order to compare forest dynamics on their range of soils, clusters of 4-5, 0.6 ha.plots were established at three sites: on shallow sandy humults, on a mosaic of deep humult and udult ultisols at and near their interface, all on sedimentaries; and on deep well-structured udult oxisols over basalt (Fig. 1). Plots were sited to avoid gaps, and this has allowed us to observe the impact of gaps on overall dynamics in the 25 years since they were established. Measurements have been taken every five years. All forests differ dramatically from those observed in the neotropics in that more than half of the canopy trees die standing. Pamela Hall (1990) has recently shown that windthrows and snapped stems account for less than 20% of tree mortality on the shallow sandy soils where the stable slopes and thick surface root matt apparently add stability, and about 35% at the other two sites. The largest canopy gaps are on the intermediate site (Fig. 2). There, clay and soft sandstone scarp slopes frequently slump after prolonged rain, leaving bare surfaces

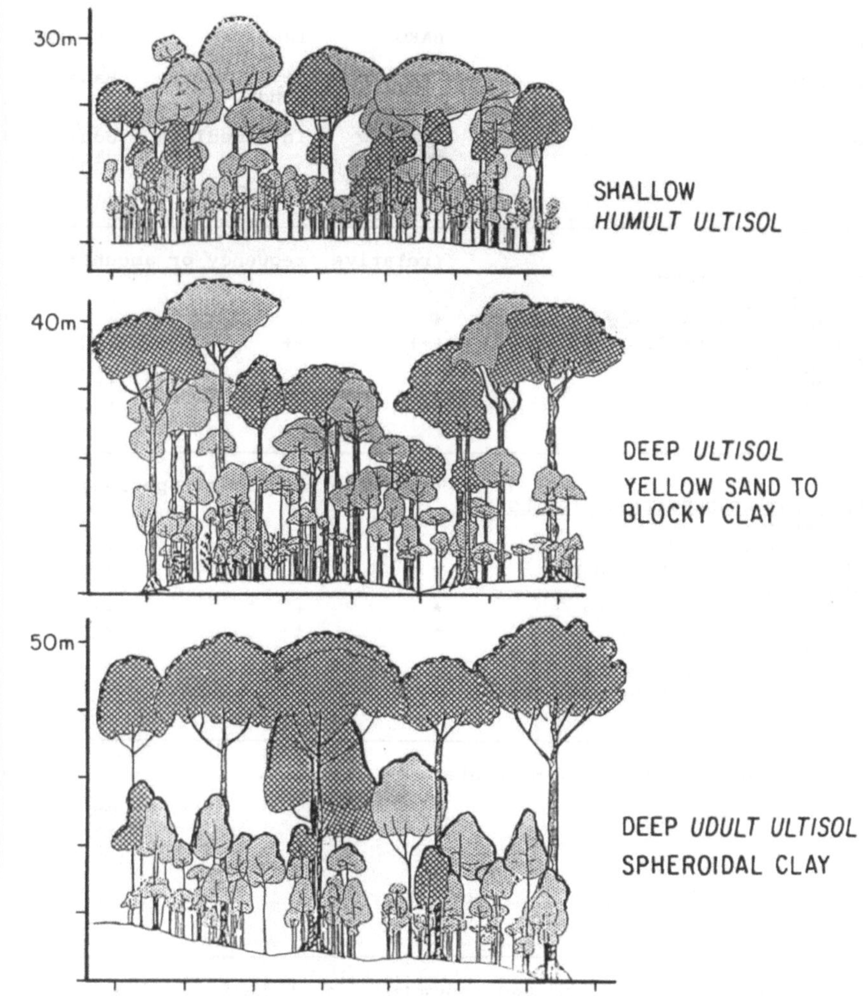

Fig. 1. 64 x 8 m profile diagrams of three mixed dipterocarp forests in northwest Borneo. From top to bottom: Bako National Park, Paleocene sandstone; Lambir N.P., Miocene sandstone and clay; Ulu Bakong, Baram, Miocene calcareous shale and clay. All trees exceeding 5 m tall included; dipterocarps cross-hatched.

SITE	BAKO	LAMBIR	MERSING
LITHOLOGY	Paleocene sandstone	Mio-pliocene sands, clay	Basalt
SOIL	Shallow humult ultisols	Intermediate	Udult ultisols
	(relative frequency or amount)		
PREVALENT GAP FORM			
Landslips	-	++	-
Emergent wind throws	*	+	+
Mortality standing	+++	++	++
MATURE PHASE CANOPY			
Density	+	+	++
Roughness	+	++	*
	(relative numbers of species)		
PREVALENT REGENERATION STRATEGY			
Pioneers:			
Short-lived	*	++	+
Building phase to canopy	-	*	+
Mature phase:			
Intolerant: Emergents	*	+	++
Intolerant: Canopy	+	+	+
Tolerant:			
Emergents	+	+	*
Canopy & subcanopy	+	+	++

- rare or absent; * few; + moderate; ++, +++ many.

Fig. 2. Canopy structure and gap characteristics of three mixed dipterocarp forests in northwest Borneo.

of up to 10 ha for recolonization *ab initio*. On the shallow sandy humults, classical pioneers such as *Macaranga*, *Trema*, and *Mallotus* are rare, small at maturity and short lived. The maximum number of pioneer species does occur at the intermediate site, although large pioneer species in such genera as *Octomeles*, *Alstonia*, and *Ptercymbium*, and fast growing intolerant climax species such as *Parashorea macrophylla* and *Dryobalanops lanceolata* only play an important part in the building phase on the clay loam oxisols over basalt. Nevertheless, the net gain in species over 20 years was less than 20% of the original species richness of the plot cluster at the basalt site, and 10% at the other two. Also, forests on the humult-udult boundary in other localities frequently overlie stable topography, yet their species richness is universally high. I therefore conclude that disturbance levels are playing a subordinate if parallel role in influencing the pattern of species richness, contrary to expectations of Connell (1978), who predicted that regular disturbance frequency would be the dominant factor in controlling species richness, as depicted in Fig. 3c.

It was a surprise to us to find no significant difference at first in mean girth growth, at any tree size, between the stand samples on these very different soils. Differences did grow over time, due to differences in the girth growth of the fastest growing individuals, climax as well as pioneer, between each stand sample. As Whitmore (1989) has explained, tropical evergreen forests include species with the climax characteristics of lack of seed dormancy and persistance of established seedlings in the understorey, yet which manifest a wide range of light responses. Among the intolerant climax genera such as *Dryobalanops*, the species with fastest girth growth are those of well-watered sites: *Dryobalanops lanceolata*, but also *Dr. rappa* in the oligotrophic peat swamps. It will come as no surprise that, unlike the pioneers, availability of water rather than of nutrients appears to determine the performance of climax species. This is also supported by the fact that the relationship between girth and height changes in these forests as a whole in relation to stature: Tall forests have more slender trees for a given girth than short forests, and this is more so the bigger the tree is. Canopy trees in tall forests, which I have suggested occur on soils with the most reliable and favorable water economies, must therefore on average grow faster in height and hence in volume than those in short forests.

Relationship of structure, gap dynamics, and species richness

Does this variation in structure and dynamics relate to species richness? The great majority of the tree species in these forests are climax species, in the sense that they principally rely on advance regeneration. Although it is possible that all emergent species require a canopy gap created by tree or major branch mortality in order to survive and grow to reproductive maturity, this seems highly unlikely in the case of many main canopy and, more particularly, subcanopy species. The majority of tree species in these forests occupy the main canopy at maturity (Ashton 1969). The greater heterogeneity of light conditions on and near the forest floor in forests on intermediate soils and on some ridges, occasioned both by canopy roughness and variability in crown density, may provide unusually diverse conditions allowing establishment of a wide variety of climax species, irrespective of gaps by canopy damage. By comparison, on shallow humult ultisols the upper canopy is diffuse but relatively uniform and the understorey is bathed in rather uniform light shade; while on deep loamy udult soils the subcanopy is dense, and the forest floor beneath is

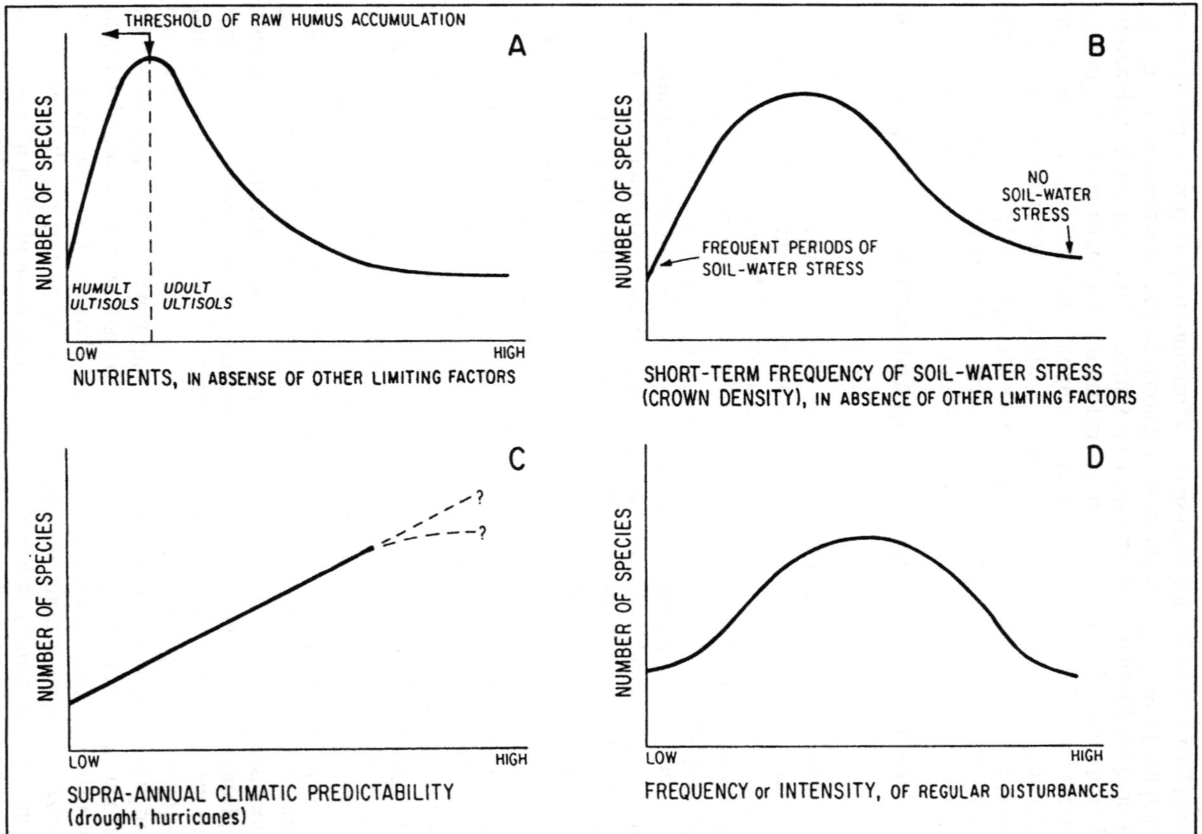

Fig. 3. Hypothetical causes of variation in tree species richness among biogeographically related evergreen forests in the lowland tropics.

uniformly shady. Emergents in these latter forests, whether of pioneer or climax species, appear to be "intolerants" which require gaps in order to grow to maturity.

On sandy humult ultisols then, more trees die standing, solitary, leaving a small gap in which advance regeneration is minimally disturbed. The greatest diversity of conditions for gap regeneration is at intermediate sites, where the soil surface may or may not bear root-matted humus, and where increased windthrows, and in some forests landslips provide bare mineral soil surfaces for colonization under a variety of light conditions. The combined effect of relatively low nutrient concentrations and likely periodic water stress appears to sustain a canopy of only moderate density even in early succession, and no species gain dominance (Fig. 3a,b). On the better structured udult soils, the dense mature phase canopy suppresses advance regneration but the death, even standing, of one emergent wide-crowned giant exposes patches of mineral soil here owing to the rapid rate of litter decomposition. A fight then immediately ensues between what advance regeneration there is, which rapidly puts on height growth, and newly established pioneers including some vines whose vast leaves, overlapping in their crowns like slates on a roof, outshade all competition: Possession is nine points of the law (Poore 1968). In such single tree gaps, the ultimate winners are either intolerant climax hardwoods such as the light red merantis (*Shorea* spp.) or kapur (*Dryobalanops lanceolata*), or dense-crowned pagoda trees (monopodial plagiotropic pioneers which contribute to the building phase). In larger gaps, particularly where advance regeneration has been damaged by treefall or landslide, succession may be postponed by a mantle of vines. As in the mature phase then, ample soil water allows a spendthrift water economy, and a dominant few suppress diversity overall. Thus, though the greatest diversity of growth strategies occurs on these well-structured udult soils, it does not lead to the greatest species richness (Fig. 3b).

Do nutrients and water therefore influence species richness entirely through their mediation of canopy structure, hence light climate for regeneration and opportunities for dominance of a few? The distinct species associations which I discussed initially cloak major gradients in generic and even familial composition, again principally correlated with mineral soil nutrient concentrations, and this suggests otherwise. The preponderance of ectotrophic mycorrhizal families, particularly Myrtaceae and Dipterocarpacae, on humult ultisols, and a tight linear correlation between some nutrient concentrations and dipterocarp species richness on these soils contrasts with the patterns of Meliaceae, Euphorbiacae and Sapindaceae which are richest on soils of highest nutrient concentration. There would appear to be ecological specialization in relation to nutrition, at least among some families, about which we so far know nothing.

In summary, I would conclude that, among equilibrium hypotheses which seek to explain the determinants of species richness, Tilman's (1982) resource-partitioning model is that which conforms most closely to the patterns observed. Connell's (1978) intermediate disturbance hypothesis is not invalidated, but the patterns it predicts seem to be somewhat parallel yet subordinate to others. It would appear - but this is where I am on the thinnest ice until more demographic, transplant and physiological research is carried out - that diversity of the regeneration niche holds a major key to species richness (Grubb 1977); while I suspect that the diversity of potential environmental requirements for reproduction may hold a second.

Management implications

At this point, before making comparisons with patterns of species richness in other regions, it would be appropriate to discuss the management implications of these observations. As a plant ecologist who has solely studied primary forests, and trees exceeding 10cm dbh, I can only summon as evidence some casual observations, supported by the very limited literature.

Currently the principal commercial products of primary Mixed Dipterocarp forests are a diversity of hardwoods of moderate to high quality and moderate but varying specific gravity. Among these, overwhelmingly the most abundant are the red merantis, comprising four sections of the dipterocarp genus *Shorea* and, in Borneo, the kapurs (*Dryobalanops* spp.) although other *Shorea* sections and dipterocarp and non-dipterocarp genera are harvested. These are exported as logs, sawwood and veneer, mostly to the east Asian industrial nations where they are used for everything from doors and window frames to forms for concrete structures.

It is probably fair to say that the Malayan Uniform System of silvicultural management (Wyatt-Smith 1963) is the only proven successful system in tropical evergreen forest. It is a form of the shelterwood system, which relies on exposing advance regeneration to prolonged periods of direct sunlight (though not completely open conditions), to which the relatively intolerant light red merantis (*Shorea* section, *Mutica*, and *S. ovalis* in peninsular Malaysia where it was developed and mostly applied), so-called on account of both their weight and color, respond by rapid increases in height growth. Mean diameter increments of 8 - 12 mm/yr and occasionally up to 20 mm/yr can be expected. I have been in a forest on udult ultisols in northwest Borneo rich in light hardwood dipterocarps, which I had believed to be primary, but which my guide had informed me was the site of his family's swidden perhaps one century earlier. This would be consistent with the rapid growth rates of mature climax and building phase pioneer light hardwoods observed in gaps in our plots on udult ultisols.

E. F. Brunig (unpublished report to the Government of Brunei) conducted a regeneration survey of tall stature mixed dipterocarp forests on deep humult ultisols at Andulau Forest Reserve in 1958, shortly after intensive logging down to 47cm dbh followed by poison-girdling of residuals down to 15 cm dbh according to the Malayan Uniform System. He found patchy but overall unsatisfactory regeneration of light hardwood dipterocarps, mostly red and yellow meranti. Anderson and Marsden (1984) surveyed trees above 10cm dbh in Andulau twenty years after logging, and found unusually low stocking of commercial species in the smaller diameter classes measured. This implies either absence of their regeneration, or mean growth rates less than 5 mm/yr dbh. Regeneration of dipterocarps and other emergents is nevertheless abundant and vigorous in these forests before logging, in the small cylindrical gaps caused by death of individual emergent trees.

On a visit to Andulau in 1988, I noted that logged areas treated by poison-girdling during the 1950's and 60's were now occupied by a dense thicket of pole-size trees, with patchy representation of climax light hardwood species. Pioneer species were abundant and diverse along old logging tracks, but were all of small stature. I also noted widespread scarification, during logging operations, of the surface root-matted raw humus. If only for conservation reasons (these species-rich forests occur on accessible undulating land near the coast, and it is questionable whether their soils could economically sustain any other crop), ecological research is urgent if what little remains of them is to be induced to regenerate. I suggest that two problems require investigation: first, the effect of the mechanical removal, during logging of the soil's humic surface on subsequent regeneration establishment and performance; and second, the light conditions required for optimal growth of the various light hardwood species on these soils.

Patterns in other forest regions

Patterns of tree species richness in broad-leaf forest have seldom been examined on this scale anywhere. Huston (1980) found a similar peaked distribution to ours in Central American forests but Gentry (1982), in extensive studies throughout the neotropics, has found that species richness increases with mean annual rainfall, with some indication that the increase levels off in aseasonal wet climates. His highest estimates from the neotropics are virtually indentical to ours. Strong (1977) has pointed out the importance of epiphyte-induced branch break in creating diverse conditions for regeneration, and hence by implication in increasing species richness in neotropical forests. Certainly, Bromeliaceae provide greatly increased epiphyte loads there compared with other regions of the lowland humid tropics, and the small- to medium-sized canopy gaps which they induce, at a constant rate and over evolutionary time, would be an optimal manifestation of intermediate disturbance (Fig. 4). In most other climates, periodic and unpredictable catastrophe in the form of supraannual climatic disturbance, particularly from water stress and hurricanes, may impose a disequilibrium on forest dynamics and increase the rate of random or selective extinctions, thereby obscuring the patterns that I have described (Fig. 3d). High predictability of climate and of disturbance regime appear to favor accumulation of species richness over time.

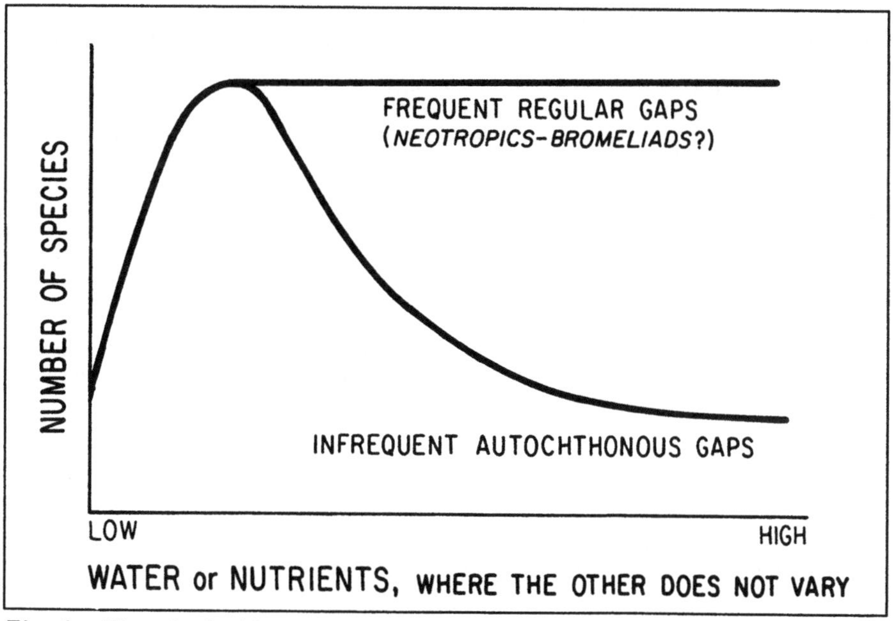

Fig. 4. Hypothetical interactions between influences on tree species richness in lowland tropical evergreen forest.

References

Anderson, J.A.R. and J.D. Marsden. 1984. Brunei resources and strategic planning study. Final report. The forest resources of Negara Brunei Darussalam. Anderson and Marsden (Forestry Consultants), Ltd. Brunei Darussalam.

Ashton, P.S. 1964. Ecological studies in the mixed dipterocarp forests of Brunei State. Oxford Forestry Memoirs 25.

Ashton, P.S. 1969. Speciation among tropical forest trees: Some deductions in the light of recent evidence. Biol. J. Linn. Soc. 1:155-196.

Ashton, P.S. 1973. Report on research undertaken during the years 1963-1967 on the ecology of mixed dipterocarp forest in Sarawak. Mimeo. 412 p.

Ashton, P.S. 1977. A contribution of rain forest research to evolutionary theory. Annals Missouri Bot. Gard. 64:694-705.

Ashton, P.S. In prep. Comparative ecological studies in the mixed dipterocarp forests of northwest Borneo. III. Patterns of species richness.

Ashton, P.S. and P. Hall. In prep. Comparative ecological studies in the mixed dipterocarp forests of northwest Borneo. II. Variation in structure and dynamics.

Ashton, P.S. and T. Panayotou. 1988. The case for multiple use management of tropical hardwood forest. Report prepared by Harvard Institute for International Development for the Internal Tropical Timber Organization.

Baillie, I.C. 1970. Further studies of drought in Sarawak, East Malaysia. J. Trop. Geogr. 43:20-29.

Baillie, I.C., P.S. Ashton, M.N. Court, J.A.R. Anderson, E.A. Fitzpatrick and J. Tinsley. 1987. Site characteristics and the distribution of tree species in Mixed Dipterocarp Forest on Tertiary sediments in central Sarawak, Malaysia. J. Trop. Ecol. 3:201-220.

Beccari, O. 1904. Wanderings in the great forests of Borneo. trs. E.H. Giglioli, ed. F.H.H. Guillemard. Constable, London.

Connell, J.H. 1978. Diversity in tropical rain forests and coral reefs. Science 199:1302-1310.

Gentry, A.H. 1982. Patterns of neotropical plant species diversity. Evol. Biol. 5:1-84.

Grubb, P.J. 1977. The maintenance of species-richness in plant communities: The importance of the regeneration niche. Biol. Rev. 52:107-145.

Hall, P. 1990. Structure, stand dynamics and species compositional change in three mixed dipterocarp forests of northwest Borneo. Ph.D. thesis, Boston University.

Huston, M. 1980. Soil nutrients and tree species richness in Costa Rican forests. J. Biogeogr. 7:147-157.

Leighton, M. and N. Wirawan. 1986. Catastrophic drought and fire in Borneo tropical rain forest associated with the 1982-83 El Niño Southern Oscillation event. In G.T. Prance, ed., Tropical Rain Forest and World Atmosphere. AAAS Symposium 101, Proceedings. Westview Press, Boulder, Colorado. Pp. 75-102.

Leslie, A. 1987. A second look at the economics of natural management systems in tropical mixed forests. Unisylva 39,155:46-58.

Morley, R.J. and J.R. Flenley. 1987. Late cainozoic vegetational and environmental changes in the Malay Archipelago. In T.C. Whitmore (ed.), Biogeographical evolution of the Malay Archipelago. Clarendon, Oxford. Pp. 50-59.

Poore, M.E.D. 1968. Studies in Malaysian rain forest. I. The forest on Triassic sediments in Jengka Forest Reserve. J. Ecol. 56:143-196.

Seth, S.K., M.A. Waheed Khan and J.S.P. Yadav. 1960. Sal mortality in Bihar. Indian For. 86:645.

Strong, D.R., Jr. 1977. Epiphyte loads, tree falls and perennial disruption: A mechanism for maintaining higher tree species richness in the tropics without animals. J. Biogeogr. 14:215-218.

Tang, H.T. and P.F. Chang. 1979. Sudden mortality in a regenerated stand of *Shorea curtisii* in Senaling Inas Forest Reserve, Negri Sembilan. Malay. For. 42:240-254.

Tilman, D. 1982. Resource competition and community structure. Princeton University Press, Princeton, New Jersey.

Whitmore, T.C. 1984. Tropical forests of the Far East. 2nd ed. Clarendon, Oxford.

Whitmore, T.C. 1989. Canopy gaps and the two major groups of forest trees. Ecology 70:536-538.

Wyatt-Smith, J. 1963. Manual of Malaysian silviculture for inland forests. Malay. For. Rec. 23.

Patterns of diversity in the boreal forest

5

JEFFREY P. THORPE

Introduction

The boreal forest is the forest biome in which temperatures are most limiting to tree growth. The North American portion of this biome stretches across Canada and Alaska (Figure 1), its limits lying parallel to the thermal isolines. Its southern boundary, at about 1300 growing degree-days above 5°C (Kauppi and Posch 1985), occurs where thermal resources become inadequate for the growth of temperate hardwoods, while its northern boundary, at about 600 degree-days, occurs where the climate becomes too severe for any tree growth.

Given the limitations imposed by temperature in this biome, and the fact that most of its present range was wiped clean by continental glaciation, it is not surprising that the boreal forest is distinctive among forest biomes in its low level of species diversity. Over most of the North American boreal forest, there are only eight species of trees: white and black spruce, balsam fir, jack pine, tamarack, trembling aspen, balsam poplar, and white birch (scientific names are listed in Appendix A).

Fig. 1. Map of Canada, showing ecological zones.

M. J. Kelty (ed.), The Ecology and Silviculture of Mixed-Species Forests, 65–79.

The three broad-leaved species are similar in having low shade-tolerance, fast juvenile growth, and prolific vegetative reproduction following disturbance. Birch and aspen tend to occur on mesic sites throughout, but birch is more important in the eastern boreal forest while aspen is concentrated in the west, where it becomes the most important tree species over large areas. Balsam poplar appears mainly on wet-mesic sites.

Jack pine is intolerant of shade, highly adapted by cone serotiny to regeneration after fire, tolerant of moisture and nutrient deficiency, and most important on coarse or thin soils. Black spruce also seeds in rapidly after fire, is intermediate in shade-tolerance, and is extremely tolerant of adverse site conditions. It is found on wet, nutrient-poor sites throughout, but also occurs on excessively dry sites in some areas, and spreads over much of the upland in the colder northern part of the boreal forest. White spruce is also intermediate in shade-tolerance, but is more site-demanding, growing mainly on mesic to wet-mesic sites, especially those rich in nutrients. Balsam fir is also found mainly on mesic to wet-mesic sites, but being poorly adapted to regeneration after fire, it is concentrated in areas protected from burning, and is a more important part of the boreal forest in the humid east where fires are less frequent. Tamarack is a relatively minor species, prominent mainly on the richer peatlands.

Wildfire is an integral part of the boreal forest (Rowe and Scotter 1973). Heinselman (1981) described the fire regime of the western boreal forest as one of large, stand-destroying fires at 50-100 year average intervals. In northern Minnesota (Heinselmann 1973) and in the Laurentian Highlands of Quebec (Cogbill 1985) the fire rotation is about 100 years. Only on the maritime margins of the biome do fire frequencies become very long. In Labrador, Foster (1985) estimated an average rotation of 500 years, while in the boreal-temperate transition forests of New Brunswick and Nova Scotia, Wein and Moore (1977, 1979) found rotations of 1000 years.

Monodominant stands are conspicuous throughout the boreal forest. In Saskatchewan, for example, about 75% of the productive forest land is covered by pure or nearly pure stands, with trembling aspen, black spruce, and jack pine accounting for most of the area (Kabzems et al. 1986). Most of these can be attributed in part to the homogenizing effect of regeneration after wildfire, for which all three species have strong adaptations.

Development of mixed stands

The most important mixed stands are those which occur on relatively productive mesic sites. In central and western Canada, these are typically dominated by trembling aspen and white spruce, with variable amounts of balsam poplar, white birch, black spruce, and balsam fir. These stands are referred to as "mixedwoods", implying mixture of broad-leaved and conifer species. In these stands, the dynamics of competition among species become important.

Trembling aspen, along with the other intolerant hardwoods, sprouts aggressively after fire and grows rapidly in juvenile stages, thus dominating young stands. Many such stands are pure aspen; about 35% of the productive forest land in Saskatchewan is placed in this cover type (Kabzems et al. 1986).

However, in some cases, white spruce and other conifers appear beneath the aspen canopy. This varies with the availability of a nearby seed source, which in extensive burns is often lacking. Some observers have noted that shallow burns, which leave some surface organic matter, favour hardwood sprouting, while hot fires which expose mineral soil are conducive to establishment of conifer seedlings (Rowe 1970, Heinselman 1981, Kabzems et al. 1986).

Because of the early dominance of aspen over spruce in these stands, and the later appearance of spruce stands from which the aspen have died out, many observers have interpreted the developmental pattern as one of succession from aspen to spruce. White spruce is the longest-lived tree species in the boreal forest, and has at least moderate shade-tolerance, so it has been considered to be the "climax" species on these sites (Moss 1953, 1955, Strong and Leggat 1981).

This view has been challenged by other ecologists. Rowe (1961) argued that white spruce does not fit the "climax" role: it is unsuccessful at regeneration under established stands, and often acts as a pioneer, invading grasslands and exposed mineral soil.

Kabzems et al. (1986) and Day and Harvey (1981) described a pattern of development in mixedwood stands in which most spruce seedlings are established in the first few years after fire, but their slow juvenile growth relegates them to an understory position. Spruce leaders penetrating the aspen canopy may be damaged by whipping of branches in the wind. By 70 to 80 years, aspen has generally levelled off in height growth, and aspen stems, which are susceptible to fungal decay, begin to break off in the wind. The longer-lived spruce continues to grow in height, and begins to emerge through the aspen canopy, gradually forming spruce-dominated stands which are still essentially even-aged.

Subsequent development of such stands depends on the establishment of new young trees in their shade. According to Rowe (1961), the only boreal tree species capable of sustained regeneration under a closed canopy is balsam fir. Fir is more shade-tolerant than the other boreal species (Logan 1969). Moreover, Place (1955) found that fir seedlings are more successful than those of spruce at establishment on undisturbed forest humus, because their long tap roots take less time to reach the mineral soil, with its more dependable moisture supply. Because of these advantages, fir seedlings are generally more numerous in advance growth than those of spruce or any other species (Dix and Swan 1971, Carleton and Maycock 1978, Doucet 1988). Fir stands in the eastern boreal forest reproduce themselves, in the absence of fire, through release of fir advance growth as the overstory trees succumb to insect and fungal attack (Westveld 1931, Morris 1948, Place 1955, Vincent 1956, Ghent 1958, Hatcher 1960, MacLean 1960). However, Rowe (1961) did not consider fir to qualify as a climax species in the western boreal forest, because of poor adaptation to the drier climate.

According to the accounts of Kabzems et al. (1986) and Day and Harvey (1981), in mixedwood stands which have reached the spruce-dominated stage, the understory consists mainly of balsam fir, as well as tolerant hardwood shrubs. Opening of the canopy through mortality of the initial generation of spruce leads to the formation of spruce-fir mixtures in which the fir component is all-aged. Other studies have shown results consistent with this pattern. Dix and Swan (1971) collected a large sample of ages from 39 stands at Candle Lake, Saskatchewan. Most ages of aspen, birch, and jack pine indicated establishment in the first 10 years after stand initiation, and those of white spruce and black spruce in the first 20 years. Only balsam fir had a wide range of ages. Similarly, in old balsam fir stands in Quebec, Cogbill (1985) found that white spruce, red spruce, and white birch formed the oldest cohorts, while fir were spread over younger ages. In these stands, it appeared that there was a high turnover in the fir component, with many trees dying before reaching 35 cm d.b.h. Replacement tended to occur in pulses, rather than continuously.

In order to test this described pattern, I analyzed some stands in which this type of succession appears to be happening, on mesic sites in Prince Albert National Park, Saskatchewan. Three of the stands had a substantial component of balsam fir, relatively unusual in the western boreal forest. In Prince Albert National Park, less than 1% of the forest area has balsam fir as a tree component (Padbury et al. 1978), while for the productive forest of Saskatchewan, fir makes up only 3% of the standing volume of timber (Kabzems et al. 1986). The study sites had apparently escaped fire for longer than usual because of their position in the lee of large lakes. For comparison, an aspen-spruce mixedwood stand, more typical of the forest on mesic sites in this region, was also analyzed by the same methods. In each stand, a 20 m by 20 m plot was laid out, diameter, height, and age were determined for all trees, and densities of seedlings and saplings were determined by sampling of subplots.

The structure of these stands is represented in Figures 2, 3 and 4 by height distributions, with the lengths of the bars representing the basal area in each height class. For ease of comparison, the distributions of ages are shown in the same way.

The youngest stand (age 100) is a typical mixedwood, with trembling aspen and white spruce roughly equal in basal area (Figure 2). The trees form a single canopy, with aspen taller on average, and more uniform in height, compared to spruce. The distribution of ages confirms the general impression of an even-aged stand in which most trees were established shortly after the last fire. All aspen ages in fact fall into a span of eight years, and even this amount of variation is probably due in part to aging errors. The spruce are more variable in age, so there was probably ongoing establishment for 10 years or more after the fire. The seedling/sapling stratum consists of dense balsam fir seedlings, and layers from these seedlings, totalling almost 40,000 stems/ha, all under breast height. None of the other species have more than a few stems in the seedling layer.

The next stand is about 190 years old (Figure 3). Within the plot sampled, there is only one old aspen, but some aspen stems can be identified in the logs lying on the ground. White spruce dominates the upper layer of the stand, with trees ranging from 20 to 32 m tall. However, this is not a continuous canopy, and there are many dead spruce on the ground as well. There is a distinct balsam fir understory with heights from 4 to 16 m, and also many dead fir stems on the ground. The age distribution reveals that the spruce are approximately even-aged, presumably dating to the first two decades after the last fire. The aspen was actually too rotten to age, but it has been shown in the oldest age class for convenience. The fir are younger than the spruce (although some of them only a few decades younger), and appear to have been established in two pulses. Probably there was some minor disturbance, early in the life of the stand, which released the first pulse of regeneration. The second pulse was probably released by the death of canopy spruce over the last several decades. Again there is a dense layer of balsam fir seedlings and saplings forming a continuous range of heights with the balsam fir trees, and no other species contributing to regeneration. Another plot, not illustrated, is about 150 years old and shows a similar stage of development.

The third stand is about 220 years old (Figure 4). Within the plot sampled, there are two old birch trees (aspen and balsam poplar stems can be identified on the

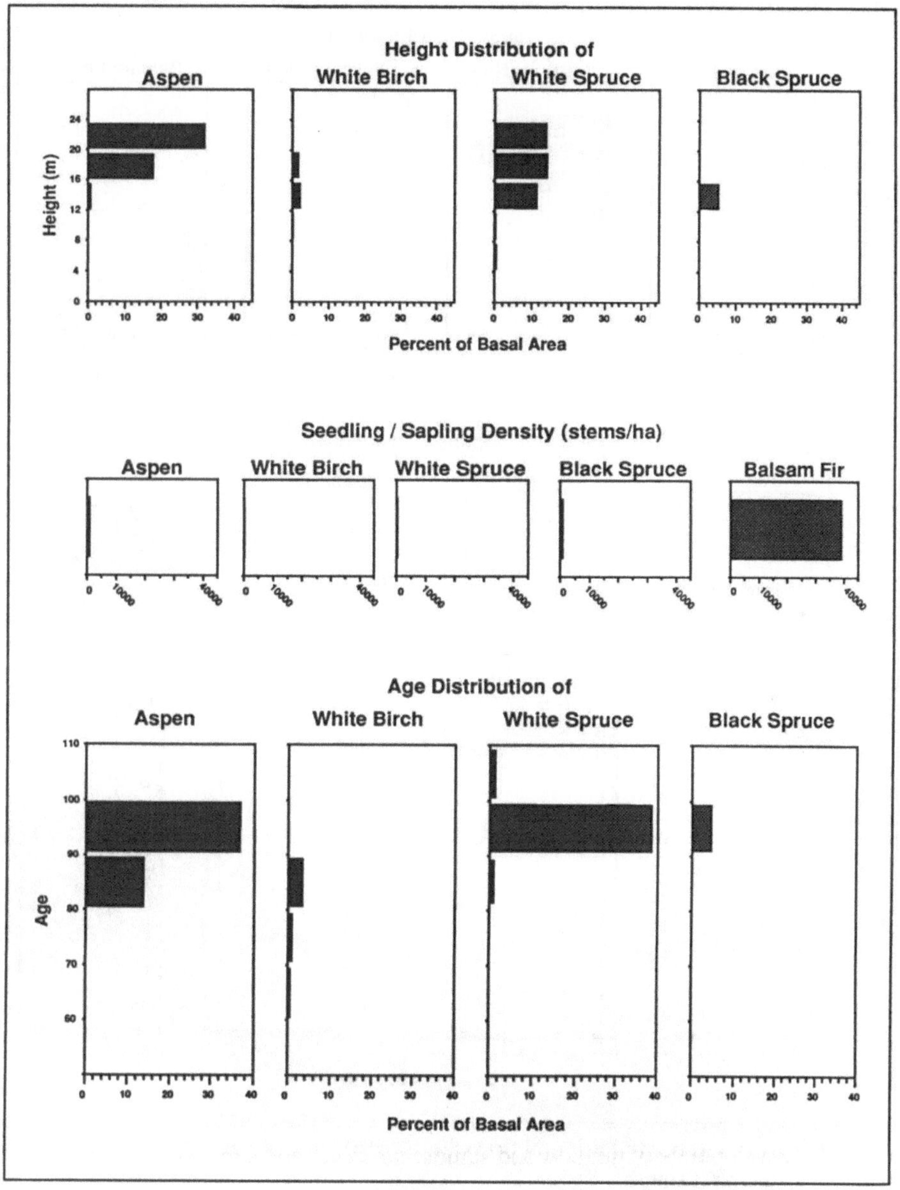

Fig. 2. Development of mixedwood stands: structure and age distribution of 100 year-old stand.

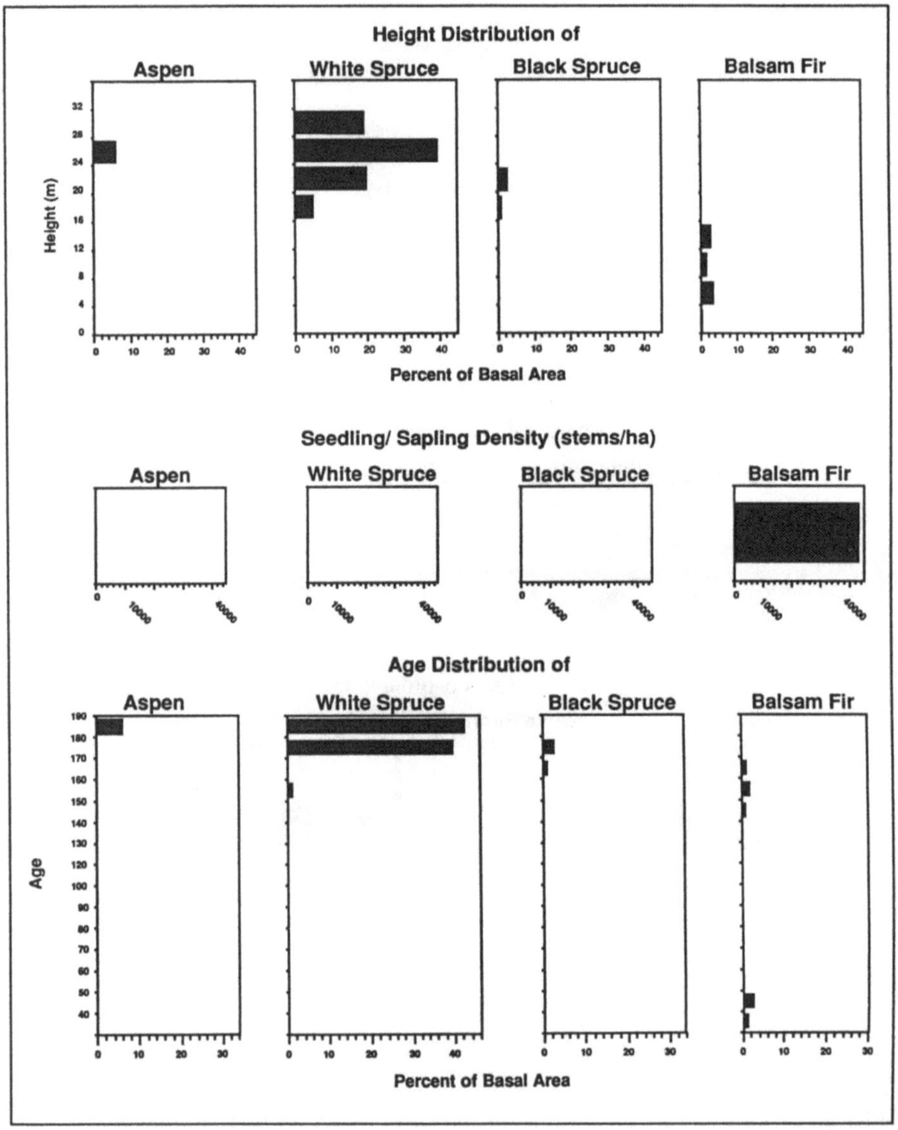

Fig. 3. Development of mixedwood stands: structure and age distribution of 190 year-old stand.

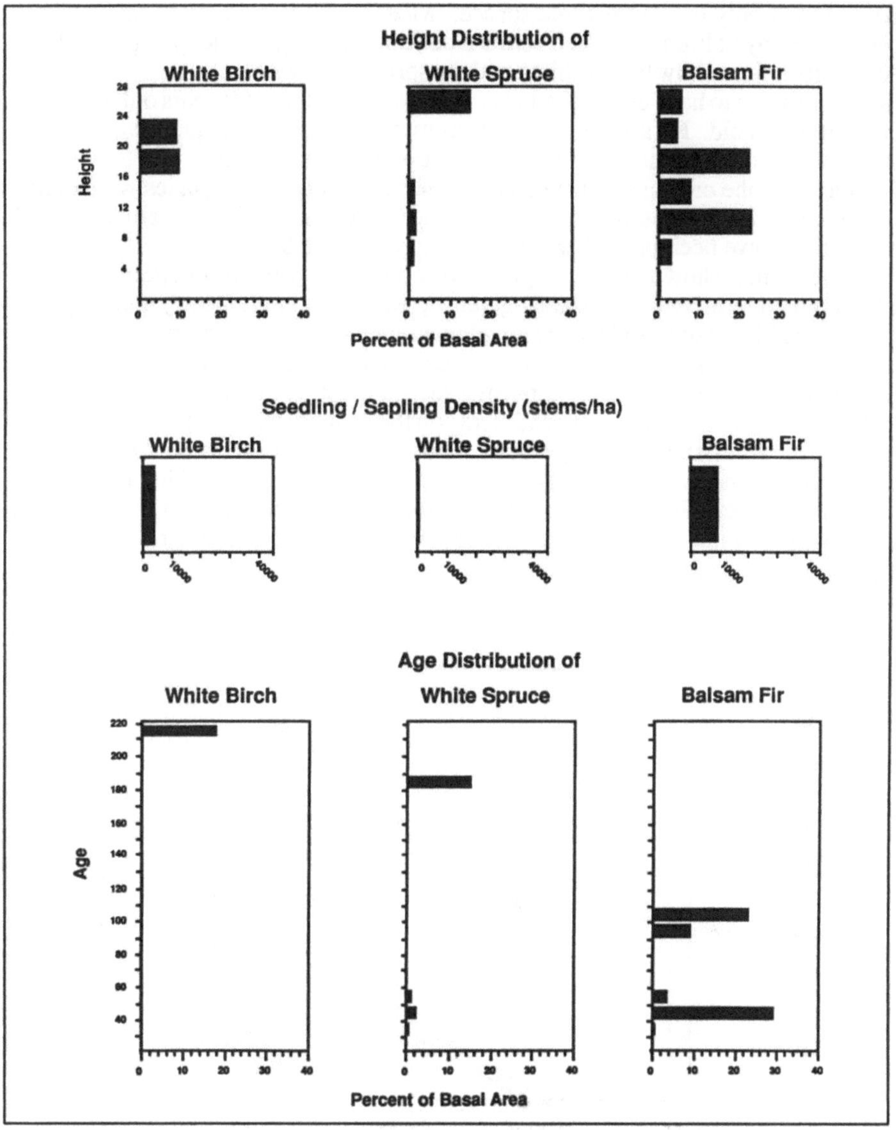

Fig. 4. Development of mixedwood stands: structure and age distribution of 220 year-old stand.

ground) and only one large white spruce. Most of the basal area of this stand is contributed by balsam fir, which forms a continuous range of heights up to 28 m. Age distributions show that the birch and the spruce are much older than the fir. The fir again appear to have established in two pulses, one 80 to 100 years old, the other 30 to 50 years old. Fir seedlings are not as numerous in this stand, because of the shade of the dense stratum of fir trees. There are a few small white spruce trees under 50 years old, the only case where any new recruitment of this species was found.

To summarize these trends, changes in the relative basal areas of spruce, fir, and hardwoods have been plotted against stand age (Figure 5).

The trends shown by these plots are consistent with the pattern of stand development in mixedwood stands given by Kabzems et al. (1986) and Day and Harvey (1981). Aspen and spruce develop as even-aged stratified mixtures, while senescence of these mixtures leads to succession to balsam fir. The long-term development of these stands remains to be seen. Fir appears to have a high mortality rate, as shown by the number of dead stems on the ground, and it is not clear whether it will ever form stands of large trees. But fir does appear to be successful in establishing dense, all-aged, self-regenerating stands, and thus taking over mesic sites in the western boreal forest. The present relative scarcity of fir in the west is apparently a result of the natural frequency of wildfire, and not of the lack of competitiveness of fir in this climate.

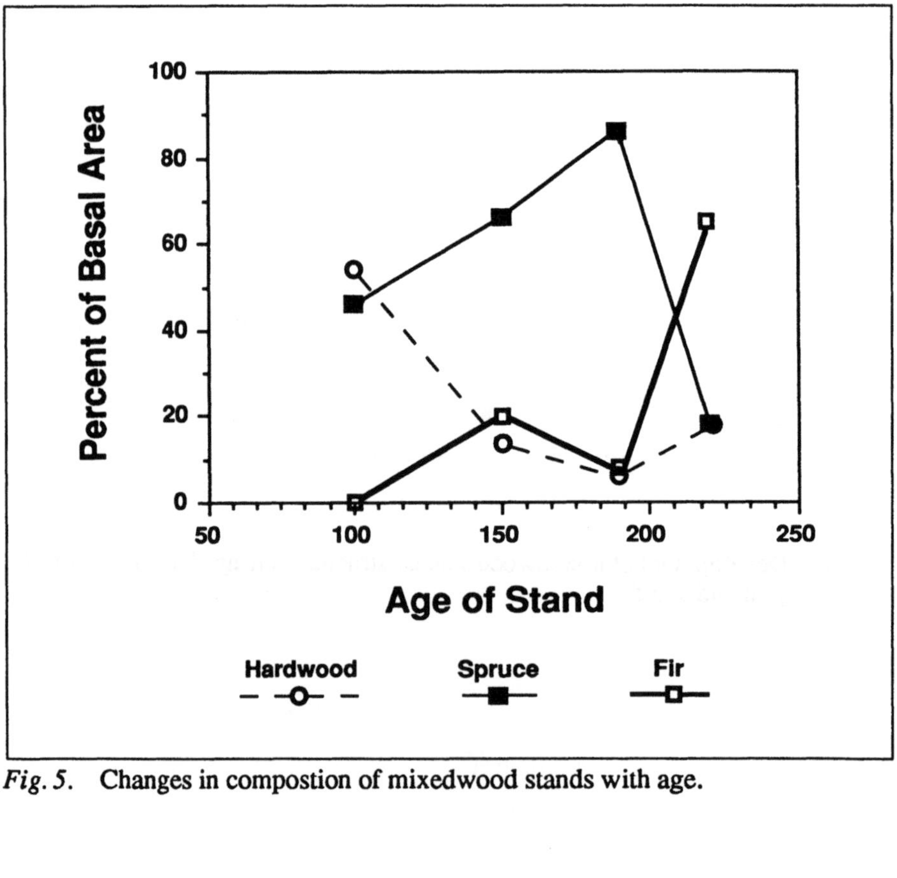

Fig. 5. Changes in compostion of mixedwood stands with age.

At the northwestern end of the boreal forest, however, balsam fir reaches its range limit, and development of mixedwood stands occurs in the absence of this species. On the floodplain of the Peace River in northern Alberta, mixedwoods are dominated by balsam poplar and white spruce, typical of wet-mesic sites. Lacate et al. (1965) found that the ages of trees in these stands suggest a development pattern similar to that described for aspen-spruce mixedwoods, with an essentially even-aged cohort of white spruce emerging as the initially dominant hardwoods die. When the spruce canopy begins to thin out, there is scattered spruce regeneration, and much shrub growth. In Alaska, Van Cleve and Viereck (1981) described similar development of balsam poplar-white spruce stands on floodplains. However, in this colder climate, there is some evidence that the mature, white spruce-dominated stage is eventually replaced by black spruce bog. This results from the thickening of the humus mat under a spruce canopy, and thus to poor drainage and formation of permafrost. Upland aspen-white spruce stands are described as developing similarly to those further east, but Van Cleve and Viereck found no stands old enough to shed light on development beyond the white spruce-dominated stage.

It is interesting to compare the development of boreal mixedwoods with patterns in the subalpine forest of the Rocky Mountains. This biome is similar in many respects to the boreal forest, and the main conifer species are closely related genetically, and similar ecologically, to boreal species. In the subalpine forest, Engelmann spruce replaces white spruce, subalpine fir replaces balsam fir, and lodgepole pine replaces jack pine. However, the broad-leaved species are much less important in the subalpine than in nearby boreal forest, especially at higher altitudes.

Bloomberg (1950) and Day (1972) analyzed age distributions in conifer stands on the east slope of the Rockies in Alberta. Lodgepole pine, which like jack pine has serotinous cones, seeds in abundantly after fire and forms even-aged stands on most sites. The spruce, which are white-Engelmann hybrids in this area at moderate altitudes, also have a wave of establishment beneath the pine in the first 10-20 years after fire. Bloomberg found little spruce regeneration after this, but Day found continued establishment, so that old stands have a single cohort of pine and a broad range of ages of spruce. This probably reflects the genetic contribution of Engelmann spruce, which is reputed to be more shade-tolerant than white spruce. Subalpine fir, which like balsam fir is poorly adapted to regeneration after fire, appears gradually, but in old stands it forms most of the understory. Fir apparently never attains great age or size. Day argued that the old spruce-fir stands are not in a climax state, but rather will continue to increase in domination by fir.

Veblen (1986), working in similar stands in the Colorado Rockies, found that fir produce most of the advance growth, althought there is also some Engelmann spruce, and fir forms the likely successor in most treefall gaps. However, he also found that fir have a higher mortality rate, as measured by numbers of treefalls relative to number of mature trees. Veblen concluded that these results are consistent with a hypothesis of coexistence, fir persisting through higher fecundity and spruce through greater longevity.

Parallels to the development of boreal mixedwoods are obvious. The boreal species are similar with respect to the fecundity and rapid turnover of fir, and the greater longevity of spruce. However, the Engelmann or hybrid spruce in the subalpine environment seem to be more successful at regeneration under an established stand, compared to white spruce in the boreal environment. In boreal mixedwoods, it appears more likely that spruce will dwindle away, leaving stands dominated by fir.

To this point, only the common kinds of mixed-species stands found on relatively productive sites have been discussed. Adverse sites in the boreal forest generally support simpler stands made up of stress-tolerant species. Jack pine stands on sand and gravel, and black spruce stands on gleysols, peatlands, and permafrost sites, present monotonous expanses of low diversity over much of the boreal forest.

However, these species also form mixed stands under some circumstances. Jack pine-black spruce mixtures, with the pine overtopping the spruce, are relatively common, often occurring on sandy sites affected by a water table in the rooting zone. As with aspen-white spruce stands, these stands have been interpreted as showing succession to a black spruce climax. It is more probable that these are even-aged stratified mixtures, in which both species have seeded in after fire, and in which the faster juvenile growth of jack pine has allowed it to overtop the spruce. In a sample of ages from such stands in Quebec, most jack pine dated to the first decade, and most black spruce to the first three decades, since stand initiation (Cogbill 1985). Kabzems et al. (1986) suggested that the jack pine component in such stands thins out at 80 to 100 years, leading to increasing dominance by the longer-lived spruce.

Black spruce also forms mixed stands with tamarack, on wet sites of intermediate nutrient regime. Black spruce again tends to be overtopped by the faster-growing, intolerant tamarack. At the maritime eastern end of the boreal forest, black spruce forms postfire stands on most sites, which undergo succession to balsam fir in the absence of further burning (Damman 1964, 1967).

Fire has a dominant effect on patterns of diversity in the boreal forest. Fire imposes uniformity on the stands formed initially. However, subsequent development takes these stands through a series of compositional stages, resulting in high diversity through time on a given site, provided the period between fires is long enough. At the landscape level, the boreal forest often appears as a mosaic of contrasting developmental stages due to fires of different dates. The result is relatively high landscape diversity, in the sense used by Romme (1982). Conversely, absence of fire can lead to low landscape diversity, as in the fir forests of humid regions, where disturbance by insect attack or blowdown leads to replacement of fir by fir.

The interaction among species in mixed stands presents more interesting silvicultural problems than in the more widespread monodominant stands. Because the conifers have generally been the more valued timber trees, herbicides and cutting have long been used to remove overtopping hardwoods and release spruce. Increasingly, however, pulping technologies have been adapted to make use of the hardwoods. As demand for hardwood has increased, some foresters have changed their policy to one of dedicating productive sites to aspen stands, and abandoning the spruce-growing potential of these sites. Perhaps there is a place here for the concept of managing these stands as stratified mixtures, as proposed by Smith (1986).

Patterns in the diversity of ground vegetation

While the ground vegetation adds many more species to the boreal flora, the diversity of the ground vegetation is still low compared to other biomes. An indication of this can be seen in the data of LaRoi (1967), who sampled 34 white spruce-balsam fir stands and 26 black spruce stands located across the whole width of the North American boreal forest. The total number of vascular species sampled was 291, while the number of species per stand ranged from 20 to 58 for spruce-fir stands and from 12 to 60 for black spruce stands.

The very low diversity of boreal tree species implies that species are wide-ranging over a variety of site types. The greater diversity in ground vegetation is often useful in differentiating site types. For example, at Fort à la Corne, Saskatchewan, jack pine stands on stabilized dunes have sparse low shrubs (*Vaccinium myrtilloides*, *Prunus virginiana*, *Rosa acicularis*) and herbs (*Maianthemum canadense*, *Oryzopsis asperifolia*), and very high lichen cover (*Cladina* spp.) (Thorpe 1990). Pine stands on level sand plains have greater cover in the shrub and herb strata, with some mesophytic species (*Amelanchier alnifolia*, *Aralia nudicaulis*) added to the xerophytes already present, and with more feather moss (*Pleurozium schreberi*, *Hylocomium splendens*) than lichen in the lowest stratum. Pine stands on interbedded sediments with seepage have dense strata of tall shrubs (*Corylus cornuta*, *Amelanchier alnifolia*, *Prunus virginiana*) and of herbs, including many moisture-demanding species (*Equisetum arvense*, *Petasites palmatus*, *Calamagrostis canadensis*).

As in the above example, the diversity of ground vegetation tends to be higher on more productive sites. Dyrenkov et al. (1981) described three types of spruce forests in northern Russia. *Vaccinium-Sphagnum* types on wet sites had 11 species of shrubs and herbs, *Vaccinium-Maianthemum* types on noncalcareous loam had 24 species, while grassy-leafy types on calcareous loams had 59 species.

The diversity of ground vegetation also tends to be higher under hardwood stands than under conifer stands. Hardwood litter is richer in bases than conifer litter, and so creates less acidic, more nutrient-rich conditions in the humus layer where most herbs are rooted. In addition, the boreal hardwoods have thin crowns, allowing relatively high light levels in the understory, resulting in the development of dense ground vegetation. Often this is dominated by a few tall shrubs, such as *Corylus cornuta* or *Acer spicatum*, but a diverse herbaceous stratum is also present.

The lowest diversity is found in dense conifer stands on low-nutrient sites. Closed black spruce-feather moss stands, which often have no more than four or five species of sparsely distributed herbs and dwarf shrubs, set an extreme of low vascular species diversity among forest communities. The least productive open black spruce stands on peat tend to have higher diversity, because of higher light levels and the appearance of the distinctive assemblage of bog species (ericaceous shrubs, sedges, insectivorous plants).

Transitions to other biomes

The North American boreal forest is bounded on the north by Arctic tundra, and on the south by grassland in the drier western portions, and by temperate deciduous forest in the more humid east (Figure 1). The transitions to tundra and to grassland occur along gradients of increasing severity for tree growth, and diversity declines along these gradients.

Within the boreal forest, landscape diversity declines northward, from the productive "mid-boreal" forest (upon which most of the above discussion is based) to the colder "high boreal" forest. In the high boreal region, simple black spruce-feather moss stands expand to a wide range of sites. Jack pine is still important on coarse soils, especially in the west, but the other species, the hardwoods, white spruce, and balsam fir become increasingly restricted to favourable sites. At Nicauba, Quebec, these are found only in the warm microclimate of hilltops (Jurdant and Frisque 1970), while in southeastern Labrador, they are found on protected concave slopes with seepage (Foster 1984). Such stands form islands of diverse ground vegetation amid the generally simple feather moss stands.

The transition to Arctic tundra occurs over a broad zone of lichen woodland, consisting of widely spaced trees, mostly black spruce, with a mat of fruticose lichens between them. Jack pine reaches its northern limit well south of the Arctic timberline, which is formed by the spruces.

The transition to semiarid grassland in western Canada is marked by the declining importance of the conifers. In the southern or "low boreal" forest in the Prairie Provinces, black spruce and tamarack become strictly confined to peatlands, and jack pine to coarse soils, while white spruce tends to retreat to moist sites. The mesic uplands become overwhelmingly dominated by trembling aspen. Some vegetation maps in fact show this zone as a deciduous forest biome. It has been argued that the conifer component here would be greater if it were not for the increasing frequency of fire as the grassland region is approached. The southern range boundaries of white spruce, black spruce, jack pine and tamarack are strikingly convergent (Zoltai 1975). At about the point that conifers disappear from the forest, grassland patches begin to appear on hilltops and south-facing slopes, gradually increasing in area southward, forming an "aspen parkland". While the diversity of tree species is very low, the mosaic of aspen stands, shrublands, grasslands, and small wetlands has very high landscape diversity. The final step in this sequence is the semiarid grassland of the Great Plains, where landscape diversity is low, except where trees appear along permanent streams.

By contrast with the tundra and grassland transitions, the transition to temperate deciduous forest occurs along a gradient of increasingly favourable conditions for the growth of a wide range of tree species. The following description is based on a literature review for eastern Canada and adjacent parts of U.S.A. (Thorpe 1986). Vegetation studies were grouped according to the temperature zones used by Chapman and Brown (1966), converted here to degree-days above 5o C:

zone 7	< 1077 degree-days
zone 6	1077 - 1307
zone 5	1307 - 1538
zone 4	1538 - 1769
zone 3	> 1769

For each of a standard series of sites in each zone, a list of tree species in approximate order of importance was compiled from the various studies having relevant data (Figure 6).

Zone 7 is "high boreal" with black spruce dominant on all sites, but balsam fir, white spruce, and white birch present in stands on mesic slopes. In zone 6, fir, white spruce, and white birch become dominant on mesic sites, and black spruce is dominant only on drier or wetter sites, while a few temperate species (white elm, black ash, white pine) appear. In zone 5, sugar maple and yellow birch (accompanied by numerous other temperate species) dominate the hilltops and upper slopes, while the boreal species are restricted to the cooler and moister lower slopes and flats. In zone 4, sugar maple dominates a wider range of mesic sites, beech becomes more important than yellow birch on these sites, and temperate species such as yellow birch and red maple become increasingly important on the lower and wetter

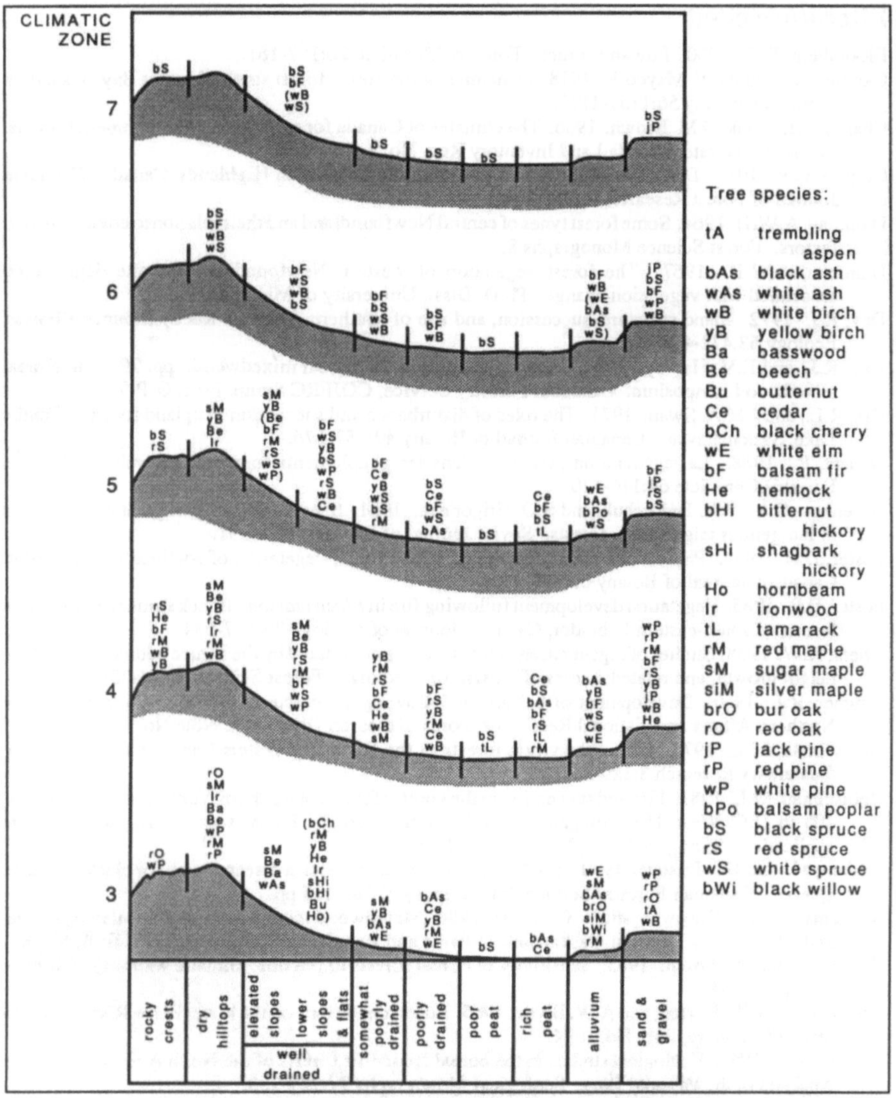

Fig. 6 Average stand composition by site along a climatic gradient, from coolest (zone 7) to warmest (zone 3), in eastern Canada. Scientific names of tree species are given in Appendix A.

sites. In zone 3, the maple-dominated stands contain many more temperate species (basswood, white ash, hickories, butternut, hornbeam), red oak becomes important on warm, dry sites, and the boreal species nearly disappear, except for black spruce in bogs.

The changes in composition along this climatic gradient provide a striking demonstration of the low diversity of the boreal forest, compared to the richness of species permitted by the environment of the temperate forest.

Literature cited

Bloomberg, W.J. 1950. Fire and spruce. Forestry Chronicle 26:157-161.

Carleton, T.J., and P.F. Maycock. 1978. Dynamics of the boreal forest south of James Bay. Canadian Journal of Botany 56:1157-1173.

Chapman, L.J., and D.M. Brown. 1966. The climates of Canada for agriculture. Environment Canada, Lands Directorate, Canada Land Inventory Rep. No. 3.

Cogbill, C.V. 1985. Dynamics of the boreal forests of the Laurentian Highlands, Canada. Canadian Journal of Forest Research 15:252-261.

Damman, A.W.H. 1964. Some forest types of central Newfoundland and their relation to environmental factors. Forest Science Monographs 8.

Damman, A.W.H. 1967. The forest vegetation of western Newfoundland and site degradation associated with vegetation change. Ph.D. Diss., University of Michigan.

Day, R.J. 1972. Stand structure, succession, and use of southern Alberta's Rocky Mountain Forest. Ecology 53:472-478.

Day, R.J., and E.M. Harvey. 1981. Forest dynamics in the boreal mixedwood. pp. 29-41 in Boreal mixedwood symposium. Canadian Forestry Service, COJFRC Symp. Proc. O-P-9.

Dix, R.L., and J.M.A. Swan. 1971. The roles of disturbance and succession in upland forest at Candle Lake, Saskatchewan. Canadian Journal of Botany 49:657-676.

Doucet, R. 1988. La régénération préétablie dans les peuplements forestiers naturels au Québec. Forestry Chronicle 64:116-120.

Dyrenkov, S.A., V.N. Fedorchuk, and S.O. Grigor'eva. 1981. Species diversity of plant communities in indigenous taiga spruce forests. Soviet Journal of Ecology 12:85-91.

Foster, D.R. 1984. Phytosociological description of the forest vegetation of southeastern Labrador. Canadian Journal of Botany 62:899-906.

Foster, D.R. 1985. Vegetation development following fire in *Picea mariana* (black spruce)-*Pleurozium* forests of south-eastern Labrador, Canada. Journal of Ecology 73:517-534.

Ghent, A.W. 1958. Studies of regeneration in forest stands devastated by the spruce budworm. II. Age, height growth, and related studies of balsam fir seedlings. Forest Science 3:184-208.

Hatcher, R.J. 1960. Development of balsam fir following a clearcut in Quebec. Canada Dept. of Northern Affairs and National Resources, Forest Research Div., Tech. Note No. 87.

Heinselman, M.L. 1973. Fire in the virgin forests of the Boundary Waters Canoe Area, Minnesota. Quaternary Research 3:329-382.

Heinselman, M.L. 1981. Fire and succession in the conifer forests of northern North America. pp. 374-405 in D.C. West, H.H. Shugart, and D.B. Botkin (eds.). Forest succession: concepts and application. Springer-Verlag, New York. 517 pp.

Jurdant, M., and G.J. Frisque. 1970. The Nicauba Research Forest: a research area for black spruce in Quebec. Canadian Forestry Service, Info. Rep. Q-X-18. 115 pp.

Kabzems, A., A.L. Kosowan, and W.C. Harris. 1986. Mixedwood section in an ecological perspective. 2nd ed. Forestry Division, Saskatchewan Parks and Renewable Resources, Tech. Bull. No. 8.

Kauppi, P., and M. Posch. 1985. Sensitivity of boreal forests to possible climatic warming. Climatic Change 7:45-54.

Lacate, D.S., K.W. Horton, and A.W. Blyth. 1965. Forest conditions on the lower Peace River. Canada Dept. of Forestry, Pub. No. 1094.

La Roi, G.H. 1967. Ecological studies in the boreal spruce-fir forests of the North American taiga. I. Analysis of the vascular flora. Ecological Monographs 37:229-253.

Logan, K.T. 1969. Growth of tree seedlings as affected by light intensity. IV. Black spruce, white spruce, balsam fir, and eastern white cedar. Canadian Forestry Service, Pub. No. 1256.

MacLean, D.W. 1960. Some aspects of the aspen-birch-spruce-fir type in Ontario. Canada Dept. of Forestry, Forest Research Div., Tech. Note No. 94.

Morris, R.F. 1948. How old is a balsam tree? Forestry Chronicle 24:106-110.

Moss, E.H. 1953. Forest communities of northwest Alberta. Canadian Journal of Botany 31:212-252.

Moss, E.H. 1955. The vegetation of Alberta. Botanical Review 21:493-567.

Padbury, G.A., W.K. Head, and W.E. Souster. 1978. Biophysical resource inventory of the Prince Albert National Park. Saskatchewan Inst. of Pedology S195. 560 pp.

Place, I.C.M. 1955. The influence of seed-bed conditions on the regeneration of spruce and balsam fir. Canada Dept. of Northern Affairs and National Resources, Forestry Branch, Bull. 117.

Romme, W.H. 1982. Fire and landscape diversity in subalpine forests of Yellowstone National Park. Ecological Monographs 52:199-221.

Rowe, J.S. 1961. Critique of some vegetational concepts as applied to forests of northwestern Alberta. Canadian Journal of Botany 39:1007-1017.

Rowe, J.S. 1970. Spruce and fire in northwest Canada and Alaska. pp. 245-254 in Proceedings Annual Tall Timbers Fire Ecology Conference, Aug. 20-21, 1970.

Rowe, J.S., and G.W. Scotter. 1973. Fire in the boreal forest. Quatemary Research 3:444-464.

Smith, D.M. 1986. The practice of silviculture. 8th ed. John Wiley & Sons, New York. 572 pp.

Strong, W.L., and K.R. Leggatt. 1981. Ecoregions of Alberta. Alberta Energy and Natural Resources, Tech. Rep. No. T/4.

Thorpe, J.P. 1986. An analysis of stand composition in relation to site in an area of spruce-fir-northern hardwoods in Maine. Ph.D. Diss., Yale University.

Thorpe, J.P. 1990. An assessment of Saskatchewan's system of forest site classification. Saskatchewan Research Council Pub. No. E-2530-5-E-90.

Van Cleve, K., and L.A. Viereck. 1981. Forest succession in relation to nutrient cycling in the boreal forest of Alaska. pp. 185-211 in D.C. West, H.H. Shugart, and D.B. Botkin (eds.). Forest succession: concepts and application. Springer-Verlag, New York. 517 pp.

Veblen, T.T. 1986. Treefalls and the coexistence of conifers in subalpine forests of the central Rockies. Ecology 67:644-649.

Vincent, A.B. 1956. Balsam fir and white spruce reproduction on the Green River watershed. Forest Research Div. (Canada) Tech. Note No. 40.

Wein, R.W., and J.M. Moore. 1977. Fire history and rotations in the New Brunswick Acadian Forest. Canadian Journal of Forest Research 7:285-294.

Wein, R.W., and J.M. Moore. 1979. Fire history and recent fire rotation periods in the Nova Scotia Acadian Forest. Canadian Journal of Forest Research 9:166-178.

Westveld, M. 1931. Reproduction on pulpwood lands in the northeast. U.S. Dept. of Agriculture, Tech. Bull. No. 223.

Zoltai, S.C. 1975. Southern limit of coniferous trees on the Canadian Prairies. Canadian Forestry Service, Information Report NOR-X-128.

Appendix A

Scientific and common names of species referred to in text

Tree species:

balsam fir	*Abies balsamea*
red maple	*Acer rubrum*
silver maple	*Acer saccharinum*
sugar maple	*Acer saccharum*
white birch	*Betula papyrifera*
yellow birch	*Betula alleghaniensis*
hornbeam	*Carpinus caroliniana*
bitternut hickory	*Carya cordiformis*
shagbark hickory	*Carya ovata*
beech	*Fagus grandifolia*
white ash	*Fraxinus americana*
black ash	*Fraxinus nigra*
butternut	*Juglans cinerea*
tamarack	*Larix laricina*
ironwood	*Ostrya virginiana*
white spruce	*Picea glauca*
black spruce	*Picea mariana*
jack pine	*Pinus banksiana*
red pine	*Pinus resinosa*
eastern white pine	*Pinus strobus*
balsam poplar	*Populus balsamifera*
trembling aspen	*Populus tremuloides*
black cherry	*Prunus serotina*
bur oak	*Quercus macrocarpa*
red oak	*Quercus rubra*
eastern white cedar	*Thuja occidentalis*
black willow	*Salix nigra*
basswood	*Tilia americana*
eastern hemlock	*Tsuga canadensis*
white elm	*Ulmus americana*

Other species:

saskatoon	*Amelanchier alnifolia*
wild sarsaparilla	*Aralia nudicaulis*
reed grass	*Calamagrostis canadensis*
reindeer lichen	*Cladina* spp.
beaked hazelnut	*Corylus cornuta*
common horsetail	*Equisetum arvense*
wild lily-of-the-valley	*Maianthemum canadense*
palmate-leaved colt's-foot	*Petasites palmatus*
feather moss	*Pleurozium schreberi*, *Hylocomium splendens*
chokecherry	*Prunus virginiana*
prickly rose	*Rosa acicularis*
blueberry	*Vaccinium myrtilloides*

Regeneration from seed under a range of canopy conditions in tropical wet forest, Puerto Rico

6

NORA N. DEVOE

Introduction: The gap-phase regeneration paradigm

There is now a generally accepted model of tropical rainforest dynamics (Swaine and Whitmore 1988). Transient canopy openings ("gaps") develop in closed forest, new trees grow in them, and the stand develops; eventually a mature forest is reconstituted. This is the essence of so-called gap-phase regeneration. The cycle is initiated with the gap, after which the stand moves through the building to the mature phase (Whitmore 1978). The gap segment is crucial because what establishes and grows then determines the floristic composition of the entire cycle, and influences composition in subsequent cycles. The gap-phase regeneration model has been applied to temperate as well as tropical forests (e.g., Jones 1945, Runkle 1979, Spies and Franklin 1989).

Many investigators examined the gap-phase phenomenon and concluded that the size of the canopy opening was the prime influence on the species composition of the regeneration therein (e.g., Baur 1968; Hartshorn 1978, Whitmore 1978, Acevedo and Marquis 1979, Brokaw 1985). Denslow (1980) reasoned that since the regeneration requirements of trees vary, early successional or pioneer vegetation can be expected to enter large gaps while later successional vegetation is confined to smaller openings. Still a third category of species was thought to have regeneration restricted to the understory.

Further refinement of the gap-phase model came with increased awareness of within-gap heterogeneity. Exposed mineral soil, areas sheltered by ground flora, and areas deeply shaded by debris may all occur in close proximity within treefall gaps. Some workers maintained that even if uniformly lit from above, these zones would be sufficiently different to separate species colonizing them (Richards and Williams 1975; Hartshorn 1980; Orians 1982; Brandani et al. 1988). In some cases, zonation was thought to result from unequal distribution of organic matter and hence nutrient fluxes, but no such inequalities were found at La Selva, Costa Rica, (Parker 1985; Vitousek and Denslow 1986) or San Carlos de Rio Negro, Venezuela (Uhl et al. 1988). Uhl et al. (1988) found no zonation in species abundance or composition by substrate conditions.

Within-gap partitioning may be as important or more important than partitioning among gaps of different sizes. Barton (1984) found that all species regenerate in all gaps with some intra-gap zonation. Transects taken perpendicular to the long axis of treefall gaps in Puerto Rican tabonuco forest showed that species locations with respect to distance from gap edges was non-random (Perez 1986). At least in the tabonuco forest, propagule supply plays a role in species zonation, with the forest understory and gap margins receiving more propagules of later successional species than gap centers (Devoe 1990). In this paper, I test the hypothesis that within-gap zonation is also related to non-uniform insolation in gaps.

M. J. Kelty (ed.), The Ecology and Silviculture of Mixed-Species Forests, 81–100.

Portions of gaps receive varying amounts of direct insolation depending upon the size and shape of the opening, the latitude, slope and aspect of the site, height of the surrounding forest, and time of year. Shade and shadow patterns have been accurately modelled from these variables (Marquis 1965; Brown and Merritt 1970; Halverson and Smith 1979; Harrington 1983). Larger gaps have greater proportions of their areas in highly exposed conditions. The generalization that large gaps are colonized by pioneer vegetation is true, but for some pioneers "large gap" means only the highly exposed regions, while for other pioneers margins of large gaps are sufficient (Barton 1984; Popma et al. 1988). Later successional species may colonize more-shaded portions of gaps at the same time that pioneers invade the centers.

Objective

I hypothesize that a mechanism for the influence of gap size upon species composition and for the zonation of species within gaps rests with the structuring of shadow patterns determining levels of light and heat reaching propagules. Species regenerating from seed along the microenvironmental gradient from fully exposed centers of large gaps to forest understory were observed. The purpose was to test the hypotheses (1) that individual gaps are non-uniform environments for plant colonization and (2) that some exposure levels are more favorable to the recruitment and growth of particular species than are others. Observations at four months and one year after gap creation were used to determine the initial rates at which species established in these environments via the seed rain and soil seed bank. This study examined regeneration only from seed; the other potential sources of regeneration, advance growth seedlings and stump sprouts, were eliminated. The importance of these other modes of regeneration is considered later in the discussion.

Methods

Study site

The study was conducted in the Luquillo Experimental Forest, northeastern Puerto Rico (18°22' N, 65°52' W). The community at the elevation range studied (330-500 m) is the tabonuco type, dominated by tabonuco (*Dacryodes excelsa*), cacao motillo (*Sloanea berteriana*), ausubo (*Manilkara bidentata*), and sierra palm (*Prestoea montana*). (Nomenclature follows Liogier and Martorell 1982). The forest is unusual among tropical forests in that these species are recognizable dominants. Compared to continental tropical forests, the biota is depauperate, with only 225 native tree species in the Luquillo Experimental Forest (Little and Woodbury 1976), many of them absent from the tabonuco type. Rainfall on the study site averages 3920 mm per year with high annual variation and slight seasonality. Mean monthly temperatures range from 21° to 24°C with mean annual relative humidity of 88% (Brown et al. 1983). The experiment was installed in a late secondary stand.

The forest light climate at the study site has extreme, stochastic spatial variability; temporal variability is diurnal and seasonal. Annual variation in daylength is more than two hours. Differences in day length of this magnitude have been shown to modify growth and development of tropical trees (Longman and

Jenik 1974). Insolation at El Verde has maxima in May and August, a slight minimum in June and a pronounced minimum in December. There are more cloudy days in summer, however, so that monthly incident radiation actually varies little, except during the winter minima. Reduced winter sunlight results from decreased angles of incidence and shorter days. Insolation reaching the forest floor beneath a closed canopy varied from about 1% of total insolation in February to about 6% in April (Odum et al. 1970).

The simulation model of Brown and Merritt (1970) was used to examine the effects of sizes and shapes of canopy openings on shadow patterns. I sought to encompass the range of light exposure levels from maximum possible irradiation, here defined as completely unshaded from 8 am to 6 pm for nine months of the year, to minimum irradiation, defined as that in the understory away from gap edges. I wanted this range of exposure values on the smallest plots that would have at least 500 m^2 in the maximally exposed condition. With the model, I calculated that the smallest opening fulfilling these conditions was 24 m north-south by 64 m east-west. Hours of direct light exposure varied clinally; the maximally irradiated area was a rectangle in the center of the plot, areas adjacent to this center received the next highest direct exposure levels and so on, with the lowest light levels in the forest understory.

Field procedures

Two areas with dimensions of 24 m north-south by 64 m east-west were cleared in March, 1987. They were separated by 54 m in the north-south direction (Fig. 1). As predicted by the model, the shadow patterns oscillated east-west and north-south during the year. The east ends of the plots were shaded in the morning before 10 am but lit in the afternoon, while the west ends were lit in the morning and shaded in the afternoon. In winter, the openings were shaded along their southern edges while the northern edges were shaded during parts of spring and summer.

Two regeneration transects were established on each plot. These extended 110 m east-west across the plots, through the openings and into the forest at either end. One-hundred and forty permanent 1 m x 1 m sampling units were randomly placed within these transects. Within each sampling unit, living tissue showing at the surface was removed by hand, so that all were uniformly free of above-ground vegetation. This cleaning in April 1987 marked the start of the experiment. No roots were excavated.

The transects were first censused in July 1987, called t_1. Vegetative reproduction was removed by uprooting. Seedlings originating within the quadrats from non-vegetative sources were tagged with numbers. Identification numbers and species of seedlings were recorded. Voucher specimens were obtained. A few species were identified only to genus in cases where the genus had numerous species. Percent cover, defined as the percentage of quadrat surface area covered by one or more layers of vegetation rooted in the same quadrat, was estimated. The height of the tallest individual (top height) on each quadrat was measured to the nearest centimeter. Average height of the vegetation on each quadrat was estimated.

The second census, called t_2, was taken in April 1988, one year after transect establishment. Untagged plants were considered new recruitment and tagged with numbers so indicating. Identification numbers assigned in the first census were checked and the condition of the plant noted as either healthy, moribund, dead, or missing. Percent cover, top height and average height were again recorded.

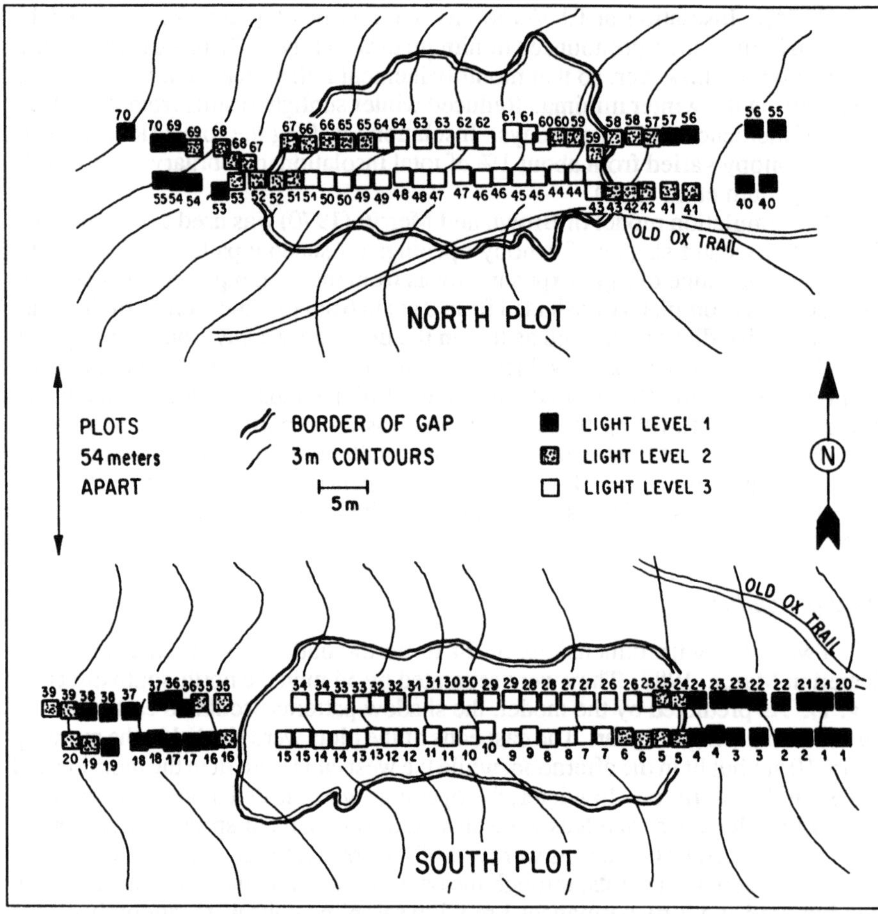

Fig. 1. Features of the study plots. Borders of gaps drawn by projecting crowns of edge trees to the ground. Sampling units (1 m x 1 m) shown twice actual size. Numbers show the two sampling units comprising each "quadrat" used in the ordination analyses. Light levels are 1=high light, 2=medium light, 3=low light.

Data analysis

A total of 3476 stems were recruited by April 1988; 92 unidentified individuals were removed from this data base prior to analysis. Separate matrices were prepared for the t_1 and t_2 samples. To reduce the number of empty cells, overall matrix dimensions, and extreme sample variation, the 140 permanent sampling units were collapsed into 70 pairs--data from each odd-numbered sampling unit were combined with that of the even-numbered one immediately following it in numerical order (see Fig. 1). Each of the 70 pairs is referred to as a "quadrat" in the discussion of data analysis. Densities or frequencies per two square meters for each species (columns) in the 70 quadrats (rows) at times t_1 (recruitment) and t_2 (recruitment minus mortality) were entered into the matrices.

Three broad categories of light level were used: 1=low, 2=medium, and 3=high. These corresponded to hours of direct light exposure as seen from the simulation model. Low light sampling units were completely shaded overhead with no nearby openings. Medium light level was assigned to sampling units with partial overhead shade or with high, complete overhead shade but pronounced light attenuation from the sides. High light sampling units were completely open overhead. (Light level values of 1.5 and 2.5 occurred for quadrats when adjacent sampling units had different light levels). Raich (1989) showed that canopy cover is inversely proportional to photosynthetic photon flux density. I therefore entered light score as a quantitative variable in a secondary matrix of quadrats (rows) by environmental attributes (columns).

Other quadrat attributes were location, slope position, and surface litter. Location referred to the plot on which the quadrat occurred (see Fig. 1). This was included as a measure of within-treatment variance, indicating how much of the variation in quadrat stocking could be attributed to random particulars of plot, such as adult tree distributions. Quadrat slope positions were read from site maps and ordered from lowest to highest. Ranks from 1=lowest (bottom of slope, west end of transects) to 70=highest (top of slope, east end of transects) were used to indicate slope position. Surface litter was scored as 1=thick (>2 cm), 2=thin (<2 cm), 3=absent, or 4=with charcoal.

Another secondary matrix consisting of species (rows) by species attributes (columns) was created for the transpose analysis of species-in-quadrats space. Attributes were life form, seed weight (data from Devoe 1990) and regeneration origin. Regeneration origin was a categorical variable indicating whether I thought colonization by the species *within and around the experimental gaps* was from 1) primarily the seed bank, 2) seed bank and seed rain more or less equally, or 3) primarily the seed rain. I also coded regeneration origin to indicate whether the species regenerated *under forest conditions* primarily via 1) the seed bank, 2) seed bank and seed rain, 3) seed rain, or 4) seed rain and advance growth, based on my observations of natural forest populations. Categories intergraded because most species exploit more than one origin.

I used Bray-Curtis polar ordination (Ludwig and Reynolds 1988) to search for trends in species composition along the environmental gradient represented by the regeneration transects (Austin 1977, 1987). Standardization by species, by site, and by both species and site (Noy-Meir et al. 1975) failed to produce results superior to those of the unstandardized analysis, which are presented here. The PC-ORD multivariate analysis package (McCune 1987) was used to ordinate the data and graph the ordinations. I used Sorensen's coefficient as the distance (dissimilarity) measure (Beals 1973). Axis endpoints were selected by the variance-regression method (Beals 1984).

The 60 and 62 most abundant species were used in the t_1 and t_2 ordinations, respectively. Kendall rank correlations (Sokal and Rohlf 1981) were used to examine relationships between the ordination axes and the environmental and response variables, and among light and the response variables. Differences in recruitment for given light levels were tested for designated species with the G log-likelihood ratio test (Sokal and Rohlf 1981).

Results

Time 1 ordination

In Figs. 2 and 3, quadrats are arranged (ordinated) in three-dimensional space based only upon similarities in species composition of regeneration, such that quadrats with least similarity in species are farthest apart. Figs. 2a and 2b show the locations of the quadrats on the ordination axes at t_1. Figs. 2c and 2d show an overlay of light scores on this quadrats-in-species-space ordination. Comparing these figures to Fig. 1, which shows the locations of the quadrats on the ground, we see that the quadrats with the same light levels are grouped. They are more similar to each other in species composition than they are to quadrats under different canopy conditions which may be physically closer together. A characteristic species assemblage was associated with each light level. At time 1, light was highly correlated with axes 1 and 2 (Table 1). That is to say, variation in species composition along axes 1 and 2 was strongly related to light level with the high light quadrats close to the origins and the low light ones near the termini (Fig. 2). Surface litter bore a similar relationship to that of light with the ordination axes. Light level was highly correlated with surface litter; the high light quadrats had low litter and low light quadrats had high litter. Axis 3 was significantly correlated with location and slope position.

Table 1. Kendall's rank correlation coefficients for the environmental and response variables with the ordination axes at t_1. Significance at the $p<.05$ level indicated by (*), at the $p<.01$ level by (**).

AXIS:	1	2	3
Light	-.557**	-.359**	-.196
Slope position	-.085	-.065	-.348**
Surface	-.559**	-.351**	-.226*
Location	.025	.037	.456**
Density	-.448**	-.046	-.086
Cover	-.562**	-.195	-.283*
Average height	.037	-.042	.038
Top height	-.414**	-.192	-.047

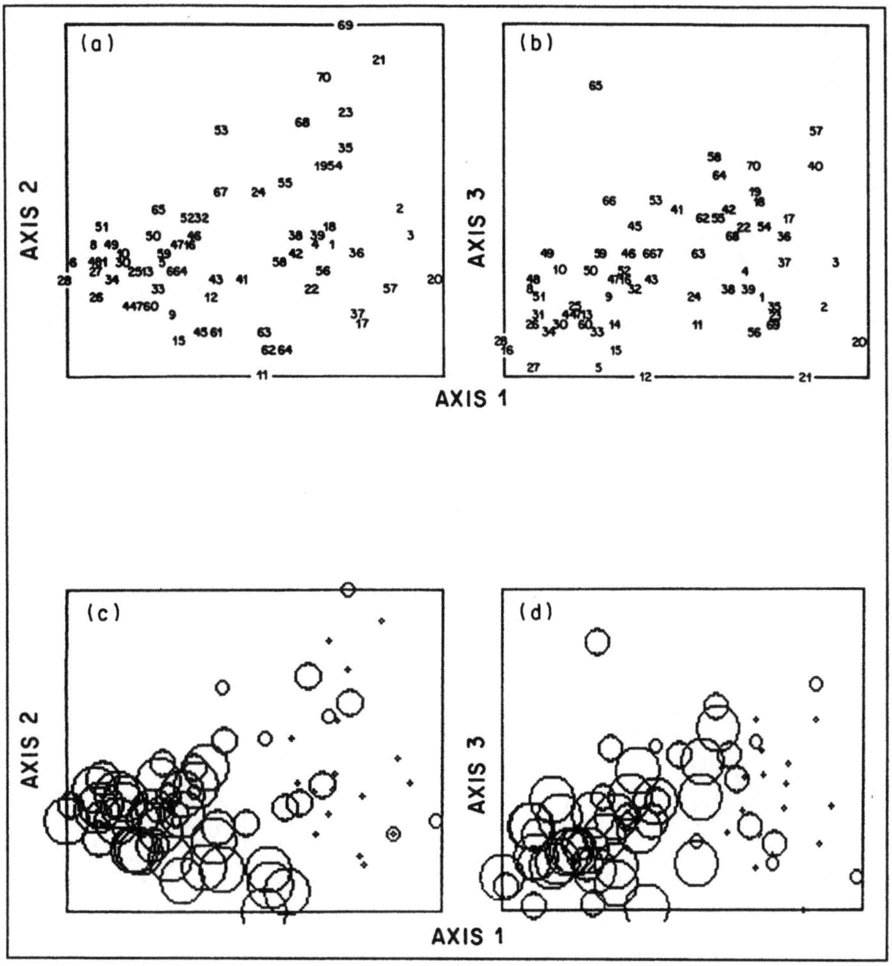

Fig. 2. Quadrats in species-space at time t_1 in relation to axes 1 and 2 (a), and in relation to axes 1 and 3 (b). The numbers in the diagrams are the quadrat numbers shown on the ground in Fig. 1. Overlays of light level on axes 1 and 2 (c) and on axes 1 and 3 (d). The sizes of the circles indicate the light levels of the quadrats (two 1 x 1 m field sampling units). The light levels in the diagrams are 1=low, 1.5, 2, 2.5, 3=high with circle sizes increasing with light level.

While quadrats were grouped by slope position in the t_1 ordination, it is questionable whether this reflected a true effect of slope position on species distribution or resulted more from patchy species distributions independent of slope. Similarly, the location effect on species composition indicated that a few species were confined to one plot or the other, probably due to biological rather than physical characteristics of the plots. Location and slope position showed no significant relationship with the response variables density, percentage cover, top height, or average height in correlation or ANOVA tests.

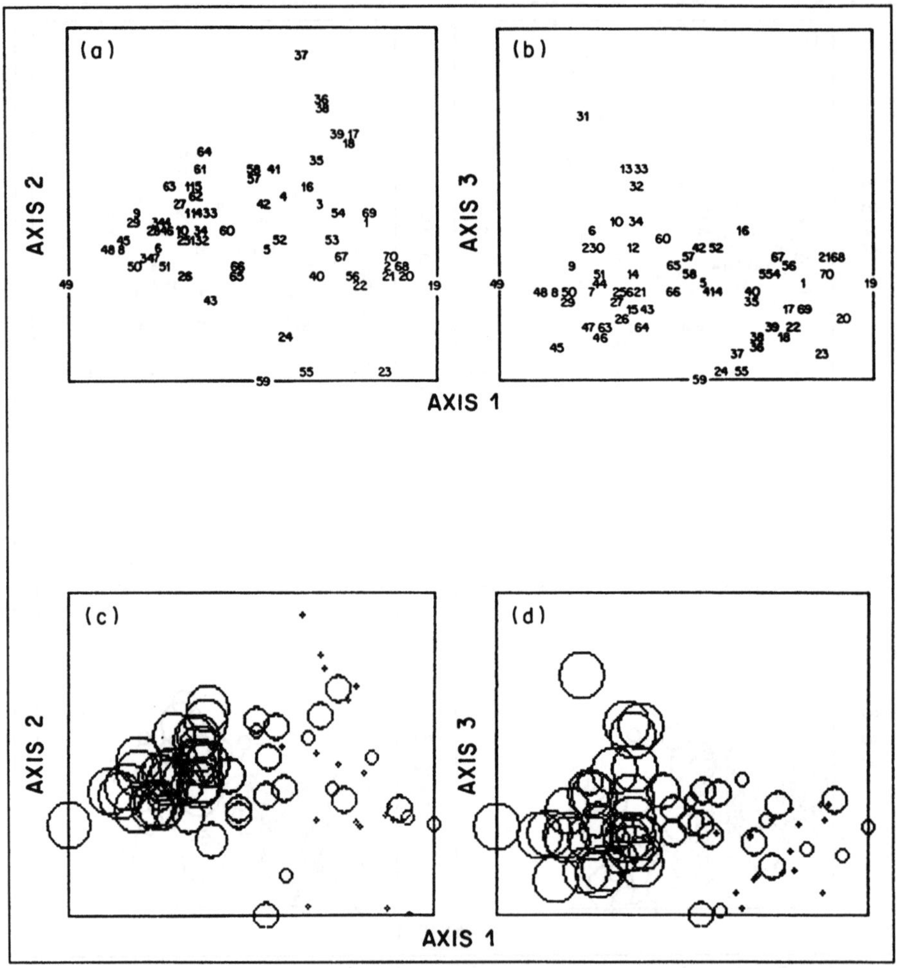

Fig. 3. Quadrats in species-space at time t_2 in relation to axes 1 and 2 (a), and in relation to axes 1 and 3 (b). The numbers in the diagrams are the quadrat numbers shown on the ground in Fig. 1. Overlays of light level on axes 1 and 2 (c) and on axes 1 and 3 (d). See Fig. 2 for explanation.

The response variables were correlated with each other and with light (Table 2). Overlays of these variables on the ordination axes showed the same pattern as the light overlay, with large values close to the origins. Regression and ANOVA T-tests showed significant effects for light at time t_1 on all the response variables except average height. Height range on the quadrats at t_1 may have been too small relative to within-treatment variation to exhibit a latent pattern.

Table 2. Kendall's rank correlation coefficients for light and the response variables at t_1. All correlations are significant at the p<.001 level except those between average height and light, density, and cover.

	LIGHT	DENS	COVER	TPHT	AVHT
LIGHT	1.000				
DENSITY	0.406	1.000			
COVER	0.513	0.451	1.000		
TOP HEIGHT	0.339	0.280	0.381	1.000	
AVERAGE HEIGHT	0.004	-0.084	0.110	0.369	1.000

Time 2 ordination

In the t_2 ordination, the endpoints for axes 1 and 3 were contrasting light levels, but both endpoints of axis 2 were in low light. Axis 1 was similar in composition to axis 1 at t_1 but axis 2 was very different and unrelated to the environmental variables. Axis 3 reversed the polarity of axis 3 at t_1 such that low light quadrats were placed near the origin and high light quadrats toward the terminus (Fig. 3). Slope position and location were no longer significantly correlated with the ordination axes, reflecting the fact that species were rarely confined to one end of a plot or to a single plot by this time (Table 3).

Table 3. Kendall's rank correlation coefficients for the environmental and response variables with the ordination axes at t_2. Significance at the p<.05 level indicated by (*), at the p<.01 level by (**).

AXIS	1	2	3
Light	-.683**	.064	.166
Slope position	-.108	.032	.044
Surface	-.635**	-.018	.137
Location	-.036	-.138	-.038
Density	-.391*	-.040	.140
Cover	-.607**	-.069	.022
Average height	-.514**	.013	.173
Top height	-.526**	.046	.201

Comparison of the two ordinations (Figs. 2a,b with Figs. 3a,b) shows that by time t_2 certain quadrats similar at time t_1 became more similar at t_2 (e.g., Q21/Q20; Q58/Q57; Q66/65) while others became less so (e.g., Q37/Q17; Q2/Q3; Q18/Q1; Q52/Q32). By t_2 site conditions had altered with increasing self-shading and biomass accumulation although the original quadrat descriptors (location, slope position, light, surface litter) were retained. Correlations between light and the response variables and among the response variables were larger at t_2 than t_1 (Table 4). At t_2 average height showed high correlation with light, an expected relationship that had not developed at t_1.

Table 4. Kendall's rank correlation coefficients for light and the response variables at t_2. All correlations are positive and significant at the $p<.001$ level.

	LIGHT	DENS	COVER	TPHT	AVHT
LIGHT	1.000				
DENSITY	0.407	1.000			
COVER	0.594	0.448	1.000		
TOP HEIGHT	0.607	0.352	0.529	1.000	
AVERAGE HEIGHT	0.593	0.361	0.560	0.809	1.000

Light had significant effect on stem density in ANOVA tests including all environmental factors at time t_1 ($F=18.5, p<.001$) and t_2 ($F=21.1, p<.001$) (Table 5). The rate of density increase had already slowed by the time of the first census. After the first census, site occupancy increased more by the growth of existing colonizers, as reflected in increases in percentage cover and height, than by the continued invasion of new individuals (Table 6).

Table 5. Mean stem densities (plants per square meter) by light level (1=low, 2=medium, 3=high).

Light Exposure Level	1	2	3
Mean Stem Density, t_1	12.1	17.4	31.4
Mean Stem Density, t_2	15.5	22.2	33.7

Table 6. Summary quadrat statistics (light levels pooled) at t1 and t2 based on measurements of the 140 sampling units.

t_1	DENSITY	%COVER	TOP HT.	AVG. HT.
MINIMUM	1.00	0.05	5.00	3.00
MAXIMUM	150.00	70.00	100.00	30.00
MEAN	21.53	11.79	17.39	6.87
STANDARD DEV	19.69	10.64	11.76	3.58
t_2	DENSITY	% COVER	TOP HT.	AVG. HT.
MINIMUM	1.00	3.00	8.00	4.00
MAXIMUM	90.00	99.99	350.00	155.00
MEAN	24.81	44.16	70.35	27.57
STANDARD DEV	17.52	36.47	65.61	26.43

Species attributes and groupings within light levels

Forty-three tree species, 13 shrub species, 21 vine species, 27 herb species, 14 grass species, and 11 fern species regenerated on the quadrats via non-vegetative origins. Species in quadrats-space were ordinated (the transpose of quadrats in species-space) and species attributes overlaid. The resulting distributions were similar at both points in time. In neither case were the ordination axes significantly correlated with any species attribute, indicating that all categories of life form, seed weight and regeneration origin occurred under all the light, slope position, and surface litter conditions on the transects.

While this result coincided with my observations, it seemed, too, that certain attributes were more common in some parts of the transects than in others. Attributes are not independent; for example, larger life forms tend to have more massive seeds (Foster and Janson 1985). In this study, species with more massive seeds and larger life forms regenerated more often from advance growth and the seed rain than from the seed bank. Although these were strong trends, the many exceptions prevented their exhibition in the transpose ordinations.

Overall stocking tended to be higher with higher light levels; however, the proportional representation of life forms varied with light level (Table 7). Trees reached their highest percentage of all stems under intermediate light, with the high light treatment showing the lowest proportion of trees. Shrubs and grasses became more abundant with increasing light level while vines reversed this trend. Only two of the 21 vine species were most abundant in low light, but one of these, *Hippocratea volubilis*, was recruited in numbers sufficiently large to determine the trend. Ferns were most abundant under low light, but their representation increased under high light from t_1 to t_2 as the tree fern *Cyathea arborea* was recruited.

Recruitment from species in the heaviest seed class was most abundant in absolute and relative terms beneath the closed canopy (Table 8). The trend was for recruitment from species with more massive seeds with decreasing light. If ferns were removed from the data, species with very small propagules would have been rare under low light.

Table 7. The distribution of life forms under treatment conditions. Stocking is the number of individuals per square meter; percentage of stems is the proportion of individuals of all life forms represented by a given life form within a given light level. Figures are based on all identified individuals, including rare species, totalling 96 species and 2935 individuals at t_1 and 3476 individuals at t_2.

	LOW LIGHT		MED LIGHT		HIGH LIGHT	
t_1	STOCKING	% STEMS	STOCKING	% STEMS	STOCKING	% STEMS
TREES	6.00	53.71	10.48	58.93	12.19	40.71
SHRUBS	.29	2.62	1.68	9.42	8.80	29.39
VINES	1.68	15.07	2.93	16.46	2.10	7.02
HERBS	1.68	15.07	1.25	7.03	3.90	13.02
GRASSES	.32	2.84	1.15	6.47	2.46	8.21
FERNS	1.20	10.70	.30	1.69	.49	1.64

(continued)

(*Table 7* continued)

t_2	LOW LIGHT STOCKING	LOW LIGHT % STEMS	MED LIGHT STOCKING	MED LIGHT % STEMS	HIGH LIGHT STOCKING	HIGH LIGHT % STEMS
TREES	7.98	52.15	13.10	56.71	12.53	38.57
SHRUBS	.44	2.87	2.90	12.55	8.69	26.77
VINES	3.73	24.40	3.98	17.21	2.14	6.58
HERBS	1.63	10.69	1.73	7.47	5.20	16.02
GRASSES	.34	2.23	1.05	4.54	2.58	7.93
FERNS	1.17	7.66	.35	1.52	1.34	4.12

Table 8. The distribution of seed weight classes of recruitment within light levels. Table based on data for 71 species of known seed weight (Devoe 1990), 2894 individuals at t_1 and 3383 at t_2.

t_1	LOW LIGHT STOCKING	LOW LIGHT % STEMS	MED LIGHT STOCKING	MED LIGHT % STEMS	HIGH LIGHT STOCKING	HIGH LIGHT % STEMS
1-9g	.49	4.42	.38	2.13	.10	.35
.1-.9g	5.76	52.10	5.08	28.88	11.00	37.34
.01-.09g	1.17	10.60	4.03	22.90	2.81	9.55
.001-.009g	.41	3.75	3.03	17.64	6.68	22.67
<.001g	3.22	29.14	5.00	28.45	8.86	30.09

t_2	LOW LIGHT STOCKING	LOW LIGHT % STEMS	MED LIGHT STOCKING	MED LIGHT % STEMS	HIGH LIGHT STOCKING	HIGH LIGHT % STEMS
1-9g	3.49	22.95	2.05	8.99	.20	.65
.1-.9g	7.07	46.55	7.57	33.22	8.51	27.16
.01-.09g	.95	6.26	4.05	17.76	3.05	9.74
.001-.009g	.39	2.57	4.33	18.97	8.59	27.44
<.001g	3.29	21.67	4.80	21.05	10.97	35.01

Regeneration from the seed bank could seldom be completely separated from that of the seed rain, although many species lacking seed bank capability could be distinguished. Under test conditions (where advance growth was eliminated), recruitment from the seed rain alone increased with decreasing light (Table 9). Under forest conditions, regeneration via the seed bank increased with increasing light (Table 10). Regeneration via advance growth showed the opposite trend. Spatial clustering of species by regeneration origin was apparent with visual inspection of the transects.

Table 9. The distribution of regeneration origins under test conditions. "SB" stands for seed bank, "SR" for seed rain; "SB + SR" denotes species that may come from either source.

t_1	LOW LIGHT		MED LIGHT		HIGH LIGHT	
	STOCKING	% STEMS	STOCKING	% STEMS	STOCKING	% STEMS
SB+SR	5.88	52.62	12.68	71.31	25.98	86.81
SR	5.29	47.38	5.10	28.69	3.95	13.19

t_2	LOW LIGHT		MED LIGHT		HIGH LIGHT	
	STOCKING	% STEMS	STOCKING	% STEMS	STOCKING	% STEMS
SB+SR	7.87	51.52	15.20	65.80	28.51	87.79
SR	7.41	48.48	7.90	34.20	3.97	12.21

Table 10. The distribution of regeneration origins under forest conditions. "SB"=seed bank, "SR"=seed rain, "AG"=advance growth.

t_1	LOW LIGHT		MED LIGHT		HIGH LIGHT	
	STOCKING	% STEMS	STOCKING	% STEMS	STOCKING	% STEMS
SB	.37	3.28	1.83	10.27	2.68	8.94
SB+SR	5.85	52.40	12.75	71.73	23.39	78.14
SR	.00	.00	.13	.70	.10	.34
AG	4.95	44.32	3.08	17.30	3.76	12.57

t_2	LOW LIGHT		MED LIGHT		HIGH LIGHT	
	STOCKING	% STEMS	STOCKING	% STEMS	STOCKING	% STEMS
SB	.24	1.60	1.55	6.71	2.85	8.77
SB+SR	7.88	51.51	15.48	66.99	25.75	79.28
SR	.02	.16	.35	1.51	.14	.42
AG	7.15	46.73	5.73	24.78	3.75	11.53

Species Richness

Number of species per light level was divided by number of square meters per light level to yield an areal measure of species richness (Table 11). For all life forms except grasses, the medium light level was most species-rich. Species richness of grasses was highest and that of trees lowest under high light. The medium light level had the highest absolute and proportional increases in species richness from t_1 to t_2.

Table 11. Species per square meter by light levels and life forms.

t_1	LOW LIGHT	MED LIGHT	HIGH LIGHT
TREES	.5122	.6829	.3729
SHRUBS	.0488	.1250	.1017
VINES	.1951	.2250	.2034
HERBS	.1220	.2000	.2034
GRASSES	.0488	.0750	.1017
FERNS	.0976	.1250	.0508
SUM	1.0245	1.4329	1.0339
t_2	LOW LIGHT	MED LIGHT	HIGH LIGHT
TREES	.5122	.8000	.3898
SHRUBS	.0488	.1250	.1017
VINES	.2195	.3750	.2712
HERBS	.1463	.2500	.2712
GRASSES	.0732	.0750	.1017
FERNS	.0976	.1750	.1017
SUM	1.0976	1.8000	1.2373

Species-specific responses

The hypothesis that light level affected the abundance of recruitment was tested for species with 15 or more individuals. The G log likelihood ratio test for goodness of fit (Sokal and Rohlf 1981) was used. The distribution of species' abundance with respect to light level was tested against a uniform distribution (null hypothesis), in which abundance was proportional only to the area within each light level (Table 12). In a few cases, the individual contribution of a particular light level was a significant departure from expectation, but the overall distribution of abundance with respect to light levels was not. In six of the 40 species tested (15%), the null hypothesis of uniform distribution could not be rejected, and "NS" (not significant) is listed in the probability column. A perfectly uniform distribution would have a G score of six or less 95% of the time. Table 12 is based on seedlings present at t_2, incorporating both mortality and recruitment.

Table 12. Recruitment greater or less than expectation for light levels 1 (low), 2 (intermediate), and 3 (high) at t_2, G score, and the probability of obtaining that score by chance.

SPECIES	GREATER IN	LESS IN	G	P
TREES				
Alchornea latifolia	2	1,3	19.31	<.001
Byrsonima spicata	2	1	7.98	<.025
Casearia arborea	2,3	1	12.60	<.005
Cecropia peltata	3	1,2	375.11	<.001
Chionanthus domingensis	3	1,2	17.57	<.001
Ficus spp.	3	1	19.55	<.001
Guarea guidonia	1,2	3	6.07	<.05
Homalium racemosum	3	1,2	7.75	<.01
Inga fagifolia	1,2	3	178.36	<.001
Laetia procera	3	1	35.62	<.001
Ocotea sintenisii	1,2	3	50.52	<.001
Prestoea montana	2	3	4.07	NS
Schefflera morototoni	2	1,3	13.15	<.005
Trema micrantha	2	1	17.08	<.001
Trichilia pallida	2	1,3	143.35	<.001
Zanthoxylum martinicense	3	1	2.77	NS
SHRUBS				
Melastomataceae spp.	3	1,2	89.17	<.001
Piper spp.	2,3	1	14.38	<.00
Psychotria berteroana	3	1,2	284.15	<.001
VINES				
Cissus sicyoides	3		3.67	NS
Heteropteris laurifolia	3	1	10.62	<.005
Hippocratea volubilis	1,2	3	127.29	<.001
Paullinia pinnata	1		21.63	<.001
Phyllanthus urinaria	3	1,2	20.74	<.001
Rajania cordata	2	1,3	25.65	<.001
Rourea surinamensis	2	3	5.55	NS
Securidaca virgata	2	1	2.74	NS
HERBS				
Gonzalagunia spicata	3	1,2	92.95	<.001
Orchid spp.			.50	NS
Pilea krugii	1	3	85.79	<.001
Sauvegesia erecta	3	1,2	211.88	<.001
Solanum rugosum	3	1,2	112.98	<.001
Solanum torvum	3	1	96.10	<.001
GRASSES				
Ichnanthus spp.	3	1,2	90.9	<.001
Pharus latifolius	2	3	9.8	<.01
Rhynchospora spp.	3	1,2	32.87	<.001
FERNS				
Adiantum latifolium	1,3	2	6.92	<.05
Blechnum occidentale	1	2,3	60.14	<.001
Cyathea arborea	3	1,2	31.22	<.001
Pityogramma calomelanos	3	1,2	20.10	<.001

Discussion

The initial species composition on the quadrats was largely maintained through the second census. Trees only moderately increased their representation, with greater numbers of low-light, large-seeded species at the second census than at the first. These species were slower to arrive on the sites than the more light-demanding, small-seeded ones, entering the understory and edge quadrats rather than the high-light quadrats. Recruitment showed segregation of species by light level; this spatial zonation was established at the outset of the experiment and maintained during the first year of regrowth.

The three light levels–understory or low light, edge or intermediate light, and gap center or high light–provided highly contrasting microenvironments for regeneration with distinct groups of associated species. Within-gap zonation was so marked as to leave no doubt that gaps are non-uniform regeneration environments. As the shadow pattern simulation model and subsequent field observation showed, gaps are not uniformly illuminated. Edges are much more often shaded than gap centers. This condition, together with factors of propagule supply, segregated species colonizing gap centers, edges, or the forest understory.

Gap size is at best only a crude indication of microenvironment and likely colonizers, since gaps of any size will encompass a range of microclimates. Use of gap size alone to describe the regeneration environment actually obscures information about how the environmental resource is partitioned by plant species, since it ignores the crucial microenvironmental variation within and around gaps. The edge condition encompasses the narrow zone on both sides of the gap margin, including areas underneath and outside the canopy cover. This microenvironment was shown to be the most favorable for tree recruitment.

Removal of felling debris at the start of the experiment did not prevent zonation of recruitment. While variation in substrate conditions may contribute to zonation as proposed by Orians (1982) and Brandani et al. (1988), it is neither the only nor the most important factor affecting species zonation in gaps. The rapid decay of litter under high light created a situation in which high light became synonymous with thin litter. This is an indirect effect of light exposure; together with its direct effects, exposure to light is the overriding influence on where and when species are recruited.

Species adapted to high light rapidly appeared in these environments. This resulted in part from their reliance on both seed rain and soil seed bank inputs to the colonizing population. No recruitment came from the soil seed bank in the low light quadrats, with the possible exception of the ferns *Blechnum occidentale* and *Adiantum latifolium.* The seed bank in the openings was probably very similar to that beneath the closed canopy at the time the plots were cut, since the whole area had a common land use history. While distribution of seeds in soil seed banks is known to be patchy (Bell 1970b; Garwood 1989), certain pioneer species are also known to be nearly ubiquitous, e.g., *Cecropia* (Cheke et al. 1979; Holthuijzen and Boerboom 1982). Much of the superiority of stocking on the high-light quadrats was due to recruitment from the soil seed bank stimulated by sudden exposure to light.

Many shade tolerant (*Dacryodes excelsa, Sloanea berteriana, Manilkara bidentata*), and intermediate (*Guarea guidonia*) tropical species are not able to maintain stocks in the soil seed bank because they lack dormancy (Marrero 1942,

1943; Moreno 1977; Ng 1980). Raich (1987) showed that germination of Malaysian rainforest species adapted to low light was inhibited in large gaps. In contrast, seeds of *Sloanea berteriana*, *Dacryodes excelsa*, and *Guarea guidonia* had germination rates of 100% in full sunlight when sown immediately upon collection and kept moist (Devoe, unpublished). Their failure to establish in large gaps is not because germination is inhibited by high light.

The probability that the relatively large seeds of shade tolerant species will arrive in the center of a gap was shown to be extremely low (this study and Devoe 1990) simply because these propagules are seldom widely disseminated. Further, if they do arrive and germinate in a newly formed gap, their establishment and survival are likely to be lower than that beneath the canopy. This is because greater insolation and consequent lower relative humidity and soil water potential enhance the risk of desiccation relative to understory conditions. Lower survival in gaps also results from increased rodent predation (Devoe 1990, Schupp 1988a,b, Sork 1987, Kasenene 1984). Opposing influences are reduction of danger of pathogen attack and mechanical damage in openings. These are important sources of mortality beneath the closed canopy that are most significant for the small seedlings arising from small seeds (Aide 1987, Augspurger 1984, Augspurger and Kelly 1984).

The harsh environment of a large, new gap *does not* favor recruitment of the large-seeded, shade tolerant timber species when recruitment is defined as germination and survival up to the stage of independence from stored seed reserves. Preliminary results of a planted-tree study (Devoe, unpublished) show that for the timber species *Dacryodes excelsa*, *Sloanea berteriana* and *Guarea guidonia*, an enhanced light regime favors more rapid growth for *established* (photosynthetically independent) seedlings. Hence, gaps favor these species at later stages in the life cycle, such as the transitions from seedlings to saplings and from saplings to poles, but gaps do not favor recruitment from seeds to seedlings for shade tolerant species. Bell (1970a) and Quarterman (1970) working at El Verde concluded that high light improved the recruitment of pioneer species while shade favored *Dacryodes excelsa*. Raich (1987) found in Malaysia that pioneer recruitment was favored by high light while shade enhanced that of later successional species.

Seed-to-seedling and seedling-to-sapling transition requirements are quite different in shade tolerant species, being first shelter and then, at least intermittently, enhanced light. In shade intolerant species, requirements at the seed-to-seedling and seedling-to-sapling-to-pole transitions are the same, namely, high light. Evidence for this conclusion comes from recruitment and height growth patterns in this study and from preliminary results of the planted-tree study, in which the pioneers *Cecropia peltata* and *Schefflera morototoni* were evaluated.

Dacryodes excelsa, *Manilkara bidentata* and *Sloanea berteriana* were sparse or absent from the seed-derived gap regeneration although the surrounding forest is composed chiefly of them. These species have maintained and increased their representation in the tabonuco forest through germination and establishment in the shade. They rely on the advance growth regeneration strategy. Advance growth strategists usually have low light compensation points and may cease height growth when light levels are too low. Death of the apical meristem occurs with prolonged suppression. At least some advance growth species can resprout repeatedly when aerial portions are damaged (Platt and Hermann 1986; Johnston and Lacey 1983); sprouting capacity is suspected for some species at El Verde although none was observed in this experiment because the advance growth was removed by uprooting.

A high frequency of low-intensity disturbance seems to enhance survival and growth of such regeneration. These species are able to respond to release following very slow growth in the understory. Their advance growth tends to be distributed throughout the forest and is usually present on the site before the partial disturbance that consolidates their dominion over it. If the advance growth had not been removed at the start of the experiment, these species probably would have been more numerous on the site from the outset.

Summary of factors influencing silvicultural choices

Seed dispersal of the most valuable timber species in the tabonuco forest of Puerto Rico is limited to short distances. Establishment of these species from seed in extremely open conditions is rare. When relying on natural regeneration, canopy openings made in harvesting must be kept small and seed trees maintained at high, uniform densities if timber species are to reproduce. Therefore, clear cutting and seed tree methods are not suited to the tabonuco forest, while shelterwood and selection methods are. The commercial timbers occur in two categories–intermediate and low light regenerators–with corresponding differences in growth rates. Because the slower growing species are extremely shade tolerant, stratified production systems allow for the most efficient use of the productive capacity of the growing stock. The choice between selection and shelterwood management should be determined by the initial age structure of the stands, knwoledge of subsequent stand development patterns, availability of professional expertise, and environmental quality considerations.

Acknowledgements

David M. Smith assisted in every phase of this work with characteristic generosity, patience and insight. Financial support was provided by the Tropical Resources Institute of Yale University and Oak Ridge Associated Universities. This publication is based on work performed in the Laboratory Graduate Participation Program under contract number DE-AC05-76OR003 between the U.S. Department of Energy and Oak Ridge Associated Universities. The Center for Energy and Environment Research of the University of Puerto Rico gave a home to these studies in cooperation with the US Forest Service Caribbean National Forest and the Institute of Tropical Forestry. Dr. George Proctor, Departamento de Recursos Naturales, confirmed the identification of many voucher specimens. Carolyn Berman supervised woods operations during felling, and a crew from the Caribbean National Forest executed the work. Alison Churchill and Tina Hubbard assisted with field measurements.

Literature Cited

Acevedo, M. and R. Marquis. 1979. A survey of the light gaps of the tropical rain forest at Lloma, Peninsula de Osa. OTS Book 78.3

Aide, T.M. 1987. Limbfalls: A major cause of sapling mortality for tropical forest plants. Biotropica 19(3):284-285.

Augspurger, C.K. 1984. Light requirements of neotropical tree seedlings: a comparative study of growth and survival. Journal of Ecology 72:777-795.

Augspurger, C.K. and C. Kelly. 1984. Pathogen mortality of tropical tree seedlings: experimental studies of the effects of dispersal distance, seedling density and light conditions. Oecologia (Berlin) 61:211-217.

Austin, M.P. 1977. Use of ordination and other multivariate descriptive methods to study succession. Vegetatio 35:165-175.

Austin, M.P. 1987. Models for the analysis of species' response to environmental gradients. Vegetatio 69:35-45.

Barton, A.M. 1984. Neotropical pioneer and shade-tolerant tree species: do they partition treefall gaps? Tropical Ecology 25:196-202.

Baur, G.N. 1968. The Ecological Basis of Rainforest Management. Forestry Commssion of New South Wales, Sydney.

Beals, E.W. 1973. Ordination: mathematical elegance and ecological naivete. Journal of Ecology 61:23-35.

Beals, E.W. 1984. Bray-Curtis ordination: an effective strategy for analysis of multivariate ecological data. Advances in Ecological Research 14:1-55.

Bell, C.R. 1970. Seed distribution and germination experiment. IN: A Tropical Rain Forest: A Study of Irradiation and Ecology at El Verde, Puerto Rico, H.T. Odum and R.F. Pigeon, eds., USAEC, Oak Ridge, TN, pp. D177-182.

Brandani, A., Hartshorn, G.S. and G.H. Orians. 1988. Internal heterogeneity of gaps and tropical tree species richness. Journal of Tropical Ecology 4:99-119.

Brokaw, N.V. 1985. Gap-phase regeneration in a tropical forest. Ecology 66:(3):682-687.

Brown, K.M. and C. Merritt. 1970. A shadow pattern simulation model for forest openings. Purdue University Experiment Station Bulletin No. 868.

Brown, S., Lugo, A.E., Silander, S. and L.H. Liegel. 1983. Research History and Opportunities in the Luquillo Experimental Forest. USFS General Technical Report SO-44. 128 pp.

Cheke, A.S., Nanakorn, W. and C. Yankoses. 1979. Dormancy and dispersal of seeds of secondary forest species under the canopy of a primary tropical rain forest in Northern Thailand. Biotropica 11:88-95.

Denslow, J.S. 1980a. Gap partitioning among tropical rain forest trees. Biotropica 12:(suppl.):47-55.

Devoe, N.N. 1990. Differential seeding and regeneration in openings and beneath closed canopy in sub-tropical wet forest. Doctoral dissertation, Yale University.

Ewel, J.J. and J.L. Whitmore. 1973. The Ecological Life Zones of Puerto Rico and the U.S. Virgin Islands. USDA Forest Service, Research Paper ITF-18.

Foster, S.A. and C.H. Janson. 1985. The relationship between seed size and establishment conditions in tropical woody plants. Ecology 66:773-780.

Garwood, N.C. 1989. Tropical soil seed banks. IN: Ecology of Seed Banks, M.A. Leck, R.L. Simpson and V.T. Parker, eds., Academic Press, New York, pp. 149-210.

Halverson, H.G. and J.L. Smith. 1979. Solar radiation as a forest management tool. A primer and application of principles. Pacific Southwest Forest and Range Experiment Station, Berkeley, CA, 13 pp.

Harrington, J.B. 1983. Solar radiation in a clear-cut strip. Petawawa National Forestry Institute, Canadian Forestry Service, Chalk River, Ontario.

Hartshorn, G.S. 1978. Tree falls and tropical forest dynamics. IN: P.B. Tomlinson and M.H. Zimmermann, eds., Tropical Trees as Living Systems, Cambridge University Press, London, pp 617-637.

Hartshorn, G.S. 1980. Neotropical forest dynamics. Biotropica 12:(2):23-33.

Holthuijzen, A.M.N. and J.H.A. Boerboom. 1982. The *Cecropia* seedbank in Surinam forest. Biotropica 14:62-68.

Johnston, R.D and C.J. Lacey. 1983. Multi-stemmed trees in rain forests. Australian Journal of Botany 31:189-195.

Jones, E.W. 1945. The structure and reproduction of the virgin forest of the north temperate zone. New Phytologist 44:130-148.

Kasenene, J.M. 1984. The influence of selective logging on rodent populations and the regeneration of selected tree species in the Kibale Forest, Uganda. Tropical Ecology 25:(2):179-195.

Liogier, H.A. and L.F. Martorell. 1982. Flora of Puerto Rico and Adjacent Islands: A Systematic Synopsis. Editorial de la Universidad de Puerto Rico, Rio Piedras. 342 pp.

Little, E.L., Jr. and R.O. Woodbury. 1976. Trees of the Caribbean National Forest, Puerto Rico. U.S. Forest Service Research Paper ITF-20, 27 pp.

Longman, K.A. and J. Jenik. 1974. Tropical Forest and Its Environment. Longman Group Ltd., London, 196 pp.

Ludwig, J.A. and J.F. Reynolds. 1988. Statistical Ecology. A Primer on Methods and Computing. John Wiley and Sons, New York, 337 pp.

Marquis, David A. 1965. Controlling light in small clear cuttings. U.S. Forest Service Research Paper NE-39 NEFES, Upper Darby, PA.

Marrero, J. 1942. Data on seed study for rapid growing species suitable for fuel. Memorandum for Record, Files of ITF, Rio Piedras, PR.

Marrero, J. 1943. A seed storage study of some tropical hardwoods. Caribbean Forester 4:99-106.

McCune, B. 1987. Multivariate Analysis on the PC-ORD System. A Software Documentation Report. April, 1987 version (revised). HRI Report No. 75. Holcomb Research Institue, Butler University, Indianapolis, Indiana.

Moreno, P. 1977. Latencia y viabilidad de semillas de arboles tropicales. Interciencia 2:298-302.

Ng, F.S.P. 1980. Germination ecology of Malaysian woody plants. Malaysian Forester 43:(4):406-419.

Noy-Meir, I., Walker, D. and W.T. Williams. 1975. Data transformations in ecological ordination II. On the meaning of data standardization. Journal of Ecology 63:779-800.

Odum, H.T., Drewry, G. and J.R. Kline. 1970. Climate at El Verde, 1963-1966. IN: A Tropical Rain Forest, H.T. Odum and R.F. Pigeon, eds., NTIS, Springfield, VA, pp. B347-418.

Orians, G.H. 1982. The influence of tree falls in tropical forests on tree species richness. Tropical Ecology 23:255-279.

Parker, G. 1985. The Effect of Disturbance on Water and Solute Budgets of Hill-slope Tropical Rainforests in Northeastern Costa Rica. Ph.D. dissertation, University of Georgia, Athens, Georgia.

Perez, I. E. 1986. Tree Regeneration in Two Tropical Rain Forests. Master of Science thesis, University of Puerto Rico, Rio Piedras, 97 pp.

Platt, W.J. and M.S. Hermann. 1986. Relationships between dispersal syndrome and characteristics of populations of trees in mixed species forests. IN: Frugivores and Seed Dispersal, A. Estrada and T.H. Fleming, eds., Dr. W. Junk, Dordrecht, Netherlands, pp. 309-321.

Popma, J., Bongers, F., Martinez-Ramos, M. and E. Veneklaas. 1988. Pioneer species distribution in treefall gaps in a Neotropical rain forest; a gap definition and its consequences. Journal of Tropical Ecology 4:77-88.

Quarterman, E. 1970. Germination of seeds of certain tropical species. IN: A Tropical Rainforest, H.T. Odum and R.F. Pigeon, eds., US Atomic Energy Commission, Oak Ridge, TN, pp. D173-175.

Raich, J.W. 1987. Canopy Openings, Seed Germination, and Tree Regeneration in Malayasian Coastal Hill Dipterocarp Forest. Ph.D. dissertation, Duke University, 167 pp.

Raich, J.W. 1989. Seasonal and spatial variation in the light environment in a tropical dipterocarp forest and gaps. Biotropica 21(4):299-302.

Richards, P. and G.B. Williamson. 1975. Treefalls and patterns of understory species in a wet lowland tropical forest. Ecology 56:1226-1229.

Runkle, J.R. 1979. Gap phase dynamics in climax mesic forests. Ph.D. dissertation, Cornell University.

Schupp, E.W. 1988a. Seed and early seedling predation in the forest understory and in treefall gaps. Oikos 51:71-78.

Schupp, E.W. 1988b. Factors affecting post-dispersal seed survival in a tropical forest. Oecologia 76:525-530.

Sokal, R.R. and F.J. Rohlf. 1981. Biometry. 2nd edition. W.H. Freeman and Company, San Francisco, 859 pp.

Sork, V.L. 1987. Effects of predation and light on seedling establishment in *Gustavia superba*. Ecology 68:1341-1350.

Spies, T.A. and J.F. Franklin. 1989. Gap characteristics and vegetation response in coniferous forests of the Pacific Northwest. Ecology 70:543-545.

Swaine, M.D. and T.C. Whitmore. 1988. On the definition of ecological species groups in tropical rain forests. Vegetatio 75:81-86.

Uhl, C., Clark, K., Dezzeo, N. and P. Maquirino. 1988. Vegetation dynamics in Amazonian treefall gaps. Ecology 69:751-763.

Vitousek, P. and J.S. Denslow. 1986. Nitrogen and phosphorous availability in treefall gaps of a lowland tropical rainforest. Journal of Ecology 74:1167-1178.

Whitmore, T.C. 1978. Gaps in the forest canopy. IN: Tropical Trees As Living Systems, P.B. Tomlinson and M.H. Zimmermann, eds., Cambridge University Press, NY, pp. 639-655.

Establishment and early growth of advance regeneration of canopy trees in moist mixed-species forest

7

P. MARK S. ASHTON

Introduction: moist mixed-species forest[1]

Climate and soil

Moist regions of temperate and tropical latitudes are largely composed of mixed aggregates of tree species. These forests can be some of the most tree-diverse ecosystems in the world, comprising species mixtures that are stratified in both vertical and horizontal space (Davis and Richards 1933, 1934, Beard 1944, Ashton 1964, Kelty 1986), and that change in composition with the progression of time via successional processes (Clements 1916, Watt 1947, Egler 1954, Drury and Nisbet 1973, Oliver 1978).

Moist regions can be broadly characterized as areas where soil water is physiologically available throughout the growing season and, in general, is not a major factor limiting plant growth. Climate and soils are the main influences on availability of soil water. The most important climatic affect is seasonality. It is expressed as the degree of fluctuation of mean monthly rainfall or temperature over a year. The nature and scale of seasonality changes latitudinally, increasing with greater distance from the equator. Moist regions in the higher latitudes are most affected by changes in temperature, with rainfall being relatively evenly distributed through the year. Largely because of plant dormancy in winter, yearly evapotranspiration rates are low compared to moist forests of more tropical climates. Regions at these temperate latitudes can therefore be moist but receive lower annual rainfall (>600 mm) than moist regions at latitudes closer to the equator.

Progressing toward more equatorial latitudes the nature of seasonality changes to one predominated by variation in rainfall distribution, with temperature change becoming more moderated. For moist regions of the higher tropical and subtropical latitudes, annual rainfall is > 2000 mm, and dry periods limited to between 2 and 4 months. In these climates a greater degree of rainfall seasonality therefore exists. The most aseasonal moist regions are predominantly in the equatorial tropics. Annual rainfall in these moist regions is very high (> 3000 mm). If dry spells occur, they do not last longer than two months and are frequently punctuated by convectional rains (Figure 1). Seasonal variation in temperature is negligible compared to moist tropical seasonal, and particularly, moist temperate regions.

Within all these moist forest regions, edaphic properties and their variation within the topography are the factors that affect water flow and retention on a site. Sites with soils where water collects and is prevented from flowing (hydric), or

[1]"Moist mixed-species forest" is defined here as those forest formations that allow many tree species to exist in mixture in time and space. Following the work of Eyre (1968), this would include mixed forests of middle latitude (conifer and broadleaf); broadleaf forests of middle latitudes; tropical rain forest; and tropical seasonal forests that are evergreen or semi-evergreen.

M. J. Kelty (ed.), The Ecology and Silviculture of Mixed-Species Forests, 101–122.

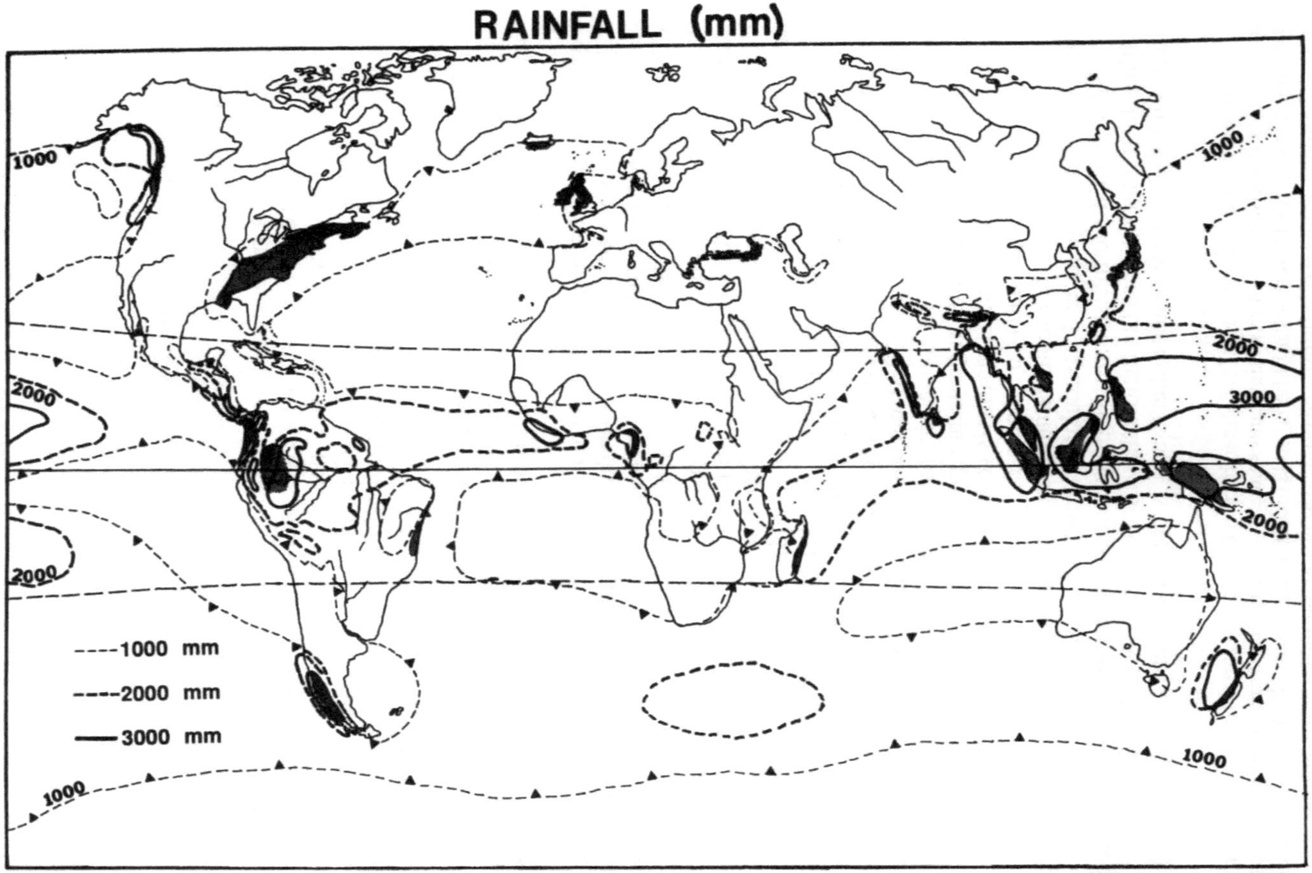

Fig. 1. Climate map depicting geographical regions that receive annual rainfalls of 1000mm, 2000mm, and 3000mm. Shading denotes those regions that have a relatively high but even rainfall distribution throughout the year with no extended dry period.

which allow water to flow too freely and cannot collect (xeric), retard mixed-species stratification within a stand. Excessive or depauperate soil nutrient accumulation can also retard mixed-species stratification. Sites with soils that allow water to flow but are also able to retain moisture, promote mixed-species stratification, provided that canopy closure is moderated by frequent small scale disturbance, therefore allowing a variable below-canopy light environment. Topographic relief, soil texture, and weathering depth, are the most important edaphic properties that influence water storage and drainage patterns.

Most forest soils within these moist regions are well weathered and deep. They are classified as either ultisols or oxisols (USDA 1975). Younger soil groups (inceptisols, entisols) have also developed where relatively recent or existing local or regional edaphic perturbations have occurred (glaciation, volcanic activity, landslides, floods).

Climatic affects therefore determine where moist mixed-species forest can exist at a regional geographic scale (Figure 2). Topography and soil determine the degree of their development at a local scale. The intensity of these factors and the interactions between them largely determine the complexity and accentuation of a stratified mixed-species stand, and the scale of their composite as a whole mosaic, within a moist forest region.

Regeneration pathways

Questions arise as to how and why so many species can co-exist together in moist mixed-species forest. Understanding the development and maintenance of tree species mixtures has direct value for the management of these moist forests and their possible emulation in plantations. This has been explained in terms of equilibrium (Ashton 1969, Grubb 1977) and non-equilibrium (Fedorov 1966, van Steenis 1969, Hubbell 1979) hypotheses.

Both equilibrium and non-equilibrium processes are influential, and play differing roles at different scales in time and space, but it is important to determine when and where each process dominates the other for each moist forest region. An aspect of this uncertainty lies in where and when equilibrating forces in a forest community can create and maintain large assemblages of tree species (Hubbell and Foster 1985). One set of mechanisms for species specialization are conditions for regeneration establishment and growth (Grubb 1977).

In moist mixed-species forest, the establishment or release of regeneration depends on some kind of disturbance to the canopy or emergent tree stratum. The type of disturbance affects the physical characteristics of the soil surface and the light environment, at which germination usually takes place. Different types of site or micro-site disturbance (or non-disturbance) affect the kind of regeneration that is established.

The original work on disturbance and patterns of regeneration in moist mixed-species forest was mostly carried out during the first half of this century (Clements 1916, Watt 1947, Egler 1954). These studies provided the basic ideas on light tolerance (Jackson 1967 a,b) and on successional development of a stand and its different phases of growth (gap, building, and mature phase after Watt 1947; stand initiation, stem exclusion, understory reinitiation, and old growth after Oliver and Larson 1990). Though these stages are somewhat arbitrary, they are useful for describing continuous changes that occur in the development of a forest.

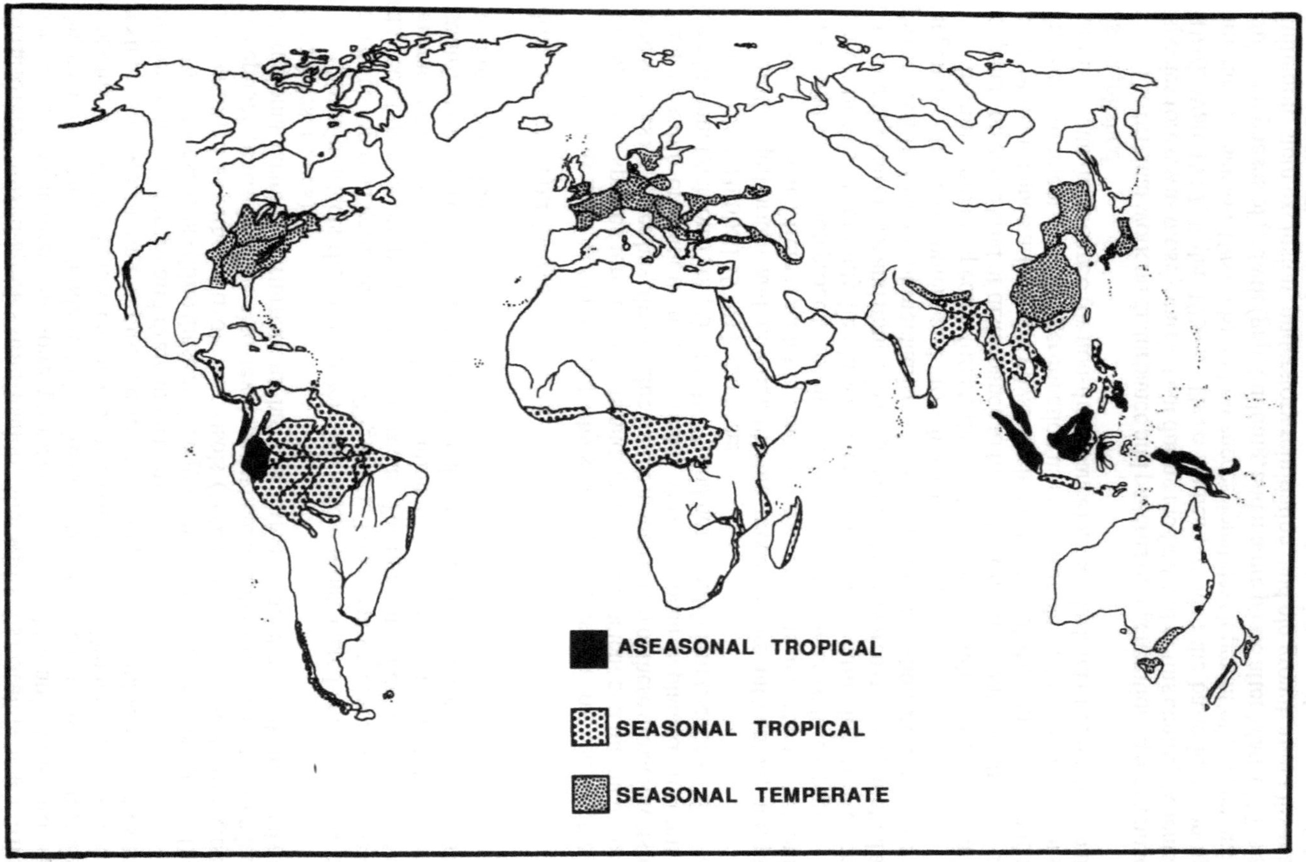

Fig. 2. Map illustrating regions of moist mixed-species broadleaf forest and their degree of seasonality.

These workers also demonstrated that some species were able to adapt to a range of contrasting micro-environments within a forest whereas others were restricted to a particular few. They showed that different species dominated a forest at the different stages of its development. Today these ideas have been refined, but have not essentially been altered (Drury and Nisbet 1973, Bazzaz 1979, Bazzaz and Pickett 1980, Bazzaz 1984, Lorimer et al. 1988). Species have been categorized as broadly belonging to a particular guild. These guilds are separated by their differences in successional status, light tolerance, and growth strategy (specialist, generalist – Bormann and Likens 1979; pioneer, late secondary, mature – Hartshorn 1978, Brokaw 1985; pioneer, small gap specialist, large gap specialist – Denslow 1980; pioneer, climax – Swaine and Whitmore 1988). In this paper guilds[2] have been defined by the successional growth phase they dominate following Watt 1947 and Whitmore 1984, and by stratum of their mature crown occupation (Figure 3).

Though many past studies have shown differences in establishment and growth between guilds (Budowski 1965, Hibbs 1982, Brokaw 1985, Spies and Franklin 1989, Devoe 1989), only some have examined the relationships between their growth morphology, physiology, and anatomy (Bjorkman et al. 1982, Fetcher et al. 1983, Bahari et al. 1985, Pearcy 1987, Popma and Bongers 1988, Strauss-Debenedetti 1989). Further, less work has examined these relationships between tree species that belong to the same guild (Canham 1988, 1989, Turner 1989), of which only some, following the classical work of Bourdeau (1954), have selected species within the same taxonomic assemblage for comparison (Walters and Field 1987, Ashton 1990, Sipe 1990, Turner 1990).

In moist mixed-species forest, the differentiation of guilds based on successional status and stratum of crown occupation appear to be at an optimum as compared to other forest formations (Figure 3).

The type, size, and frequency of disturbance that occurs within moist mixed-species forest allow for this differentiation. The early successional guilds (which dominate gap and building phases) rely on buried or windblown seed that germinate and establish on a site after the occurrence of disturbance. Excluding this, the majority of guilds within moist mixed-species forest rely on advance regeneration[3] in at least some form for establishment and growth (see Figure 3). This suggests that the ability of a forest to develop stratified mixtures is in large part dependent on a disturbance regime that allows for the perpetuation and differentiation of the advance regeneration niche. This appears to be an important characteristic that separates the complexity of moist mixed-species forest from other kinds of forest.

Accepting this view, canopy trees that rely on advance regeneration and that dominate the mature phase of a forest are of critical importance to retaining the structural integrity of a managed moist mixed-species forest. Most of these canopy trees would be categorized as belonging to the dominant mature phase regeneration guild (Figure 3). Though their reproduction, growth, and development is seemingly similar, many species within this guild can co-exist together in the same forest, often as con-generic groups.

---

[2]A "guild" is defined here as a collection of tree species that hold common silvical characteristics.

[3]"Advance regeneration" is defined as that regeneration which exists as seedlings or saplings beneath the canopy of a forest before a disturbance creates or releases new canopy growing space; or to that regeneration which, after being damaged by a disturbance, arises by sprout or coppice growth from surviving roots or stumps.

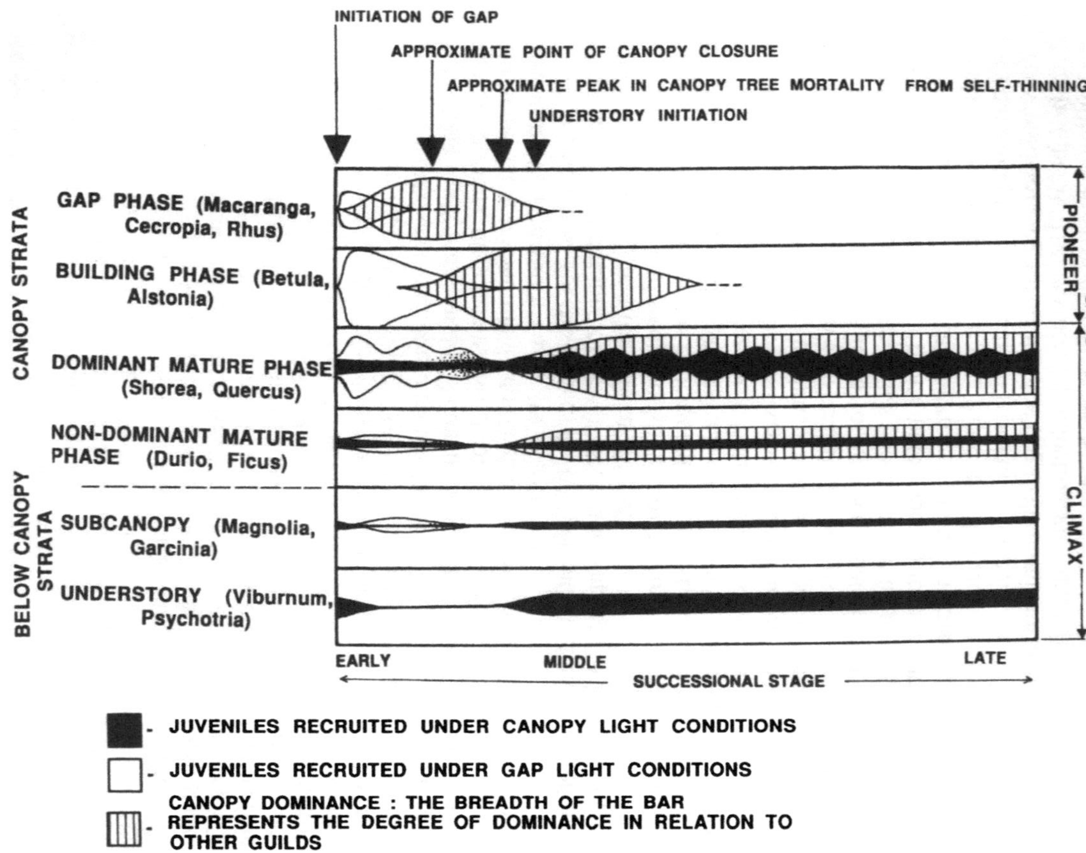

Fig. 3. Generalized diagram depicting recruit frequency and canopy dominance of guilds over the different successional stages of stand development for moist mixed-species forest. (Note the periodic recruitment of seedlings for tree species that dominate the mature phase of the canopy).

The remaining portion of this paper is devoted to elucidating the niche partitioning and co-existence of tree species that dominate the canopy of the mature phase of moist mixed-species forest.

Discussion: Regeneration establishment of dominant mature phase canopy tree species

Biological and economic importance

Moist mixed-species forests have a mature phase canopy stratum dominated by particular groups of species that often form close taxonomic assemblages at the generic level (Table 1). Several important families that comprise these genera are noticeably pan-temperate (Aceraceae, Fagaceae), pantropical (Leguminosae, Meliaceae), and some dominate particular moist geographic regions (Bignoniaceae - Central America; Dipterocarpaceae - South and Southeast Asia; Juglandaceae - North America; Flindersiaceae, Myrtaceae - Australasia).

In the mature phase of a forest's development the crowns of these canopy genera modify the effects of radiation and rainfall, creating a buffer against the extremes, and promoting a diversity of micro-environments within which other plants and animals can live. These canopy trees are also indirectly responsible for protecting the soil of a forest that, when exposed, is often susceptible to nitrogen leaching and particulate erosion (Bormann and Likens 1979). They also give the site protection that enables a watershed to provide the quality and stable yield of water used downstream by cities and crops.

Many of these tree species produce the quality hardwood and softwood timbers used for interior panelling, flooring, framery, cabinetry, turnery, and ornamentation. These timbers are exploited solely from moist mixed-species forest. This includes many of the common recognizable timbers such as greenheart, luan, maple, mahogany, meranti, oak, rosewood, satin, and walnut. The crowns and physical structure of these trees also provide support and create a micro-environment that allows many of the sub-canopy plant species to produce other goods and services that are extracted from moist mixed-species forest. Examples are fruit trees, latex-bearing trees, many condiments and medicinals, and plants which can be repeatedly harvested for cord or sugar.

Because these canopy tree genera have been categorized as belonging to the same guild, by implication, they therefore have in common key silvical characteristics. In particular their modes of reproduction (flowering, pollination, and fruiting), and regeneration establishment (germination, survival, growth, and release) have many similarities (Table 2). The most important silvical characteristic common to these tree species is the site specialization and establishment of their regeneration. Understanding this is critical toward the development of regeneration methods for the management of moist mixed-species forest.

Site specialization

How and why can tree species co-exist within moist mixed-species forest? Evidence suggests differences in species composition can exist across the topography due to edaphic factors: soil moisture (Bourdeau 1954, Curtis 1959), and nutrient status (Ashton 1964, Whittaker 1966, Whitmore 1982, Baillie and Ashton 1983); biotic factors, including density dependence of host-specific seed predation (Janzen 1971), seedling herbivory (Becker et al. 1985), and micro-faunal and floral

	TEMPERATE	
NORTH AMERICA	**EUROPE**	**NORTHEAST ASIA**
ACERACEAE	ACERACEAE	ACERACEAE
Acer	*Acer*	*Acer*
FAGACEAE	FAGACEAE	FAGACEAE
Castanea	*Castanea*	*Castanea*
Fagus	*Fagus*	*Castanopsis*
Quercus	*Quercus*	*Fagus*
		Lithocarpus
JUGLANDACEAE		*Quercus*
Carya		
Juglans		
	TROPICAL	
SOUTH & CENTRAL AMERCICA	**WEST & CENTRAL AFRICA**	**SOUTH & SOUTHEAST ASIA**
BIGNONIACEAE	LEGUMINOSAE	CLUSIACEAE
Tabebuia	*Cynometra*	*Mesua*
	Dalbergia	
ELAEOCARPACEAE ?	*Macrolobium*	DIPTEROCARPACEAE
Sloanea	*Pericopsis*	*Dipterocarpus*
	Pterocarpus	*Dryobalanops*
LAURACEAE ?		*Shorea*
Nectandra	MELIACEAE ?	
Ocotea	*Entandrophragma*	LEGUMINOSAE
	Khaya	*Intsia*
LEGUMINOSAE		*Koompasia*
Dalbergia		*Milletia*
Dipteryx		*Parkia*
Inga		*Pericopsis*
Macrolobium		
Mora		**AUSTRALASIA**
Pterocarpus		
		ELAEOCARPACEAE ?
MELIACEAE ?		*Elaeocarpus*
Cedrella		
Guarea		FLINDERSIACEAE ?
Swietenia		*Flindersia*
MYRISTICACEAE ?		LAURACEAE ?
Virola		*Bielschmiedia*
		Cryptocaria
	TEMPERATE	
FAGACEAE		FAGACEAE
Notofagus		*Notofagus*
		MYRTACEAE
		Eucalyptus

Table 1. Families and important genera of tree species that are canopy dominants within the mature phase of moist mixed-species broadleaf forest. Question marks denote those families from the more seasonal moist tropics that do not show true mast fruiting.

REPRODUCTION

- Pollination vectors are either wind (temperate) or small insects (Hymenoptera -- tropical)
- Seed is with storage tissue
- Seed is dispersed by gravity (often aided by wind or animals)
- Fruiting time can be at regular annual intervals with abundant seed each fruiting, or more or less supra-annual with distinctly different amounts of seed at each fruiting (mast fruiting)
- Seed show no classical seed dormancy though delayed germination may occur, particularly in seasonal regions

ESTABLISHMENT AND GROWTH

- Seed requires partial shade protection for germination and early survival
- Seedlings require an increase in light (as compared to understory conditions) for their establishment and growth
- Seedling survival and establishment is usually site specific, according to particular biotic, microclimatic and edaphic characteristics

Table 2. Silvical characteristics of canopy tree species belonging to genera assemblages that dominate the mature phase of moist mixed-species broadleaf forest. These characteristics should be interpreted broadly. Exceptions will exist.

symbioses (Janos 1988, Newberry et al. 1988); and by climatic factors, including relatively predictable patterns of forest micro- and macro-climate after disturbance (Clements 1916, Egler 1954, Drury and Nisbet 1973, Denslow 1987, Poulson and Platt 1988, 1989, Smith and Ashton in prep), or temperature difference attributed to elevation (Whittaker 1967, Wali and Krajina 1973) (Table 3).

These workers have emphasized that these site factors differentiate species establishment and growth of different guilds (Bazzaz 1979, 1984). Few studies have investigated differences between species belonging to the same guild (Sipe 1990). This suggests to some authors that no or little difference exists and, that if differences are present, they are inconsequential (Pickett 1980, Pickett and White 1985).

This paper proposes that canopy tree species that would be categorized as belonging to the dominant mature phase guild show patterns in regeneration establishment and early growth that are contrary to this viewpoint. Recent studies suggest that species belonging to the same genus have different adaptations to a dominating biotic, edaphic, climatic, or physiographic factor. Each species was shown to have adaptations best suited to a specific topographic location which possesses the combination of site conditions that optimizes its survivorship and performance, each of these in turn representing segments along gradients of those site factors which its genus most responds. These adaptations are reflected in the

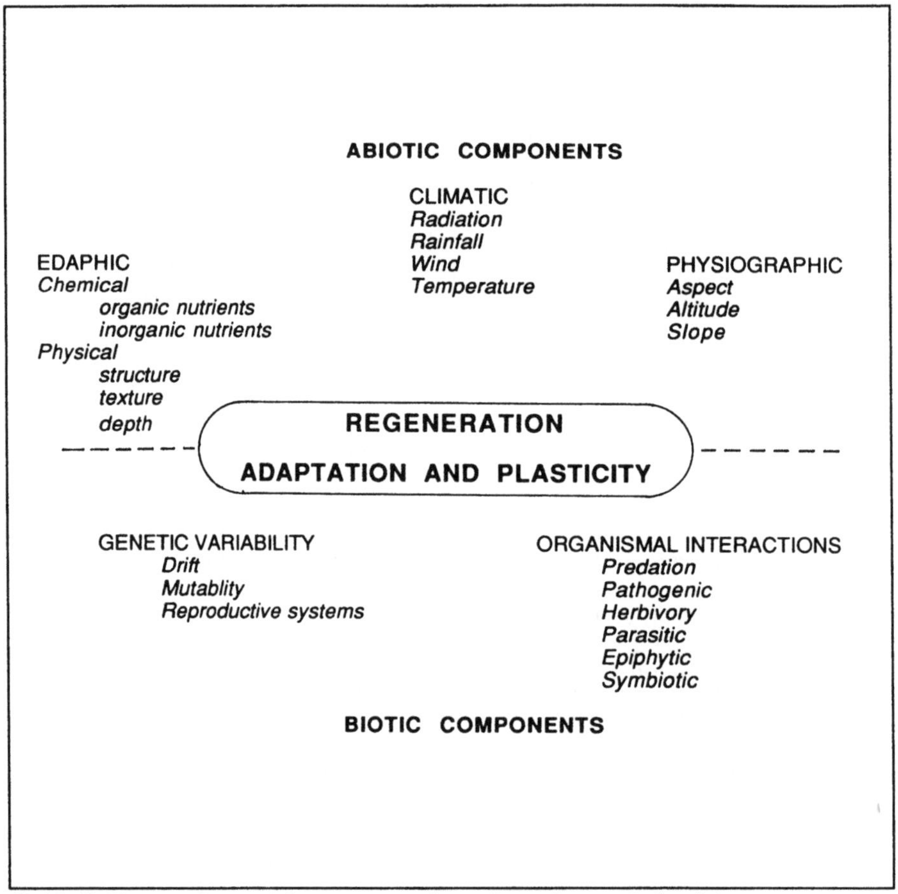

Table 3. Factors determining regeneration adaptation and plasticity.

different tolerance levels that each species had toward the site factor gradients. Abiotic and biotic factors that determine adaptation and plasticity (tolerance) have been listed in Table 3.

In general, the studies also showed that a particular site factor was primarily responsible for differentiating species within a genus, though this did not preclude other factors having secondary interactions which could allow further specialization of their regeneration niches.

Two studies have been summarized as examples of these ideas and hypotheses, each of which is based on work currently being done by the author (Ashton 1990, in prep). The cases described are designed to demonstrate that species within each genus have particular tolerance adapations and plasticities at different scales and levels of seedling development. The first describes the canopy tree genus *Shorea* from moist mixed-species forest in the aseasonal tropics. The second example is of the canopy tree genus *Quercus*, from moist mixed-species forest in a seasonal temperate region.

An example of adaptation and plasticity along a light continuum: *Shorea* section *Doona* (Dipterocarpaceae)

Description of the moist forest region

The geography, geology, climate, and soils of the southwest upland hills of Sri Lanka have been described by de Rosayro (1959), Moorman and Panabokke (1961) and Cooray (1967). In summary, the topography is one of a series of parallel ridges and valleys that lie along an east-west axis. The rock types are khondalites and charnokites of metamorphic origin, overlain by tropical ultisols on the lower slopes and valleys, with both tropical entisols, inceptisols and ultisols on the steeper ridges (USDA 1975). The elevational differences between valley and ridge do not generally exceed 100 m, with a mean elevation of approximately 600 m above sea level.

The mean annual rainfall is between 3750 and 5000 mm. Most rain falls during the southwest monsoons (April to July) and the northeast monsoons (September to January). The mean monthly temperature of the area ranges between 25 and 27°C. The forest has been classified after the two dominant mature phase canopy genera as a *Mesua-Shorea* community (de Rosayro 1942). Its floristic composition has been described by de Rosayro (1954) and Gunatilleke and Gunatilleke (1981, 1983).

Site characterization of Shorea *section* Doona

Shorea is perhaps the most important genus in the Dipterocarpaceae, a family of trees that dominate the canopy of moist mixed-species forest-types in south and southeast Asia. Four *Shorea* section *Doona* species (duns) occupy different parts of the forest topography. The lower slopes and valleys, where small rivers and perennial streams occur, have *S. megistophylla*. *Shorea trapezifolia* predominantly occupies the deep soils of the midslopes and ridges but can also occur on the bottomlands. *Shorea disticha* can also occur with *S. megistophylla*, as well as on the ridges with *S. worthingtonii*. It predominates on the midslopes but does not occur with *S. trapezifolia* (Figure 4). *Shorea worthingtonii* is restricted to the ridges or steep slopes that are shallow, rocky and thin to bedrock.

Physiological, anatomical and morphological adaptations

Shorea megistophylla seedlings are the most light-demanding of the four *Shorea* (Figure 5). Optimum net photosynthesis is in full sun. Their leaves are the most sun-adapted, having thick blades and cuticles and a large leaf size. Though seedling water-use is more efficient[4] than the other *Shorea* species in full sun, seedlings have dramatically higher transpiration rates. They therefore require a free supply of water. Best seedling growth occurs when exposed to periods of direct sunlight that last longer than 3 hours and where soil water is freely available. This can occur in the centers of approximately circular gap-sites greater than 450 m^2, that are located on the lower slopes and valleys.

Shorea disticha is intermediate in light tolerance. Its optimum net photosynthesis is at a light intensity of 50% of full sun. Seedlings also grow well under an array of different light conditions and, because of this, suggests it has a wide light

[4]A measure of water-use efficiency is the ratio between optimum net photosynthesis and transpiration at that optimum net photosynthesis.

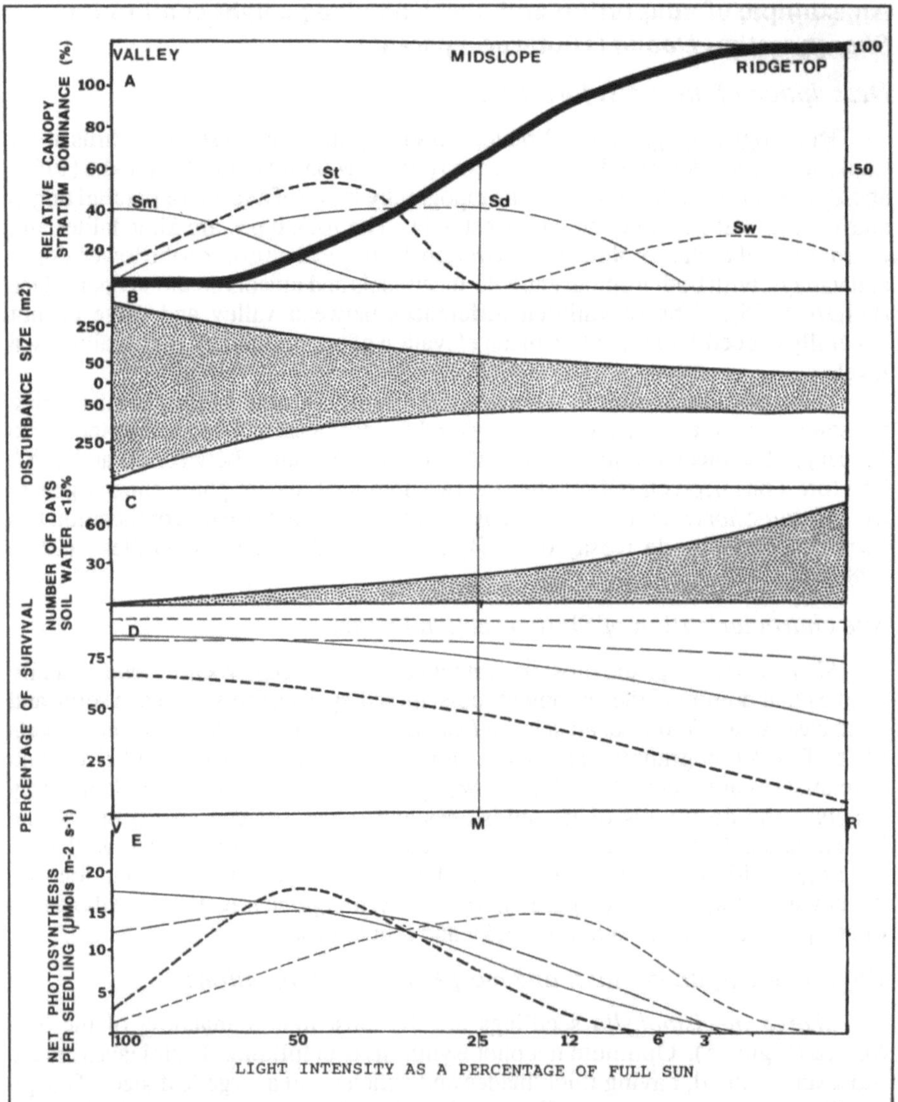

Fig. 4. Diagram depicting: A) distribution of the four *Shorea* species across the topographic gradient; B) natural disturbance size as a measure of light intensity across the topographic gradient; C) low soil-water as a measure of drought across the topographic gradient; D) survival as a measure of drought tolerance of *Shorea* species across the topographic gradient; survival was calculated two years after initial establishment; E) relative index of net photosynthesis of seedlings at different light intensities as a measure of light tolerance of *Shorea* species; the index was based on the product of appropriate mean net-dry leaf weight and photosynthetic rate (µmols m^{-2} s^{-1}) for each species at different light intensities. Sm = *S. megistophylla*; St = *S. trapezifolia*; Sd = *S. disticha*; Sw = *S. worthingtonii*.

tolerance. This is demonstrated by its greater palisade layer thickness and plasticity between seedling leaves from shade and sun conditions, as compared to other *Shorea* species. Best growth occurs when seedlings are exposed to between 1.5 and 3 hours of direct sunlight. This can occur in the center of gap-sites that vary in area between 200 and 450 m^2.

The most shade-demanding species is *S. worthingtonii* (Figure 5). Its optimum net photosynthesis is at a light intensity of 20% of full sun. In shade, seedlings are water-use efficient, with leaves that are small, with thick blades and cuticles, and with a low number of stomata. At understory light levels (< 3% of full sun) seedlings allocate a higher proportion of dry weight to roots than to stem compared to the other *Shorea* species. Under these low light levels, seedling root development allows for greater capacity of nutrient and water uptake. This root allocation phenomenon, together with its leaf characteristics, make *S. worthingtonii* the most drought tolerant of the *Shorea* species under shade. However, in full sun, seedlings have thin cuticles making their leaves susceptible to dehydration. *Shorea worthingtonii* therefore is susceptible to pronounced dieback and mortality in these light conditions. Best growth occurs in the center of gap-sites of about 200 m^2. In smaller gaps it grows better than other *Shorea* species.

Shorea trapezifolia is similar in light tolerance to *S. disticha*. Seedlings have a poor water-use efficiency because of a high transpiration. This is largely because of leaf anatomy. Seedling leaves have a thin blade and cuticle with a high number of stomata. These attributes, together with epidermal and palisade cell dimensions, reveal a low plasticity, with little anatomical ability to adapt to micro-sites other than those to which it is best suited. This, together with poor root allocation at low light levels, make it the most susceptible of the *Shorea* species to water stress. Best seedling growth occurs when periods of direct sun last between 1.5 and 3 hours (which can occur at the intermediate gap sizes of 200 to 450 m^2), and where soil water is freely available.

Site factor gradients across the forest topography

In the forest, differences in light intensity and periodicity of direct sunlight appear to occur as a gradient that is related to the type and size of disturbance across the topography. This gradient is hypothesized to be the primary factor responsible for the distribution of these four *Shorea* species. Valley bottoms and midslopes with deep soils are more prone to large disturbances that are caused by multiple tree-falls. This disturbance type most likely occurs during the monsoon season when rain-soaked soils and wind interact to make trees more susceptible to windthrow. Ridgetops and midslopes that have rocky or thin soils allow a more stable forest to develop because tree roots are able to penetrate the cracks within the bedrock. On these sites the forest is more prone to smaller disturbances created by the death of single trees that die standing. These deaths can often be caused by lightning strikes or, perhaps by trees that succumb to disease and rot because of old age. Although this pattern of trees showing greater wind stability on the more exposed ridgetops may be counter-intuitive, there is also evidence for the same pattern in tropical moist forests in Puerto Rico for similar reasons – better rooting in the ridgetop soil with rocks near the surface (Basnet 1990).

This disturbance-size gradient can be directly related to a light gradient, with larger disturbances in the low lying sites having brighter light regimes than those smaller disturbances on the ridgetops. This encourages the establishment and

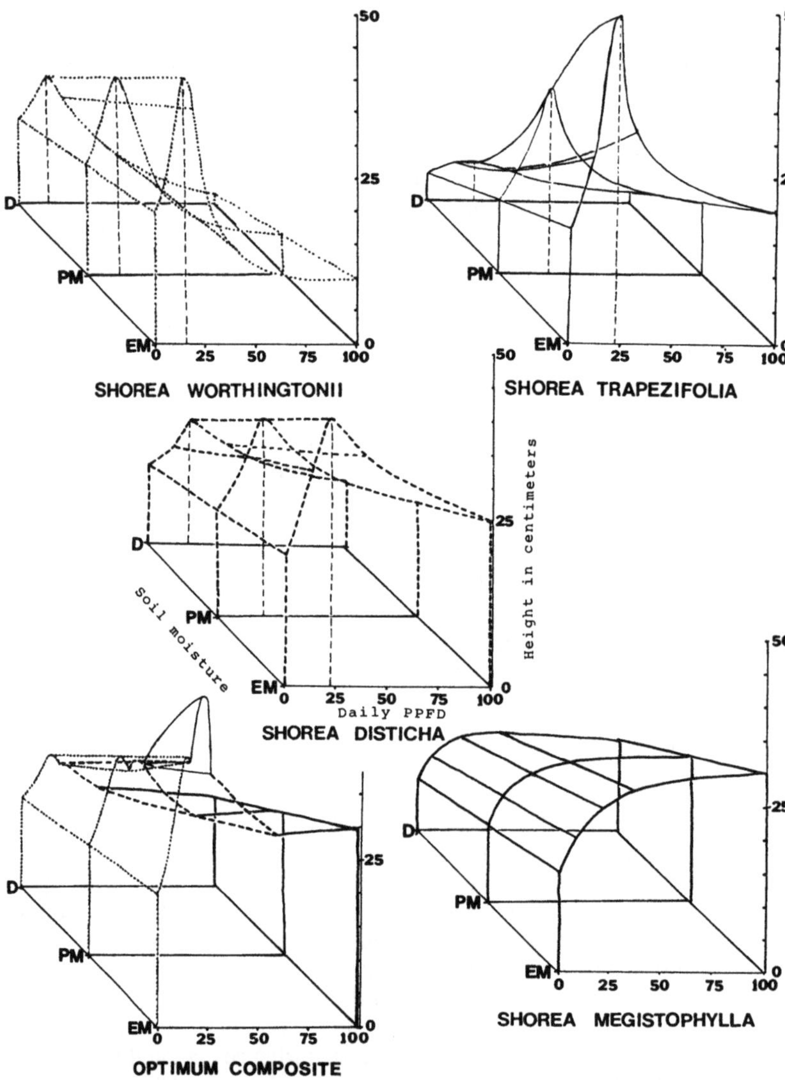

Fig. 5. Schematic diagrams of height growth after 800 days across different regimes of soil moisture and light for each of the *Shorea* species. For the soil moisture regimes D = droughty; PM = periodically moist; EM = ever-moist. Light is measured as daily photosynthetic photon flux density (PPFD) as a percentage of full sun. The bottom left-hand diagram illustrates the optimum growth performance "surface" of the four combined *Shorea* species.

growth of the most light-demanding *Shorea* species in the valley and the most shade-demanding or tolerant on the ridges.

A secondary factor affecting *Shorea* species distribution is a soil moisture gradient that exists across the topography from valley to ridgetop. During unseasonably dry periods, which can occur at irregular, supra-annual intervals, thin ridgetop soils are more water-stressed than lower lying sites. This favors the survival of drought-tolerant *Shorea* species on the ridges (see Fig. 4 for summary).

An example of adaptation and plasticity along a soil moisture continuum: *Quercus* section *Erythrobalanus* (Fagaceae)

Description of moist forest soils

This moist temperate mixed-species forest is located in the eastern upland physiographic region of southern New England, USA. The topography is undulating, ranging from 170 to 415 m above sea level. The bedrock is of metamorphic origin, consisting of a mixture of granite, gneiss and schist. This is generally overlain by glacial till soils that range from moderate to well drained, acidic, stony loams that would be classified as inceptisols (USDA 1975). Local variations in parent materials, texture, alluvial deposits, and slope can produce poorly drained and excessively drained sites as well. The regional climate is hot and humid in summer (mean temperature 20°C) with cold winters (mean temperature -4°C), with a mean annual rainfall (1100 mm) distributed fairly evenly throughout the year.

The forest is classified as Central Hardwood-Hemlock-Pine (Westveld et al. 1956). The pre-settlement forests of this region were diverse in composition and structure, played upon by a natural disturbance regime that includes hurricanes, other windstorms and ice storms, fire, and insect and pathogen epidemics (Raup 1966). Human disturbances have accentuated these patterns (Bormann and Likens 1979). During periods of permanent settlement (1700-1900) this region was nearly completely cleared for agriculture. Only small patches of forest remained in the most inaccessible areas. After 1900 much of the land was abandoned and secondary natural forest developed.

Site characterization of Quercus *section* Erythrobalanus

Quercus is the most commercially important genus in the Fagaceae, a family of trees that dominate the canopy stratum of the mixed-species forests of eastern North America, the central uplands of Mexico, as well as north and east Asia, and Europe. In southern New England upland regions, three different species of *Quercus* section *Erythrobalanus* (the red oaks) occupy different parts of the forest topography, in mixture with up to three species of section *Leucobalanus* (white oaks).

The lower slopes and valleys of the the forest have *Q. rubra*. On the shallow to bedrock ridges *Q. velutina* occurs. Both species have wide overlapping site ranges and can frequently, though not predominantly occur on each other's sites. *Quercus coccinea* occurs on the ridges and upper slopes of the topography. Often it occurs as a co-dominant on deep soil ridges in mixture with *Q. rubra* and *Q. velutina* (Figure 6). In other instances, it occurs on ridgetop sites as a co-dominant with *Q. velutina*, and on still other ridge sites, it can occur as the sole dominant *Quercus* species in section *Erythrobalanus*.

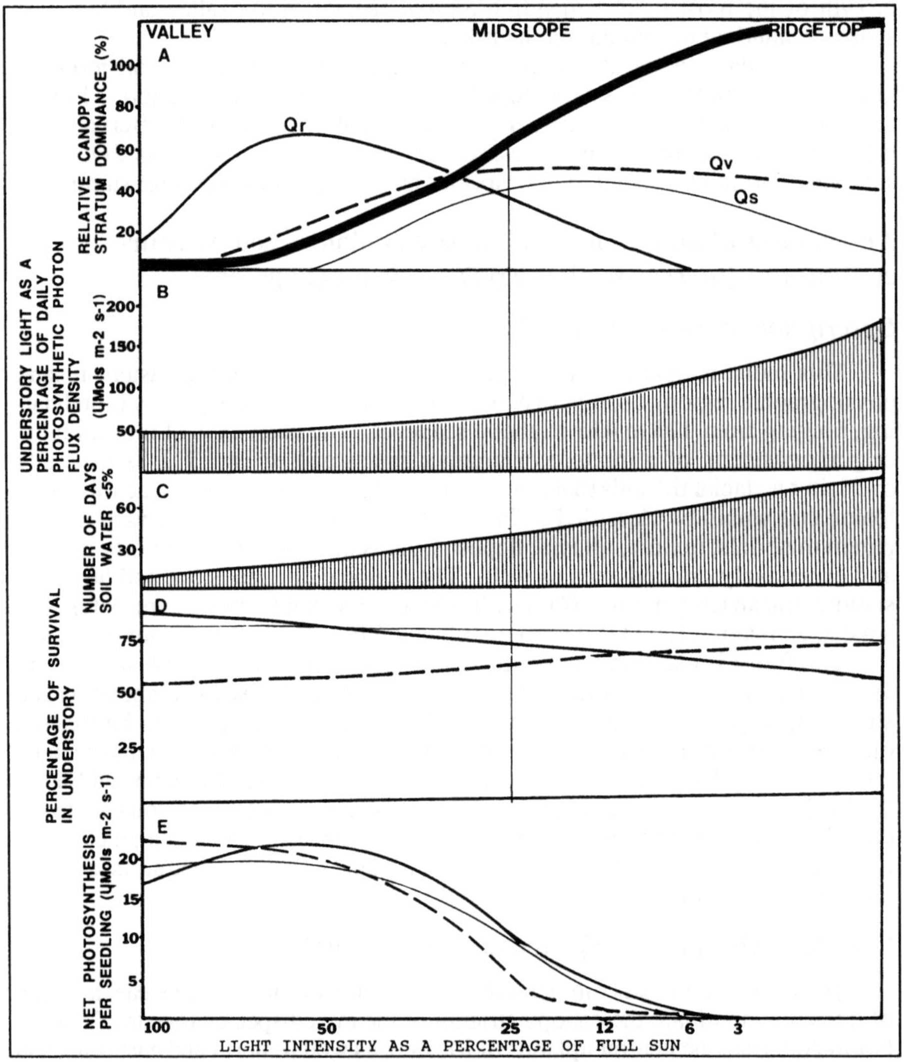

Fig. 6. Diagram depicting: A) distribution of the three *Quercus* species across the topographic gradient; B) natural groundstory light intensity beneath a closed canopy across the topographic gradient; C) low soil-water during the growing season (April - October) as a measure of drought across the topographic gradient; D) survival as a measure of drought tolerance of *Quercus* species across the topographic gradient; E) relative index of net photosynthesis of seedlings at different light intensities as a measure of light tolerance of *Quercus* species; the index is based on the product of the appropriate mean net-dry leaf weight and photosynthetic rate (μmols m^{-2} s^{-1}) for each species at different light intensities. Qc = *Q. coccinea*; Qr = *Q. rubra*; Qv = *Q. velutina*.

Physiological, anatomical, and morphological adaptations

Experiments show *Q. rubra* seedlings to be the most shade-tolerant of the three, though optimum net photosynthesis is at full-sun light intensity. Seedling leaf blade and cuticle are the thinnest, and blade surface area is the largest of the *Quercus* species. Studies also demonstrate that seedlings allocate a greater proportion of their dry weight to stem growth and leaf area and less to root development compared to the other *Quercus* species. Both anatomical and morphological attributes suggest that seedlings of *Quercus rubra* are more susceptible to water stress. Best growth occurs in ever-moist soils, where seedlings are exposed to periods of direct sunshine that is periodic, and that lasts on average for about six hours a day. When seedlings are exposed to full sun all day, leaves appear chlorotic, and height growth is retarded. In moist understory light conditions, seedlings are able to survive better than other *Quercus* species.

Quercus velutina is the most light-demanding, with optimum net photosynthesis at full-sun light intensity. The leaf blade and cuticles of seedlings are the thickest compared to other *Quercus* species. Seedling leaves are also much more pubescent, with a lower number of stomata. Leaf size can also become markedly reduced in dry soils. Compared to other *Quercus* species, seedlings allocate a greater proportion of their dry weight to tap-root development than to promoting stem growth or leaf area. These root, leaf size, and other anatomical attributes make *Q. velutina* the most drought tolerant. At light levels below that of full-sun, seedlings are susceptible to dieback. At low light intensities, which are equivalent to those beneath a fully stratified forest canopy at ground level, seedlings do not survive. Best growth occurs in moist soils, where seedlings are exposed to periods of direct sunlight that last longer than six hours. In drier soils, but in the same light conditions *Q. velutina* achieves better growth than other *Quercus* species.

Seedlings of *Q. coccinea* are also light-demanding, with net optimum photosynthesis at full sun intensity. Leaf blade and cuticle thickness are similar to that of *Q. rubra*. Leaf size and number dramatically decline on progressing from moist to dry soils. Like *Q. velutina*, a greater proportion of seedling dry weight is allocated to roots than to stem, compared to *Q. rubra*. Unlike *Q. velutina* more of the dry root weight is allocated to fine root development than to tap-root. Best growth occurs in moist soil, where seedlings are exposed to periods of direct sunshine similar to that of *Q. velutina*. Similarly, at light levels lower than full sun, it is susceptible to dieback and mortality. Anatomical and morphological characteristics suggest it to be adapted to drier sites than *Q. rubra* but less than that of *Q. velutina*.

Site factor gradients across the forest topography

Differences in soil moisture across the topography appear to be the primary gradient responsible for *Quercus* species differentiation. In the valley, soils remain moist throughout the growing season, but on the ridgetops pronounced droughts can occur. In particular, moisture stress to seedlings is more severe in understory conditions, because of root competition with the overstory and greater soil-water losses through canopy evapo-transpiration. This favors more drought-tolerant *Quercus* species on ridgetops and drier slopes.

A secondary factor affecting species distribution is a light gradient that occurs across the topography beneath the forest overstory. These below-canopy light levels can be related to differences in degree of canopy stratification. Greater availability of soil-water in valley soils promote greater canopy stratification when compared to more drought-prone soils of the ridgetops. Light regimes in the understory during the growing season are lower in the valleys than on ridgetops. This promotes the survival of more shade-tolerant *Quercus* species in the understory of the valley.

Compared to site factor gradients affecting *Shorea* species, differences in the size and type of disturbance appears not to create a distinct light gradient across the topography. *Quercus* seedling leaf-out is closely synchronized, and occurs before canopy leaf-out in spring. All species enjoy similar moist sunny conditions, and accumulate much of their photosynthate for their shoot extension in early summer. Disturbances are also more catastrophic and less predictable. This suggests equilibrating forces are less evident and might explain why patterns in the niche partitioning of *Quercus* species regeneration overlap considerably. This is accentuated by the ability of section *Erythrobalanus*, unlike *Shorea*, to hybridize. Both hybridization, and a wider species tolerance overlap, suggest strategies to accomodate more unpredictable forest disturbance regimes and climate (see Figure 6 for summary).

Summary

These examples demonstrate that species classified as late successional, and belonging to the same regeneration guild, have specific morphological, physiological, and anatomical characteristics that differentiate them from their congeneric associates. These factors have allowed each species to adapt to a specific ecological niche within the same whole forest environment, and suggests that moist mixed-species forest can maintain large taxonomic assemblages of late-successional canopy tree species through predictable, specialized and/or species-specific mechanisms for establishment. It can be hypothesized therefore that, for a particular moist forest region, different canopy genera can co-exist within the same landscape by differentiating along primary gradients of different site factors (Figure 7).

These studies also document relationships, particularly pertaining to tolerance adaptations and their plasticities, between species of the same guild. They demonstrate that these relationships are similar to those between species of different guilds, but that they are not as simple (linear) or as ecologically noticeable.

Results also indicate regeneration establishment is an important period for determining site specialization of a species. Patterns in regeneration establishment of these assemblages appear to reflect forest processes that are in equilibrium whereby, after a natural disturbance, a forest will revert back to its former structure and composition. These results imply that the degree of equilibrium-promoting processes within a mixed-species forest may be associated with the species diversity of its canopy stratum. Greater climatic (and hence, disturbance) predictability may promote a more refined degree of specialization across a gradient, allowing more species to co-exist within the generic assemblage, at least among species which are abundant. Moist mixed-species forests in less predictable climates can be hypothesized to promote fewer generic assemblages, with species within each assemblage having less specialized, more overlapping adaptations, with greater abilities to change (plasticity). This, in part, has also been suggested by Runkle (1989), who proposed that temperate forests have reduced options for phenological differentiation, because winter induces a more synchronized growth response from regeneration. Another example might be moist mixed-species tropical forest in Central America which, though tree species diverse, is not dominated by particular canopy

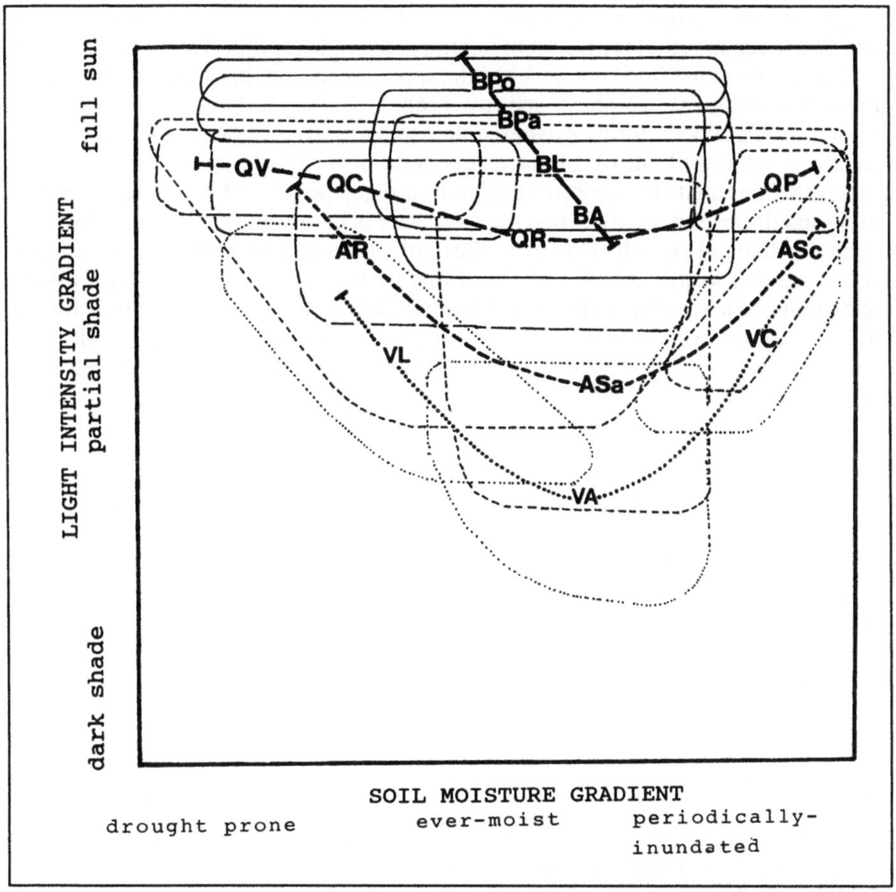

Fig. 7. A hypothetical diagram depicting generic assemblages across soil moisture and light regimes for moist mixed-species forest of southern New England. Each genus dominates a particular gradient as shown by the axis orientation of the assemblage. Boundaries for each species represent their tolerance plasticity. Generic assemblages belonging to several guilds are represented, of which the maples and oaks would be classified as belonging to the dominant mature phase.

Species codes are as follows:

- Canopy and Subcanopy Stratum Assemblages

  Birches - BPo = *Betula populifolia* (gray birch), BPa = *B. papyrifera* (paper birch), BL = *B. lenta* (black birch), BA = *B. alleghenensis* (yellow birch)

  Red oaks (section Erythrobalanus) - QV = *Quercus velutina* (black oak), QC = *Q. coccinea* (scarlet oak), QR = *Q. rubra* (northern red oak), QP = *Q. palustris* (pin oak)

  Maples - AR = *Acer rubrum* (red maple); ASa = *A. saccharum* (sugar maple); ASc = *A. saccharinum* (silver maple)

- Understory Stratum Assemblages

  Blackhaw viburnums - VL = *Viburnum lentago* (nannyberry); VA = *V. alnifolia* (hobblebush); VC = *V. cassinoides* (wild raisin).

tree genera. Workers have shown these forests to be more strongly influenced by disturbance regimes that are frequent but unpredictable (Hubbell and Foster 1986).

Shelterwood methods are perhaps the most conducive to obtaining regeneration of dominant mature-phase canopy trees of moist mixed-species forest. In essence, shelterwood cutting creates minor gaps (partial overstory disturbances) that allow establishment of advance regeneration but minimizes the growth of species that dominate the canopy stratum of the gap phase and building phase of a young stands development. When the overstory is harvested, the regeneration can respond with rapid growth. However, the application and development needs to be further refined, so that creation of minor disturbances can emulate natural disturbance regimes across topographic gradients. This is particularly apparent for the moist aseasonal tropics, where indiscriminant logging can dramatically alter the future structure and productivity of a forest (whether the logging is excessively heavy or overly conservative).

Acknowledgements

The ideas in this paper are largely based on studies that were carried out as part of my doctoral work at Yale University. I would like to acknowledge advice received from David Smith, Graeme Berlyn, Bruce Larson, and my father Peter Ashton. I would also like to thank Matt Kelty for advice and help received from editing.

References

Ashton, P.S. 1964. Ecological studies in the mixed dipterocarp forests of Brunei State. Oxford Forestry Memoirs 25.

Ashton, P.S. 1969. Speciation among tropical trees: some deduction in light of recent evidence. Biological Journal of the Linnean Society 1:155-196.

Ashton, P.M.S. 1990. Seedling response of *Shorea* species to moisture and light regimes in a Sri Lankan rain forest. Ph.D. Thesis, Yale University, New Haven, CT.

Baillie, I.C. and P.S. Ashton. 1983. Some soil aspects of the nutrient cycle in the mixed dipterocarp forests of Sarawak, east Malaysia. In: "Tropical Forest Resources and Management" (S.L. Sutton, T.C. Whitmore and A.C. Chadwick, eds.). Blackwell Scientific Publications, Oxford.

Bahari, Z.A., Pallardy, S.G., and W.C. Parker. 1985. Photosynthesis, water relations and drought adaptations in six woody species of oak-hickory forests in central Missouri. Forest Science 31:557-569.

Basnet, K. 1990. Studies of ecological and geological factors controlling the pattern of Tabonuco forests in the Luquillo Experimental Forest, Puerto Rico. Ph.D. Thesis, Rutgers University, New Brunswick, NJ.

Bazzaz, F.A. 1979. The physiological ecology of plant succession. Annual Review of Ecology and Systematics 10:351-371.

Bazzaz, F.A. 1984. Dynamics of wet-tropical forests and their species strategies. In: "Physiological Ecology of Plants of the Wet Tropics (E. Medina, H.A. Mooney and C.A. Vazquez-Yanes, eds.), pp. 233-243. Dr. W. Junk, The Hague.

Bazzaz, F.A., and S.T.A. Pickett. 1980. The physiological ecology of tropical succession: a comparative review. Annual Review of Ecology and Systematics 11:287-316.

Beard, J.S. 1944. Climax vegetation in tropical America. Ecology 25:127-158.

Becker, P., L.W. Lee, E.D. Rothman, and W.D. Hamilton. 1985. Seed predation and coexistence of tree species; Hubbells's models revisited. Oikos 44:382-390.

Bjorkman, O., M. Ludlow, and P. Morrow. 1982. Photosynthetic performance of two rainforest species in their habitat and analysis of their gas exchange. Carnegie Institute of Washington Year Book 71:94-102.

Bormann, F.H., and G.E. Likens. 1979. Patterns and Processes in a Forested Ecosystem. Springer-Verlag, New York.

Bourdeau, P. 1954. Oak seedling ecology determining segregation of species in Piedmont oak-hickory forests. Ecological Monographs 24:297-320.

Brokaw, N.V.L. 1985. Gap-phase regeneration in a tropical forest. Ecology 66:682-687.

Budowski, G. 1965. Distribution of tropical rain forest tree species in the light of successional process. Turialba 15:40-42.

Canham, C.D. 1988. Growth and canopy architecture of shade-tolerant trees: the response of *Acer saccharum* and *Fagus grandifolia* to canopy gaps. Ecology 69:786-795.

Canham, C.D. 1989. Different responses to gaps among shade-tolerant tree species. Ecology 70:548-550.

Clements, F.E. 1916. Plant succession. An analysis of the development of vegetation. Carneigie Institute of Washington, Publication No. 242.

Cooray, P.G. 1967. An introduction to the geology of Ceylon. Spolia Zeylanica 31:1-314.

Curtis, J.T. 1959. The Vegetation of Wisconsin. University of Wisconsin Press, Madison, WI.

Davis, T.A.W. and P.W. Richards. 1933. The vegetation of Moraballi Creek. Part I. Journal of Ecology 21:350-384.

Davis, T.A.W. and P.W. Richards. 1934. The vegetation of Moraballi Creek. Part II. Journal of Ecology 22:106-155.

Denslow, J.S. 1980. Gap partitioning among tropical rainforest trees. Biotropica 12 (Suppl):47-59.

Denslow, J.S. 1987. Tropical rainforest gaps and tree species diversity. Annual Review of Ecology and Systematics 18:431-451.

de Rosayro, R.A. 1942. The soils and ecology of the wet evergreen forests of Ceylon. Tropical Agriculture 98:78-80; 153-175.

de Rosayro, R.A. 1954. A reconnaissance of Sinharaja rain forest. Ceylon Forester 1:68-74.

Devoe, N.N. 1989. Differential seeding and regeneration in openings and beneath closed canopy in sub-tropical wet forest. Ph.D. Thesis, Yale University, New Haven, CT.

Drury, W.H., and I.C.T. Nisbet. 1973. Succession. Journal of the Arnold Arboretum 54:331-368.

Egler, F.E. 1954. Vegetation science concepts in initial floristic composition. A factor in old field vegetation development. Vegetatio 4:412-417.

Eyre, S.R. 1968. Vegetation and Soils. Edward Arnold, London.

Fedorov, A.A. 1966. The structure of the tropical rain forest and speciation in the humid tropics. Journal of Ecology 54:1-11.

Fetcher, N., B.R. Strain and S.F. Oberbauer. 1983. Effects of light regimes on the growth, leaf morphology and water relations of seedlings of two species of tropical trees. Oecologia 58:314-319.

Grubb. P.J. 1977. The maintenance of species-rich communities: the importance of the regeneration niche. Biological Reviews 52:107-145.

Gunatilleke, C.V.S. and I.A.U.N. Gunatilleke. 1981. The floristic composition of Sinharaja - a rain forest in Sri Lanka with special reference to endemics and dipterocarps. Malay Forester 44:386-396.

Gunatilleke, C.V.S. and I.A.U.N. Gunatilleke. 1983. A forestry case study of the Sinharaja rain forest in Sri Lanka. In: "Forest and Watershed Development and Conservation in Asia and the South Pacific" (L.S. Hamilton, ed.). Westview Press, Colorado.

Hartshorn, G.S. 1978. Tree falls and forest dynamics. In: "Tropical Trees as Living Systems (P.B. Tomlinson and M.H. Zimmerman, eds.), pp. 617-638. Cambridge University Press.

Hibbs, D.E. 1982. Gap dynamics of hemlock-hardwood forest. Canadian Journal of Forest Research 12:522-527.

Hubbell, S.P. 1979. Tree dispersion, abundance and diversity in a tropical deciduous forest. Science 203:299-1309.

Hubbell, S.P. and R.B. Foster. 1986. Canopy gaps and the dynamics of a neotropical forest. *In*: "Plant Ecology" (M.J. Crawley, ed.), pp 51-76. Blackwell Scientific Publications, Oxford.

Jackson, L.W.R. 1967a. Effect of shade on leaf structure of decidous tree species. Ecology 48:498-499.

Jackson, L.W.R. 1967b. Relation of leaf structure to shade tolerance of dicotyledonous tree species. Forest Science 13:321-323.

Janos, D.P. 1988. Mycorrhiza applications in tropical forestry: are temperate-zone approaches appropriate? In: "Trees and Mycorrhiza" (F.S.P. Ng, ed.), pp. 133-188. Forest Research Institute, Malaysia, Kuala Lumpur.

Janzen, D.H. 1971. Seed predation by animals. Annual Review of Ecology and Systematics 2:465-492.

Langenheim, J.H., C.B. Osmond, A. Brooks and P.J. Ferrar. 1984. Photosynthetic responses to light in seedlings of selected Amazonian and Australian rainforest tree species. Oecologia 63:215-224.

Lorimer, C.G., L.E. Frelich, and E.V. Nordheim. 1988. Estimating gap origin probabilities for canopy trees. Ecology 69:778-785.

Moorman, F.R. and C.R. Panabokke. 1961. Soils of Ceylon. Tropical Agriculture 117:4-65.

Newberry, D.M, I.J. Alexander, D.W. Thomas and J.S. Gartlan. 1988. Ectomycorrhizal rain-forest legumes and soil phosphorus in Korup National Park, Cameroon. New Phytologist 109:433-450.

Oberbauer, S.F. and B.R. Strain. 1986. Effects of canopy position and irradiance on the leaf physiology and morphology of *Pentaclethra macroloba* (Mimosaceae). American Journal of Botany 73:409-416.

Oliver, C.D. 1978. The development of northern red oak in mixed species stands in central New England. Yale University School of Forestry and Environmental Studies, Bulletin No. 91. 63 p.

Oliver, C.D., and B.C. Larson. 1990. Forest stand dynamics. McGraw-Hill, New York. 467 p.

Pearcy, R.W. 1987. Photosynthetic gas exchange responses of Australian tropical forest trees in canopy, gap and understory micro-environments. Functional Ecology 1:169-178.

Pickett, S.T.A. 1980. Non-equilibrium coexistence of plants. Bulletin of the Torrey Botanical Club 107:134-145.

Pickett, S.T.A., and P.S. White (eds.). 1985. The ecology of natural disturbance and patch dynamics. Academic Press, London.

Popma, J. and F. Bongers. 1988. The effect of canopy gaps on growth and morphology of seedlings of rain forest species. Oecologia 75:625-632.

Poulson, T.L. and W.J. Platt. 1988. Light regeneration niches. Bulletin of the Ecological Society of America 69:264.

Poulson, T.L., and W.J. Platt. 1989. Gap light regimes influence canopy tree diversity. Ecology 70:553-555.

Sipe, T. 1990. Gap partitioning among maples (*Acer*) in the forests of central New England. Ph.D. Thesis, Harvard University, Cambridge, MA.

Smith, D.M., and P.M.S. Ashton. Patterns in the early growth and establishment of canopy tree regeneration across strip cuts in southern New England. (In preparation).

Spies, T.A., and J.F. Franklin. 1989. Gap characteristics and vegetation response in coniferous forests of the pacific northwest. Ecology 70:543-545.

Strauss-Debenedetti, S.I. 1989. Responses to light in tropical Moraceae of different successional stages. Ph.D. Thesis, Yale University, New Haven, CT.

Swaine, M.D., and T.C. Whitmore. 1988. On the definition of ecological species groups in tropical rain forests. Vegetatio 75:81-86.

Turner, I.M. 1989. A shading experiment of some tropical rain forest tree seedlings. 1:383-389.

Turner, I.M. 1990. Tree seedling growth and survival in a Malaysian rain forest. Biotropica 22:146-154.

U.S.D.A. Soil Conservation Service. 1975. Soil taxonomy. A basic system of soil classification for making and interpreting soil surveys. Agricultural Handbook No. 436. USDA Soil Conservation Service, Washington, D.C.

van Steenis, C.G.G.J. 1969. Plant speciation in Malesia with special reference to the theory of non adaptive saltory evolution in speciation in tropical environments. Biological Journal of the Linnean Society 1:97-133.

Walters, M.B., and C.B. Field. 1987. Photosynthetic light acclimation in two rainforest *Piper* species with different ecological aptitudes. Oecologia 72:449-456.

Wali, M.K., and V.J. Krajina. 1973. Vegetation - environment relationships of some Sub-boreal Spruce zone ecosystems in British Columbia. Vegetatio 26:237-381.

Watt, A.S. 1947. Pattern and process in the plant community. Journal of Ecology 35:1-22.

Westveld, M., R.I. Ashman, H.I. Baldwin, R.P. Holdsworth, J.H. Lambert, H.J. Lutz, L. Swain, and M. Standish. 1956. Natural forest vegetation zones in New England. Journal of Forestry 54:332-338.

Whitmore, T.C. 1982. On Pattern and process in forests. *In*: "The Plant Community as a Working Mechanism (E.I. Newman, ed.). Blackwell Scientific Publications, Oxford.

Whitmore, T.C. 1984. Tropical rain forests of the Far East. Second edition. Clarendon Press, London.

Whittaker, R.H. 1966. Forest dimensions and production in the great Smoky Mountains. Ecology 47:103-121.

Whittaker, R.H. 1967. Gradient analysis of vegetation. Biological Review, Cambridge Philosphocal Society 42:207-264.

Productivity of Mixed-Species Stands

III

Comparative productivity of monocultures and mixed-species stands

8

MATTHEW J. KELTY

Introduction

Ecological theory suggests that there is a potential productivity advantage to be gained by designing managed forest stands to contain more than one tree species. The basis for this advantage, as noted by Ewel (1986) and Vandermeer (1989) is rooted in fundamental niche theory—two or more species must use resources differently if they are to coexist on a site. Differential resource use among species suggests that the species in a mixture may utilize the resources of a site more completely than any single species would be able to do, leading to greater overall productivity. However, the link between differential resource use among species and greater total resource use does not necessarily exist in all cases. For example, it is possible that a mixture of species may simply subdivide the total resource base that one highly efficient species may completely use on its own. In addition to niche separation, which potentially applies to all species in mixture, certain specific combinations of species may exist in which one species may directly benefit from the presence of another.

The concept of even a potential increase in production in mixed-species stands has not generally been incorporated into forestry practice. In situations where timber yield is the primary objective of management, a clear trend exists toward favoring monocultures of the most productive species having suitable wood quality. Where species mixtures are favored, the reasons generally include one or more of the following: protection from disease and insect outbreak, resistance to wind damage and other abiotic stress, risk reduction and compensatory growth (i.e., if one species succumbs to any factor, other species may survive and respond with increased growth), maintenance of landscape aesthetics, and conservation of native plant and animal species. These objectives are quite logically seen as overriding productivity considerations in many situations, and lower yields are often considered to be a necessary sacrifice that accompanies the use of mixtures. However, there is little information directly linking productivity and number of species in forest stands. If this relationship were better understood, it might be possible to design specific mixtures to have higher yields than monocultures, or to achieve some of the objectives listed above while minimizing any decreases in yield. This paper focuses on the questions: 1) are there situations in which mixed-species stands have greater total stand yields than monocultures of the component species, and 2) if so, under what conditions, or with what sets of species is this likely to occur?

The data required to answer these questions would best be obtained from experiments in which monocultures of each of two component species plus mixtures of varying species proportions are all grown under the same site conditions, over a rotation length logical for those species. The time and space requirements for such experiments with forest stands have almost entirely precluded their use. However, similar studies with herbaceous plant species are logistically much simpler, and have been pursued in considerable detail by plant

M. J. Kelty (ed.), The Ecology and Silviculture of Mixed-Species Forests, 125–141.

population ecologists and agronomists, using annual crop species. These have identified many of the principles of yield comparisons quite well, and the application of these principles to forest stands will be considered. Also, a number of forest studies will be reviewed which employ less complete experimental designs but still provide a starting point for understanding yield patterns in mixtures of tree species.

General approach to production comparisons with herbaceous species

Two basic experimental designs (the "additive" and "substitutive" designs) have been used in studies of productivity and competition with herbaceous species (Harper 1977). Although neither is free of problems associated with their artificiality (Vandermeer 1989), the substitutive design has been the one more commonly used for yield comparisons of mixtures and monocultures. In studies of this type, also referred to as "replacement series" experiments, a series of plots is grown under identical site conditions, each having been sown at the same total plant total density, but with varying proportions of two constituent species. Monocultures of each of the two species are also grown at the same density. Production is assessed by measuring total biomass yield, usually at the end of one growing season. (The additive design differs in that the density of one species is held constant throughout all plots, and a second species is added at varying densities. This design is often used to study the effect of a weed species on a crop planted at a standard density, but it has been used for some forest mixture studies as well.)

The effects of combining two species in a replacement series can then be analyzed by comparing the yield of each species in mixture with its yield in monoculture (Harper 1977). For any particular mixture of species A and B, a relative yield (RY) of each species, and a relative yield total (RYT) can be calculated by:

$$\text{RY(species A)} = \frac{\text{yield of species A in mixture}}{\text{yield of species A in monoculture}}$$

$$\text{RYT} = \text{RY(species A)} + \text{RY(species B)}$$

An expected value of RY and RYT can be calculated, assuming the two species use resources in identical ways (Figure 1a). For example, in a 50:50 mixture of species A and B, the expected RY of each species would be 0.5. The expected value of RYT for all plots is always 1.0. An actual RYT value greater than 1.0 for any mixture indicates that either significant niche separation or a direct beneficial relationship exists among species, and suggests that a potential productivity advantage may exist for mixtures compared to monocultures. RYT values less than 1.0 indicate an antagonistic relationship between the species in mixture. However, it is necessary to compare absolute (rather than relative) yield values to identify the highest yielding stand in these experiments. If one of the component species is much more productive in monoculture than the other, mixtures can have RYT's exceeding 1.0, and yet not exceed the yield of the more productive species in monoculture (Figure 1b).

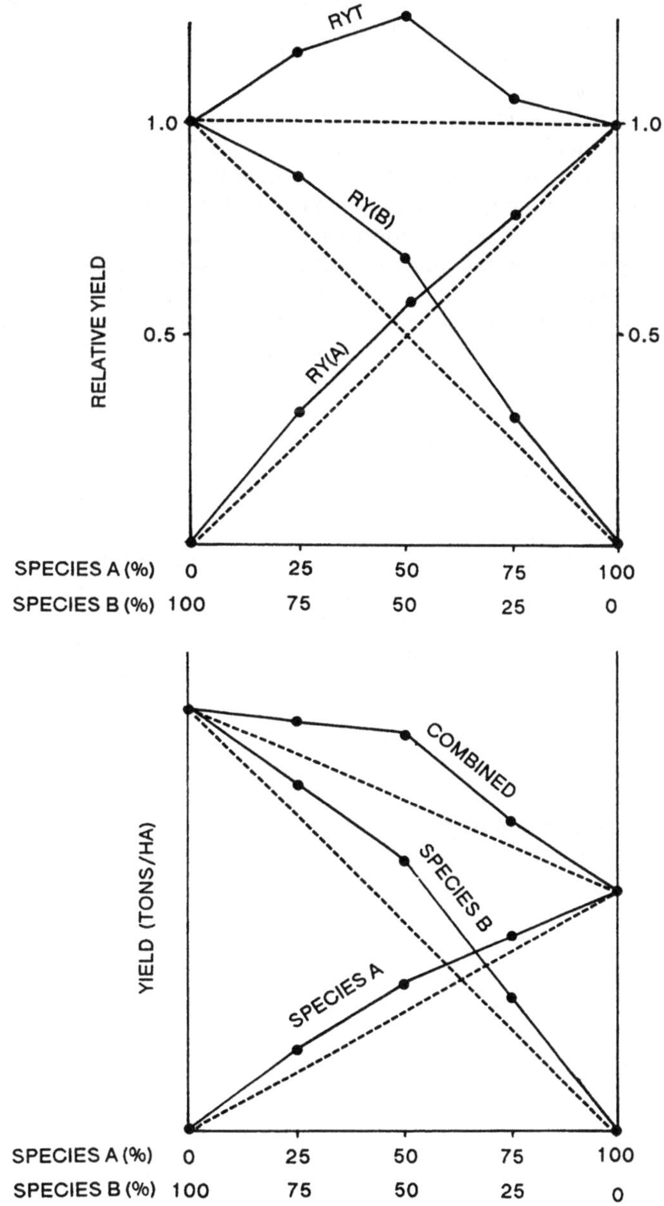

Fig. 1. Results of hypothetical replacement series experiment, expressed as (a) relative yield and (b) absolute yield. Dashed lines represent expected yield in intra- and inter-specific interactions were equivalent. Solid lines represent experimental yields.

A limitation of this design is that a specific replacement series must have a given planting density, site quality, and rotation length. Yield comparisons among mixtures and monocultures would be expected to vary if these factors were altered. However, studies often do not use multiple replacement series to incorporate variation in these factors.

In a survey of yield studies of mixtures, Trenbath (1974) concluded that situations where the yield of mixtures was significantly greater than that of the highest yielding monoculture or significantly less than that of the lowest yielding monoculture were not common. Harper (1977) noted that most studies combined crop species that were not particularly chosen for niche separation, so that this result should not be surprising. A number of replacement series studies have been designed to isolate the factors that are associated with the cases in which mixtures did produce greater yields than monocultures; these have combined closely related species (or different varieties of the same species) with only minor differences in growth characteristics (Ellern et al. 1970, Trenbath and Harper 1973, Hill and Shimamato 1973). These studies have demonstrated that mixtures can outyield monocultures when the species or varieties have:

1. differences in height, form, or photosynthetic efficiency of foliage;
2. differences in phenology, such as timing of foliage production and duration of photosynthetic activity;
3. differences in root structure, particularly depth of rooting.

These characteristics all cause interspecific competition to be less intense than intraspecific competition, with the key to the reduction in competition often being a spatial stratification of foliage or roots. Species having important differences in characteristics which allow them to coexist in mixture with high productivity are said to have good "ecological combining ability" (Harper 1977).

Some combinations of species may have greater yield in mixtures because of a commensalistic or mutualistic relationship between the species, in which one or both species directly benefit from the presence of another. Studies have concentrated on mixtures involving nitrogen-fixation by one species, in which the other species benefits from increased nitrogen availability. Many intercropping schemes in use in tropical agriculture involve a legume as one of the species to attempt to achieve this effect. The direct benefits appear to result from the nitrogen-fixing species increasing the nitrogen available to the companion species through root exudation, production of nitrogen-rich tissues (fine roots and foliage) which release nutrients upon decomposition, or possibly a direct transfer of nitrogen between species via mycorrhizae. However, increases in production could also simply be a result of niche separation in nitrogen nutrition, since nitrogen-fixing species do not draw as heavily upon soil nitrogen. For the most part it is difficult to determine the relative importance of these mechanisms in field experiments, but it is clear that direct benefit is important in at least some cases (Vandermeer 1989).

In a review of the ecological principles of intercropping for agriculture, Vandermeer proposed that the factors described above could be grouped into two principles that serve as "a core to understanding how intercrops function" (1989, p. 12). He called these the "competitive production principle" (in which two species have reduced competition in mixture compared to monocultures, thereby utiltizing resources more efficiently), and the "facilitative production principle" (in which one species positively affects the growth of another in mixture). The discussion of potential advantages of mixtures in forest production that follows will make use of this division, with the principles referred to here simply as "competitive reduction" and "facilitation".

Competitive reduction

Studies of how different tree species interact have tended to divide the areas of competition into two types: competition for light and competition for soil resources (both nutrients and water). This is largely because the methods of study are quite different, with light being more extensively studied simply because crowns are more accessible than roots. Although there clearly are interactions between crown and root competition, it is still a reasonable way to structure the discussion of competition reduction.

Reduction of crown competition—theoretical considerations

Much of the potential niche separation among tree species with respect to crown competition and productivity involves characteristics that control both the quantity of light intercepted (crown structure and density) and the efficiency of use of intercepted light for photosynthesis (leaf anatomy and physiology). These characteristics are often used to classify species as sun- or shade-adapted (shade-intolerant or shade-tolerant, respectively, in the forestry literature), recognizing that a gradient exists among species. Light-use efficiency differs among species along this gradient. The foliage of shade-adapted species reaches the light compensation point (the point at which photosynthesis meets respiration requirements) at lower light levels than that of sun-adapted species. However, shade-adapted species become light-saturated at lower light levels and with lower maximum photosynthetic rates than sun-adapted species. The anatomical and biochemical adaptations underlying these patterns are not fixed for each species (Boardman 1977). Leaves developing in different light levels within the crown of a single tree have different characteristics, such that a vertical gradient in acclimation exist from sun leaves to shade leaves. However, the range of acclimation is limited for each species; no single species has high efficiency at both extremes of light intensity (Boardman 1977).

At the canopy level, shade-adapted species tend to have greater total amounts of foliage per unit land area and intercept more light than canopies of sun-adapted species, because their efficiency at low light levels allows more foliage to survive in deeper crown layers (Assmann 1970, Monsi et al. 1973).

The clearest way of using mixtures to make more complete use of available light is by combining species that develop a stratified canopy with a sun-adapted species in the high-light environment in the overstory with a shade-adapted species in the low light of the understory. Trenbath (1981) and Vandermeer (1989) described two ways in which a mixture with this canopy structure can exceed the monoculture of sun-adapted species in photosynthetic capacity:

1) A sun-adapted species uses a relatively low proportion of available light, because of the inability of its foliage to survive in the lower, shaded portions of the canopy. Beneath a complete canopy of the sun-adapted species, sufficient light would still be available for a shade-adapted species to grow; if such an understory were added, the total photosynthetic capacity of the mixed canopy would exceed that of the sun-adapted monoculture. In this type of mixture (Mixture A in Fig. 2), the increase results from greater total

light interception by the added foliage in the understory, rather than by a change in light-use efficiency.

2) Stands with a somewhat different canopy structure may show an increased efficiency in light use as well as an increase in total light interception (Mixture B in Fig. 2). Foliage in the lower crown levels of a monoculture of sun-adapted species survives at a very low net photosynthetic rate – at or near the compensation point. Foliage of a more shade-adapted species would be more efficient in light use at these lower light levels. If stand structure were controlled such that the understory species replaced the foliage of the sun-adapted species at these light levels, total efficiency of light use would be increased, in addition to an increase in the total amount of light intercepted.

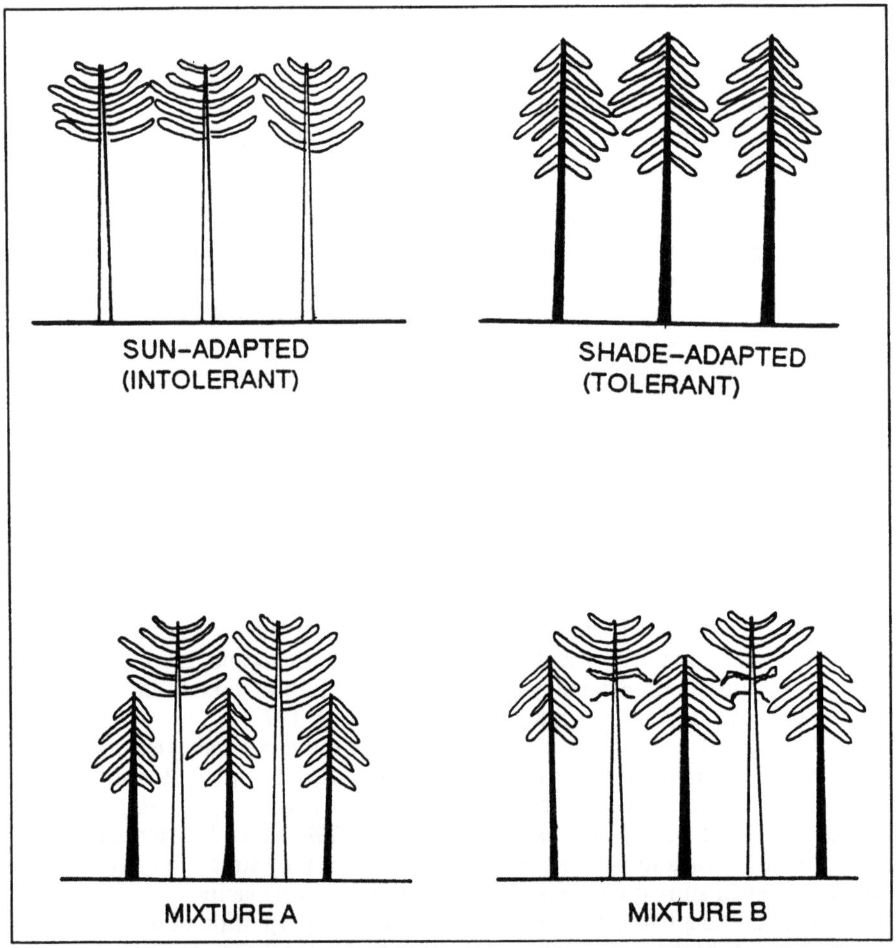

Fig. 2. Profile diagram of two different stand structures in mixtures of two species. Mixture A has complete vertical stratification between species, with an overstory canopy equivalent to that of the monoculture of the sun-adapted species. Mixture B has a more widely spaced overstory, with foliage of the shade-adapted species replacing that of the overstory species at mid-canopy levels.

The comparison of a mixture with the shade-adapted monoculture is not as clear. The canopy of the shade-adapted monoculture would likely intercept approximately the same quantity of light as a mixture, since the depth of the canopy would be controlled in both cases by the shade species. The potential advantage of the mixed canopy depends upon the range of sun and shade adaptations among species being greater than that of sun and shade leaf developmental acclimation within any one species. A mixed canopy would have greater photosynthetic capacity if the foliage of the sun-adapted species could utilize high intensity light at the top of the canopy more efficiently than sun leaves of the shade-adapted species.

Most considerations of sun- and shade-adaptations deal only with foliage-level characteristics. Givnish (1988) has pointed out that the traditional compensation point deals only with carbon balance in foliage and does not define the point at which the leaf can contribute to energy gain of the whole plant. Additional costs must be deducted from photosynthesis, including night respiration of foliage; branch, stem, and root maintenance respiration; and construction costs of leaves and supporting structures. Sun and shade adaptations may be importantly affected by species differences in respiration rates and construction costs of various branch and crown architectures, rather than simply differences in foliage characteristics.

The basic stratified canopy structure in mixtures tends to develop naturally because sun-adapted species generally have greater rates of juvenile height growth than shade species. This reduces competition since the species occupy different niches by capturing light at different intensities and locations within the canopy. The spatial arrangement of trees of different species must be fine-grained (i.e., trees must be adjacent to trees of a different species) in order for reduction of competition to occur. The establishment of different species in monospecific blocks will tend to minimize the effects of combining species in a stand. The different structures illustrated in Figure 2 can be attained by controlling density and species composition during stand establishment and thinning. The main point being made in noting this distinction is to suggest that the most productive canopy structure may not be a complete overstory canopy above an understory, but rather a more widely spaced overstory with understory trees that are not completely suppressed. This would be similar to the pattern of widely spaced emergents growing above a main canopy, which often occurs naturally in mixed stands. Of course in practice, there is no reason not to combine three or more species in canopy structures that take advantage of both of these forms of canopy stratification.

Reduction of crown competition—empirical evidence

Forestry experiments of the replacement series design would be useful in testing whether these crown competition reduction factors actually have significant effects on stand production. The ones that most closely resemble that design are a set of plantation experiments described by Assmann (1970). These plantations were established in the late 19th century on various sites in Germany and Switzerland, using the principal timber species of the region. Data were generally available for only one mixed stand, so the effects of variation in species proportion could not be studied; also, in some cases there was a monoculture of only one of the component species present for comparison. However, these studies have considerable value in that they are long-term comparisons of stands growing on adjacent plots, with detailed measures of biomass yield of stems or (in some cases) stems plus branches.

In each case, canopies developed a stratified structure with the less shade-tolerant species forming an upper canopy above the more tolerant species. The species combinations studied were as follows (with the upper canopy species listed first): Scots pine (*Pinus sylvestris*) and Norway spruce (*Picea abies*); Scots pine and beech (*Fagus sylvatica*); sessile oak (*Quercus petraea*) and beech; larch (*Larix decidua*) and beech; Norway spruce and silver fir (*Abies alba*). In the first three species combinations, only monocultures of the overstory species were available for comparison, whereas data for both monocultures were available for the latter two.

These consistently showed that mixed stands had greater yields than monocultures of the less tolerant upper canopy species and (where comparisons were possible) greater than the more tolerant species monoculture as well. The importance of canopy stratification in these yield results was demonstrated in two additional studies described by Assmann. In pure stands growing adjacent to one another, Norway spruce outyielded beech in stemwood biomass by 10 to 100%, depending upon site characteristics. Mixtures proved to be more productive than pure spruce stands if beech remained in the understory; however, if spruce was thinned to allow beech to develop into a single canopy stratum together with spruce, total mixed-stand yield was less than that of pure spruce. Similar results were obtained with oak/beech mixtures, where oak was thinned to increase height growth of understory beech and eliminate canopy stratification.

Two additional European plantation studies (Prokop'ev 1976, Poleno 1981) showed similar results. Both compared mixtures of Scots pine and Norway spruce in which pine formed an overstory above spruce, and concluded that mixtures had greater yields than monocultures of either species. The study by Poleno was the most complete of any of those described above; it included plots that differed in stand age, species proportions, and site qualities.

A study of mixed stands of akamatsu (*Pinus densiflora*) and hinoki (*Chamaecyparis obtusa*) in Japan made similar comparisons. These stands were found to develop stratified canopies by age 30, with akamatsu forming an upper canopy layer above hinoki (Kawahara and Yamamoto 1982). Total stem volume per hectare was found to be greater in mixed stands than pure stands of either species, with the maximum volume occurring in stands with 35% akamatsu to 65% hinoki (Kawahara and Yamamoto 1986).

Several additional studies of natural stands provide evidence concerning yield comparisons, although the direct comparisons are less complete than for the studies described above. In the northeastern U.S., stands composed of oak (*Quercus*), maple (*Acer*), and birch (*Betula*) overstories with understories of the tolerant conifer eastern hemlock (*Tsuga canadensis*) had higher yields than adjacent stands with the same overstory species composition but lacking hemlock (Kelty 1989). The hemlock yield was essentially additive to that of the overstory. Comparisons with yield tables for hemlock suggested that the hardwood/hemlock mixture would also have greater yields than a pure hemlock stand. In a study of mixed stands of Douglas-fir (*Pseudotsuga menziesii*) and western hemlock (*Tsuga heterophylla*) in the northwestern U.S., in which the more tolerant hemlock grew in a lower canopy position, Wierman and Oliver (1979) compared basal area yields with yield tables for both species. These indicated that mixed stands would have greater yields than pure stands of either species, with the difference being greater in comparison with pure Douglas-fir.

Patterns of basal area yields in natural stands in which conifers grow as emergents above a main canopy of diverse hardwood species have been studied by Ogden (1986) and Enright (1982). In New Zealand, Ogden found that the emergent kauri (*Agathis australis*) appeared to have minimum competition with associated species. The main-canopy species had the same average basal area density, regardless of the density of the *Agathis* component, which varied over a wide range among different stands. Ogden referred to this phenomenon as "additive basal area." Enright observed the same pattern in rain forests in Papua New Guinea, where the emergent conifer *Araucaria hunsteinii* grows above a main canopy of hardwood species.

Reduction of root competition

Compared with studies of crown stratification and competition, much less work has been done concerning reduction of root competition in mixtures. The principle that root stratification can be associated with increased total production has been demonstrated with herbaceous species (Ellern et al. 1970, Trenbath and Harper 1973). Different tree species have different root structures and rooting depths (Spurr and Barnes 1980). This factor is considered very important in silvicultural decisions favoring mixtures in European silviculture. This is related mostly to differences in depths of the main woody root structures, which affects the stability of a stand against windthrow. This is one of the reasons that deeper rooting fir and beech are interplanted with shallow-rooted spruce (Burschel and Huss 1987).

However, it is not clear whether differences in depths of main rooting structures correspond to functional separation in nutrient or water uptake. Trees with deep root systems do have feeder roots which can mine water and nutrients from deeper soil layers, but the comparative distribution patterns in pure and mixed-species stands and on different soil types are not known. In natural stands of oak, maple, and birch in the northeastern U.S., Lyford (1980) found that the oaks had deeper taproots and woody laterals than associated species. However, the deep oak laterals produced higher order laterals that grew upward and elaborated networks of fine roots, with the result that the non-woody feeder roots of all the species were concentrated near the surface of the mineral soil and in the forest floor. Thus, the existence of significant reduction of competition for nutrient and water uptake is not clear for this case. Other studies indicate that a trend exists toward greater concentration of fine roots in the uppermost layer of soil with increasing stand age (Vogt et al. 1981, Berish and Ewel 1988), which may indicate a minimal functional separation of roots, except during the establishment phase.

One study in the humid tropics in Costa Rica (Berish and Ewel 1988) has directly examined the relation between species diversity and root development. A comparison was made between successional vegetation plots containing more than 100 plant species and a monoculture of the tree species *Cordia alliodora*. The monoculture at age 2.5 years had developed the same root biomass and surface area as the diverse successional plots at 5 years, and had a similar vertical root distribution pattern. In this case, the tree monoculture exploited the soil as completely as the diverse stand, and had the same total biomass productvity (J. Ewel pers. comm.)

Even if not separated spatially, roots of different species in mixtures may utilize soil resources more efficiently if they take up different forms of nutrients. This may occur with nitrogen, with different species taking up greater proportions of ammonium or nitrate (Waring and Schlesinger 1985), but no studies link this characteristic with production increases in forest mixtures.

Phenology

Differences in phenology among species in a mixed stand may reduce competition in crowns or roots temporally rather than spatially. Differences in timing of foliage production and duration of photosynthetic activity are important in some agricultural intercropping systems, and may be important in forests as well. One situation that can occur in forests but not in annual crops is that in which one species of a mixture is deciduous and the other evergreen. Depending upon the timing of the leafless period in relation to climatic limitations on the growing season, the evergreen species might have a period of growth with little competition for light, water, or nutrients. This may be especially effective if the deciduous species were in an upper canopy position above the more tolerant evergreen, thus having phenological differences complement foliage stratification effects. This may be the case in studies of deciduous hardwoods above shade-tolerant, evergreen hemlock (Kelty 1989).

Effects of site quality

Although site quality has not been explicitly considered in this discussion, it clearly is important in controlling yield patterns in mixtures. Mixtures which achieve some measure of competitive reduction are likely to have increased yields only if the supply of the resource for which competition is reduced is limiting to production. The studies described above in which mixed stands had stratified canopies all took place in moist climates on sites where soil moisture was not critically limiting. In these situations, the ability to capture light tends to be the limiting factor, since the understory would not compete intensely with the overstory for soil moisture. On drier sites, understories are more likely to reduce the growth of overstory trees through root competition (Dale 1975, Kelty et al. 1987). In contrast, combinations of species that efficiently utilize moisture or nutrients would be expected to show yield advantages only where those factors are limiting, such as sites that are dry, inherently infertile, or excessively leached.

Facilitation

The principal way in which facilitation mechanisms have been used to affect production directly in both forestry and agriculture is by increasing the nitrogen available to the crop species. Other potentially important forms of facilitation mainly involve the reduction of loss of yield to pests, rather than affecting the production process itself; these will not be considered here. There are two ways in which species mixtures can be used in silviculture to improve nitrogen availability. In the first, mixtures are used to increase litter decomposition rates, thereby increasing the nitrogen available for tree uptake, without changing the total nitrogen levels in the soil-plant system. In the second, tree species that fix atmospheric nitrogen through symbiotic association with bacteria are used to increase total nitrogen levels in the system. Both of these approaches will be considered.

Increasing litter decomposition rates

In regions with cold climates where forests are dominated by conifers, tree growth can become limited by low available nitrogen concentrations, even though total nitrogen levels are adequate. Litter decomposition rates are quite low in these kinds of stands, because low temperatures are limiting to microbial activity and because of the resistant nature of the conifer litter, particularly its high carbon:nitrogen ratio. Litter production can exceed decomposition in these forests, so that nitrogen becomes stored in organic form in the forest floor.

This situation is common in European forests because conifers are favored for management, both where they are naturally dominant and where native hardwood stands have been converted to conifers. A number of silvicultural techniques have been developed to overcome this problem by increasing nitrogen mineralization rates. Reducing stand density through thinning allows more light to reach the forest floor, resulting in higher temperatures and faster rates of decomposition. Prescribed burning also increases mineralization rates, although it also volatilizes some of the nitrogen. These methods can be used with conifer monocultures. However, another alternative for achieving the same result is the use of mixed conifer-hardwood stands; this has become a standard prescription in European forestry (Matthews 1989). These mixtures are designed to have a hardwood species with high foliar nutrient content included with the principal conifer crop species to decrease the carbon:nitrogen ratio of the litter, thereby increasing microbial activity. In Germany, the problem of slow litter decomposition in pure spruce stands is one of the reasons for including beech with spruce in young stands as mature spruce monocultures are regenerated (Burschel and Huss 1987). In Scandinavia and the Soviet Union, birches are often added to spruce and pine plantations for this purpose. Although it is not necessary to use a nitrogen-fixing tree species to achieve this effect, species such as grey alder (*Alnus incana*) are recommended for these mixtures because they have foliar nitrogen concentrations that are often much higher than those of other hardwoods (Mikola et al. 1983).

One of the problems with these mixtures is that of crown competition from the hardwoods, which often have much lower commercial timber value than the conifers. This problem is greatest with pioneer hardwood species which have rapid juvenile height growth. In these cases, the hardwood species is included at the minimum proportion at which an appreciable effect on litter decomposition rate is achieved, so that they do not suppress the conifer crop species. In some cases, it works well to keep the hardwoods segregated in multiple rows. For example, in a study in the boreal region of the Soviet Union, Prokop'ev (1978) found that 6 to 8 rows of pine alternating with 2 rows of birch were best in keeping birch from overtopping the pines.

Increasing total nitrogen with nitrogen-fixing species.

In some forests, stand productivity is limited by low total nitrogen levels. This occurs on soils that are inherently infertile, or on better soils that have been depleted of nutrients following agricultural use, destructive logging practices, or the repeated practice of litter collection. In these situations, increases in nitrogen availability in mixtures depend upon the use of a nitrogen-fixing tree species; the ones most commonly used are legumes, such as *Acacia*, *Leucaena*, and *Albizia*, which are symbiotically associated with *Rhizobium* species, or genera such as *Alnus* and *Causarina* associated with *Frankia* species. The use of these tree species in mixed stands is generally of interest only where the nitrogen-fixing species is not of principal interest for timber production purposes because of poor wood quality or stem form. If the nitrogen-fixing species is acceptable in these regards, it is often used in pure plantations, as is increasingly done in Southeast Asia (Domingo 1983).

One situation involving nitrogen-fixing species in mixture is that of Douglas-fir and red alder (*Alnus rubra*) in the northwestern U.S. On many soils of the region, nitrogen is an important limiting factor to the growth of Douglas-fir, a highly valuable timber species commonly planted in monoculture. Red alder is a native, early successional species with much lower timber value. Two studies of this species combination have compared plots in pure Douglas-fir plantations with plots in the same plantations where red alder has been added through natural establishment (Binkley 1984) or planting (Miller and Murray 1978). These studies are of the additive design, since the Douglas-fir is at a constant density in all plots, and total tree density is higher in the mixed plots.

In comparisons on infertile sites, the facilitative mechanism of increased nitrogen availability has been demonstrated in the elevated nitrogen levels in the soil and in the Douglas-fir foliage in mixed stands compared to the Douglas-fir monoculture (Tarrant 1961, Binkley 1984). Yield comparisons in these studies showed the combined effects of facilitation and increased competition in mixtures. The higher density in mixed stands resulted in higher Douglas-fir mortality, but the average size of survivors was greater. At the stand level, the Douglas-fir component of mixed stands was equal to that of the pure Douglas-fir stand in terms of stem volume (Miller and Murray 1978) or total biomass (Binkley 1984). With the alder component of the mixture added in, total yield of the mixture was approximately double that of the pure Douglas-fir stand in both cases.

In a similar comparison on a fertile site, the facilitative effect appeared to be absent; nitrogen concentrations in Douglas-fir foliage were equally high in mixtures and pure stands. Yield comparisons showed that the average size of Douglas-fir was lower in the mixed stand compared to the monoculture on this site; at the stand level, total Douglas-fir biomass was lower in the mixed stand compared to the pure Douglas-fir stand. The production of the alder component made up for this decrease, so that total stand biomass of the mixture and the pure Douglas-fir were equal (Binkley 1984). Thus, the use of mixtures on fertile sites appears unpromising.

One problem with these mixtures, even on appropriate sites, is that red alder often outgrows Douglas-fir in height in early stages of stand development. In the study by Miller and Murray, the desired effect of significant increases in Douglas-fir height and diameter growth occurred only for the relatively small proportion of Douglas-fir trees that escaped early suppression and eventually developed into emergent crown positions above the alder. They suggested that best results would be obtained if a small number (50 to 100 trees/ha) of uniformly distributed dominant red alders were incorporated into Douglas-fir plantations to provide adequate nitrogen levels but minimize the crown suppression problem. Another alternative is to choose a different nitrogen-fixing species which competes less severely. Sitka alder (*Alnus sinuata*) has been studied as such an alternative, since it is a small tree (maximum height of about 8 m), and it tends to be overtopped by Douglas-fir more quickly during early stand growth on many sites (Harrington and Deal 1982). Although Sitka alder fixes less nitrogen than red alder, it was found to give comparable increases in Douglas-fir stand productivity and nitrogen content of litterfall when growing in mixed stands (Binkley et al. 1984). In this mixture, Sitka alder was half the height of the dominant Douglas-fir at stand age of 23 years.

The problem of competition of the nitrogen-fixing species with the more valuable crop tree species is not unique to the Douglas-fir/red alder mixture, but is one of the key factors controlling decisions to use nitrogen-fixing species in mixed-species silvicultural systems generally (Gordon 1983). For example, this is the case when mixtures are used to increase growth rates of plantations of black walnut (*Juglans nigra*), a highly valuable timber species of the central U.S. Several nitrogen-fixing tree and shrub species, including black alder (*Alnus glutinosa*) and autumn-olive (*Elaeagnus umbellata*), cause increased growth of walnut when planted in mixture, with the growth being proportional to the nitrogen-level increases in the soil (Campbell and Dawson 1989). The choice of species is governed by the intensity of crown competition given by the nitrogen-fixing species as well as the amount of nitrogen added; in this case, the shrub species autumn-olive is favored in both regards.

One method used to minimize competition has been to abandon tree/tree and tree/shrub mixtures in favor of the addition of herbaceous legumes such as lupin (*Lupinus* spp.) to forest stands. Lupins have been used in pine stands in Germany since 1910 on sites degraded by annual removal of pine litter (Mikola et al. 1983), and are now successfully employed in New Zealand, Australia, and Europe both for stand establishment and for underplanting beneath established stands (Turney and Swethurst 1983).

In the cases described above, the nitrogen-fixing tree species had comparatively low (or no) value as a crop, so there was little interest in complete yield comparisons of mixtures and monocultures. Studies concentrated on increases in the growth of the principal crop species in mixture compared with that species in monoculture; the monoculture of the nitrogen-fixing species was generally not available for comparison. However, Debell et al. (1989) have described studies of fuelwood plantations in which maximum total biomass yield was a management objective, and a replacement series experimental design including both monocultures was used. For those reasons, their study is of particular interest for the present discussion. In earlier studies, DeBell et al. (1985) had found that growth of *Eucalyptus saligna* fuelwood plantations in Hawaii was limited by low nitrogen levels on some sites, and mixtures of *Eucalyptus* with several leguminous tree species had greater total stand yields on those sites. On sites with lower rainfall and where nitrogen availability was higher, mixtures had lower yields than *Eucalyptus* monocultures (DeBell et al. 1987)

A replacement series experiment was carried out on one of the low fertility sites with high rainfall to compare growth of the most promising of these legumes, *Albizia falcataria*, in mixture with *Eucalyptus* (DeBell et al. 1989). Plot density was held constant at 2500 trees/ha, with different plots having the following proportions of *Eucalyptus*:*Albizia*—100:0, 89:11, 75:25, 66:34, 50:50, 34:66, 0:100. Total aboveground dry weight was measured after 4 years, when the largest trees were 12 to 18 m in height and 14 cm dbh. Biomass yields are shown in Figure 3; results are included for two kinds of pure *Eucalyptus* plots—ones which had received essentially the same treatment as other plots, and ones which had received added nitrogen fertilizer.

The biomass yield of *Albizia* was approximately proportional to its planting density in each plot (e.g, the yield of *Albizia* in the 50:50 mixture was about one-half of the *Albizia* monoculture). Biomass yields of the *Eucalyptus* component of the mixture were consistently greater than that of the *Eucalyptus* monoculture (relative yield of *Eucalyptus* was greater than 1.0), which indicates that facilitation

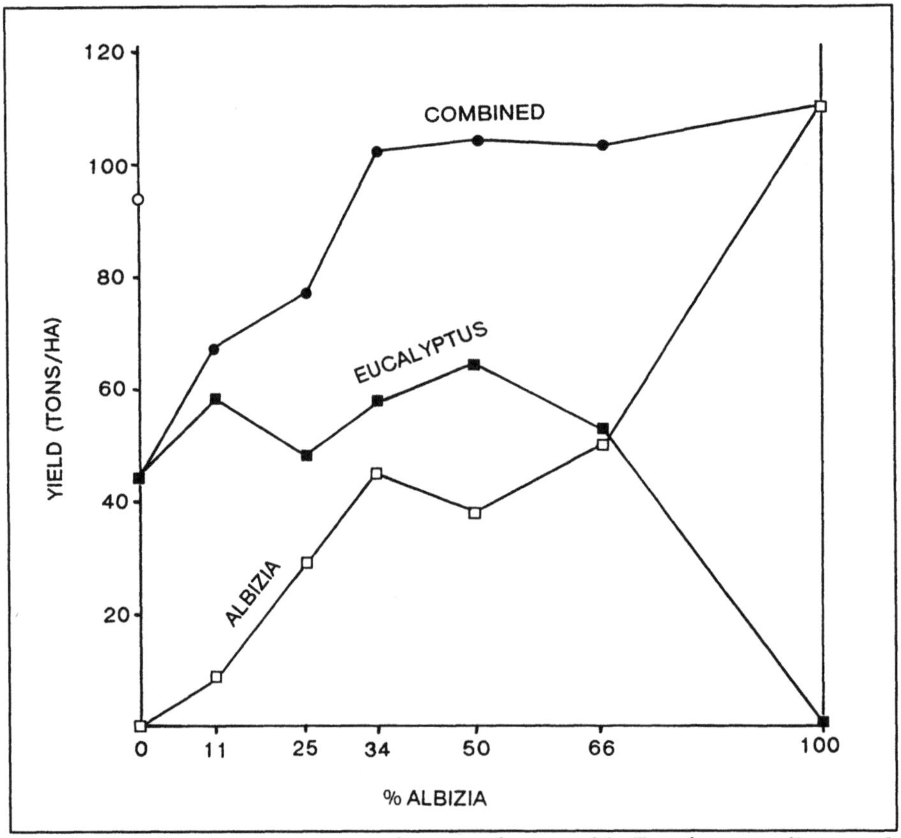

Fig. 3. Results of replacement series experiment with *Eucalyptus saligna* and *Albizia falcataria* in Hawaii, from DeBell et al. (1989). Yield expressed as total aboveground dry weight at stand age of 4 years. Plots received N,P,K-fertilization during first year, and only P,K-fertilization in subsequent years. Open circle shows yield of pure Eucalyptus that received continued N,P,K-fertilization after first year.

was occurring, since no level of reduction in competition could cause this to occur (Vandermeer 1989). Nitrogen concentrations in *Eucalyptus* foliage and soil in the mixed stands with more than 34% *Albizia* were equal or greater than those of pure fertilized *Eucalyptus* plots. This, together with the growth response of the *Eucalyptus* monoculture to nitrogen fertilizer, indicates that the anticipated mechanism of a general increase in ecosystem nitrogen levels was the cause of *Eucalyptus* production increases in mixed stands.

The relative yield totals for all mixtures were greater than 1.0 (i.e., greater than expected if no competitive reduction or facilitation was positively affecting yields). However, total absolute biomass yields were not significantly different for pure *Albizia* and the three mixtures with the higher proportions of *Albizia* (66%, 50%, and 34% *Albizia*). Because *Albizia* had much greater yield than *Eucalyptus* in pure stands on this site (2.5 times greater), the marked effects of facilitation were not great enough to cause mixtures to outyield pure *Albizia*.

The final choice of which stand type to use in a silvicultural system depends, of course, upon management objectives. *Albizia* has wood of much lower density than *Eucalyptus* (reducing its efficiency of use for fuelwood) and is problematic for harvesting because of its multi-stemmed form. Therefore, DeBell and coworkers concluded that the mixture of 66% *Eucalyptus*:34% *Albizia* would likely be the best for economic production of firewood—giving the maximum total yield with the minimum component of *Albizia*. Yields of pure *Eucalyptus* stands could approach this level only with the added investment of intensive nitrogen fertilization. It should be remembered, though, that yield patterns leading to this decision may change with different rotation lengths or planting densities.

Conclusions

The principles of competitive reduction and facilitation that have been used to explain patterns of yield in mixtures of herbaceous species appear to apply in mixtures of tree species as well. Although the forestry evidence is meager compared to that from agricultural intercropping studies, enough exists to indicate a general pattern in yield studies of forests. Relative yield totals of mixtures in these studies frequently exceed 1.0, indicating that the effects of competitive reduction or facilitation cause mixtures to exceed the production expected if yield were simply the average of monoculture yields of the component species. In these cases, the yield of the mixture will certainly exceed that of the lower yielding monoculture, but not necessarily that of the higher yielding monoculture. It may well be that forests are more likely than herbaceous stands to show significant effects of species interactions on production. The greater lifespan and larger stature of trees allows more opportunity for reduced competition via spatial separation in roots and canopy (J. Ewel, pers. comm.). The greater lifespan may also allow more opportunity for the facilitative effect of nitrogen-fixing species to occur, through the increase of soil nitrogen levels from continual litter production and decomposition.

It is clear, however, that mixtures will not always exceed the production of the highest yielding monoculture; in fact, this may not even be a common situation. In some cases, tree species simply may not differ enough to reduce competition significantly. In others, niche separation among species in mixtures may only result in a partitioning of the same total resource that a highly efficient species could fully utilize in monoculture, as seen in the the Costa Rican example (Berish and Ewel 1988). In yet other cases, significant beneficial interactions may occur, but they may not be of great enough magnitude to cause mixtures to outyield the monoculture of a highly productive species, as in the *Eucalyptus/Albizia* example from Hawaii (DeBell et al. 1989). The greatest evidence for mixtures having a higher yield than either monoculture exists for stands with stratified canopies in moist temperate forests; this may be due partly to the greater frequency of research in these kinds of forests.

It is clear that there are major gaps in our knowledge of the connection between tree species diversity and yield. Only a small fraction of the possible types of species combinations have been studied, and most of the studies have examined two-species mixtures; the effects of increasing diversity further to tens of species or even more are not known.

However, the studies reviewed here make it possible to identify the general set of conditions under which increased yields in mixtures relative to monocultures are likely to occur. There is no evidence to indicate that a mixture of randomly selected species would generally outyield a monoculture of the most productive component species. Species that are used in mixtures must have good ecological combining ability—significant differences in growth characteristics that will reduce competition or foster facilitation. Furthermore, the species interactions must increase efficiency of use of a resource that is a limiting factor to productivity. Thus, a mixture with a sun-adapted overstory above a shade-adapted understory is not likely to have higher yields on sites where soil moisture rather than light is limiting to production. Similarly, a mixture of a legume with a non-nitrogen-fixing tree species is unlikely to show increased yields on soils with adequate nitrogen levels. The matching of the ecological combining ability of species with the limiting factors of a site is of primary importance.

Even though mixtures in many cases may not have higher production than monocultures of the most productive species, studies of the kind reviewed here still have a broad application in silvicultural planning for both planted and naturally established stands. The patterns observed in these studies make it possible to design mixed stands which have the highest yield possible while providing some of the benefits that are associated with certain mixtures of species, including stand stability, reduction of risk of total crop loss, resistance to pests and diseases, reduction of fertilizer needs, increased wildlife habitat diversity, and improved landscape aesthetics.

References

Assmann, E. 1970. The principles of forest yield study. Pergamon Press, Oxford. 506 p.

Berish, C.W., and J.J. Ewel. 1988. Root development in simple and complex tropical successional ecosystems. Plant and Soil 106: 73-84.

Binkley, D. 1983. Ecosystem production in Douglas-fir plantations: interactions of red alder and site fertility. For. Ecol. Manage. 5: 215-227.

Binkley, D., J.D. Lousier, and K. Cromack, Jr. 1984. Ecosystem effects of sitka alder in a Douglas-fir plantation. For. Sci. 30: 26-35.

Boardman, N.K. 1977. Comparative photosynthesis of sun and shade plants. Annual Rev. Plant Physiol. 28: 355-377.

Burschel, P., and J. Huss. 1987. Grundriss des Waldbaus: Ein Leitfaden fur Studium und Praxis. Verlag Paul Parey, Hamburg and Berlin. 352 p.

Campbell, G.E., and J.O. Dawson. 1989. Growth, yield, and value projections for black walnut interplantings with black alder and autumn olive. North. J. Appl. For. 6: 129-132.

Dale, M.E. 1975. Effects of removing understory on growth of upland oak. USDA For. Serv. Res. Pap. NE-321. 10 p.

DeBell, D.S., C.D. Whitesell, and T.H. Schubert. 1985. Mixed plantations of *Eucalyptus* and leguminous trees enhance biomass production. USDA For. Serv. Res. Pap. PSW-175. 6 p.

DeBell, D.S., C.D. Whitesell, and T.B. Crabb. 1988. Benefits of *Eucalyptus-Albizia* mixtures vary by site on Hawaii Island. USDA For. Serv. Res. Pap. PSW-187. 6 p.

DeBell, D.S., C.D. Whitesell, and T.H. Schubert. 1989. Using N2-fixing *Albizia* to increase growth of Eucalyptus plantations in Hawaii. For. Sci. 35: 64-75.

Domingo, I.L. 1983. Nitrogen fixation in Southeast Asian forestry: research and practice. Pp. 295-315, in Gordon and Wheeler (1983).

Ellern, S.J., J.L. Harper, and G.R. Sagar. 1970. A comparative study of the distribution of roots of *Avena fatua* and *A. strigosa* in mixed stands using a 14C labelling technique. J. Ecol. 58: 865-868.

Enright, N.J. 1982. Does *Araucaria hunsteinii* compete with its neighbors? Austral. J. Ecol. 7: 97-99.

Ewel, J.J. 1986. Designing agricultural systems for the humid tropics. Annual Rev. Ecol. Syst. 17:245-271.

Givnish, T.J. 1988. Adaptation to sun and shade: a whole-plant perspective. Aust. J. Plant Physiol. 15: 63-92.

Gordon, J.C. 1983. Silvicultural systems and biological nitrogen fixation. Pp. 1-6, in Gordon and Wheeler (1983).

Gordon, J.C., and C.T. Wheeler, eds. 1983. Biological nitrogen fixation in forest ecosystems: foundations and applications. Martinus Nijhoff/Dr. W. Junk Publisher, The Hague. 342 p.

Harper, J.L. 1977. Population biology of plants. Academic Press, New York. 892 p.

Harrington, C.A., and R.L. Deal. 1982. Sitka alder--a candidate for mixed stands. Can. J. For. Res. 12: 108-111.

Hill, J., and Y. Shimamato. 1973. Methods of analysing competition with special reference to herbage plants. I. Establishment. J. Agric. Sci. 81: 77-88.

Kawahara,T., and K. Yamamoto. 1982. [Studies on mixed stands of akamatsu (*Pinus densiflora*) and hinoki (*Chamaecyparis obtusa*) (I) Productivity and decomposition rate of organic matter.] J. Jap. For. Soc. 64: 331-339. (in Japanese, with English summary)

Kawahara, T., and K. Yamamoto. 1986. [Studies on mixed stands of akamatsu (*Pinus densiflora*) and hinoki (*Chamaecyparis obtusa*) (III) Stem volume of mixed stands.] J. Jap. For. Soc. 68: 327-332. (in Japanese, with English summary)

Kelty, M.J. 1989. Productivity of New England hemlock/hardwood stands as affected by species composition and canopy structure. For. Ecol. Manage. 28: 237-257.

Kelty, M.J., E.M. Gould, Jr., and M.J. Twery. 1987. Effects of understory removal in hardwood stands. North. J. Appl. For. 4: 162-164.

Lyford, W.H. 1980. Development of the root system of northern red oak (*Quercus rubra* L.). Harvard Forest Pap. No. 21. 30 p.

Matthews, J.D. 1989. Silvicultural systems. Clarendon Press, Oxford. 284 p.

Mikola, P., P. Uomola, and E. Malkonen. Application of biological nitrogen fixation in European silviculture. Pp. 279-294, in Gordon and Wheeler (1983).

Miller, R.E., and M.D. Murray. 1978. The effects of red alder on growth of Douglas-fir. In: D.G. Briggs, D.S. DeBell, and W.A. Atkinson (eds.), Utilization and management of alder. USDA For. Serv. Gen. Tech. Rep. PNW-70, p. 283-306.

Monsi, M., Z. Uchijima, and T. Oikawa. 1973. Structure of foliage canopies and photosynthesis. Annual. Rev. Ecol. Syst. 4: 301-327.

Ogden, J. 1986. Additive basal area, regeneration, and self-thinning in populations of kauri (*Agathis australis*) growing in mixed conifer-angiosperm forest in New Zealand. Unpubl. manuscript, Auckland Univ., Auckland, N.Z.

Poleno, Z. 1981. Vyvoj smisenych porostu. [Development of mixed forest stands.] Prace VULHM 59: 179-202. (in Czech, with English summary)

Prokop'ev, M.N. 1976. [Mixed plantings of pine and spruce]. Lesnoe Khozyaistvo 5: 37-41. (in Russian, with English summary)

Prokop'ev, M.N. 1978. [The creation of mixed pine/birch plantations]. Lesnoe Khozyaistvo 1: 53-57. (in Russian, with English summary)

Spurr, S.H., and B.V. Barnes. 1980. Forest ecology. Third ed. John Wiley and Sons, New York. 687 p.

Trenbath, B.R. 1974. Biomass productivity of mixtures. Adv. Agron. 26: 177-210.

Trenbath, B.R. 1981. Light-use efficiency of crops and the potential for improvement through intercropping. In: Willey, R.W., (ed.), Proceedings of the international workshop on intercropping. ICRISTAT, Hyderabad, India, p. 141-154.

Trenbath, B.R., and J.L. Harper. 1973. Neighbour effects in the genus *Avena*. I. Comparison of crop species. J. Appl. Ecol. 10: 379-400.

Turvey, N.D., and P.J. Smethurst. Nitrogen fixing plants in forest plantation management. Pp. 233-259, in Gordon and Wheeler (1983).

Vandermeer, J. 1989. The ecology of intercropping. Cambridge University Press, Cambridge. 237 p.

Vogt, K.A., R.L. Edmonds, and C.C. Grier. 1981. Seasonal changes in biomass and vertical distribution of mycorrhizal and fibrous-textured conifer fine roots in 23- and 180-year-old subalpine *Abies amabilis* stands. Can. J. For. Res. 11: 223-229.

Waring, R.H., and W.H. Schlesinger. 1985. Forest ecosystems: concepts and management. Academic Press, New York. 340 p.

Wierman, C.A., and C.D. Oliver. 1979. Crown stratification by species in even-aged mixed stands of Douglas-fir--western hemlock. Can. J. For. Res. 9: 1-9.

Exploring the possibilities of developing a physiological model of mixed stands

9

MICHAEL B. LAVIGNE

Introduction

Mixed stands are difficult to manage silviculturally because each species has a different growth habit and mixed stands frequently grow on the most productive sites (Smith 1986). These same characteristics make it difficult to predict the development of mixed stands, but knowledge of the growth of these stands is essential for designing silvicultural practices. The high potential productivity of many mixed stands make it important to address these problems.

One approach to predicting the development of mixed stands is constructing physiological models. Physiological models are based on our knowledge of the autecology of the species involved, and our ability to describe the growth environment. Consequently these models use parameters that have some biological meaning, in comparison to more empirical approaches. These models can be relatively simple or very complex (Landsberg 1986). Our understanding of physiological ecology has increased to the point where the behaviour of trees in response to their environment can reasonably be described in mathematical terms and the use of these models can improve forestry practices (Landsberg 1986; Landsberg and McMurtrie 1985). The requirement of more empirical approaches to growth prediction for long-term data bases, such as comprehensive permanent sample plot programs, can be circumvented by physiological models, but other data about physiological processes and their responses to site and climate are required for each species whose growth is to be predicted. Physiological models can be used to estimate responses to silvicultural treatments that have never been performed in the woods and responses to unprecedented changes in the environment such as that which might be caused by changing climate.

Physiological models for predicting the growth and development of mixed stands have not been produced yet. However, models of the growth of pure stands based on physiological processes have been constructed. In this paper, the current state of the art in physiological models of pure stands is assessed to determine what changes and additions are required to develop physiological models of mixed stands. It appears to be feasible to develop physiological models of mixed stands but further development of procedures for estimating light interception, and a better paradigm for height growth will be required to improve the confidence in model predictions.

Development of physiological models for mixed stands

Foundation for physiological models

The physiological processes responsible for the uptake and release of carbon and energy are dealt with explicitly in models that take the carbon budget approach (Figure 1). In the carbon budget approach, net primary production (NPP) is the difference between the annual net photosynthetic production (P), which is gross

M. J. Kelty (ed.), The Ecology and Silviculture of Mixed-Species Forests, 143–161.

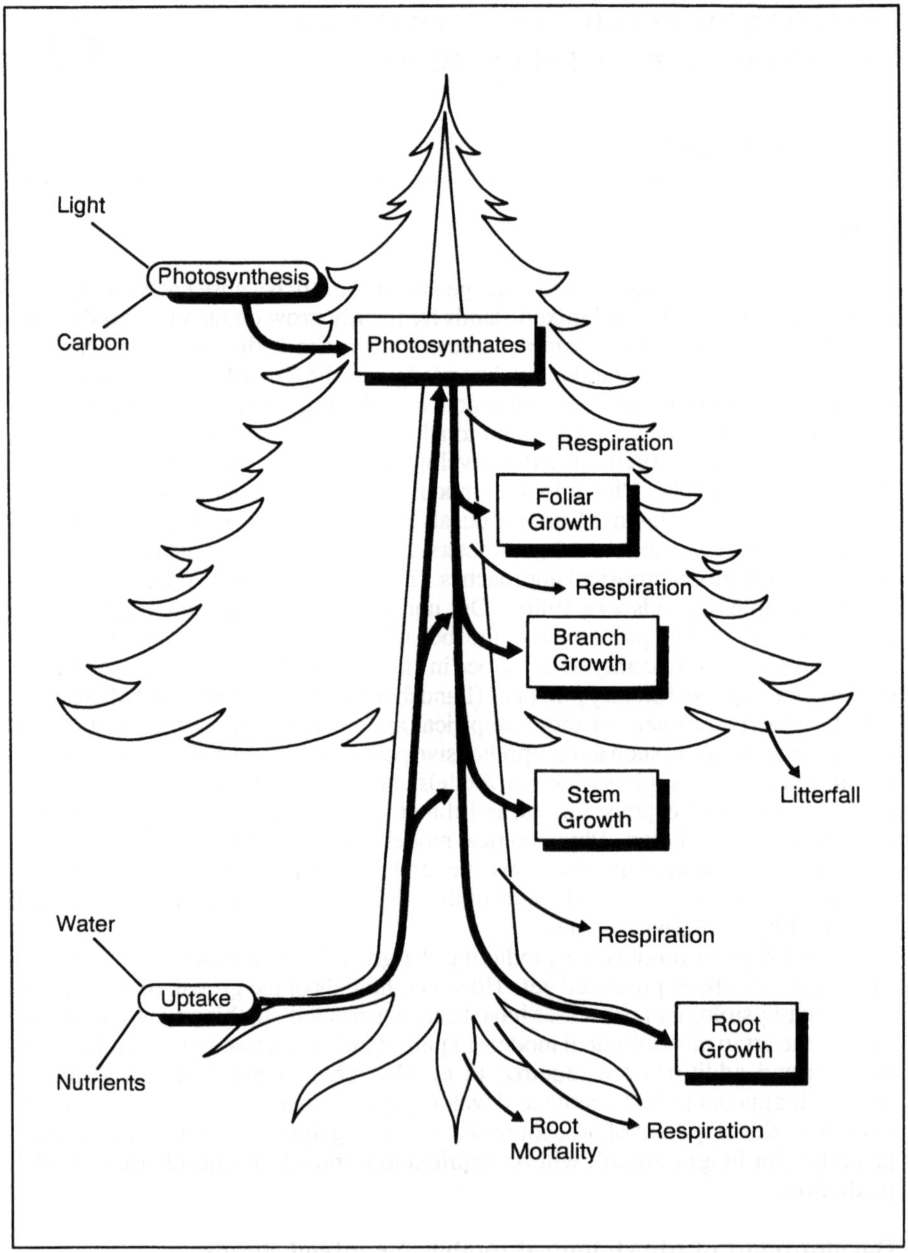

Fig. 1. **The flow of materials into and out of a tree according to the carbon budget approach.**

photosynthesis less foliar respiration, and the annual total respiration of the whole plant excluding foliage (R).

$$NPP = P - R \quad (1)$$

Growth (G) is net primary production minus losses of biomass by shedding, herbivory, and other causes (S).

$$G = NPP - S \quad (2)$$

Annual growth is usually estimated by physiological models. It is normal to assume that there are no net annual changes in storage carbohydrates to contribute to the carbon balance, and so the annual cycle of stored carbohydrates is not incorporated in models.

Net primary production is allocated to each component in physiological models. Normally translocation is not modelled, and therefore allocation is treated phenomenologically. Allocation must be included in models to update the sizes of the photosynthetic apparatus and the respiring tissues, and to estimate the increasing quantity of economically valued forest products. The values of allocation coefficients primarily indicate the strategy of a species, including some responses to site conditions.

The processes responsible for acquiring water and nutrients are included implicitly in many models by taking into account the dependence of photosynthesis and allocation on the supplies of nutrients and water. Root activity rate is included in some models as a variable determining allocation. In some cases a model was constructed for a particular site. These models do not incorporate the site dependent aspects of water and nutrient acquisition in any way.

Numerous process models based on the carbon balance approach have been constructed in recent times. Noteworthy among the published models are those by Yahata et al. (1979), McMurtrie and Wolf (1983), Jarvis and Leverenz (1983), Mohren et al. (1984), Mohren (1987), Valentine (1985, 1988, 1990), Mäkelä (1986, 1988, 1990a), Mäkelä and Hari (1986), Landsberg (1986), West (1987), Sievänen et al. (1988), and Ludlow et al. (1990). Each model differs from the others in the manner in which it deals with some of the components of this basic framework. These models are the basis for the review of physiological models that follows.

Photosynthesis

Photosynthetic production by a crown depends on the quantity of photosynthetically active radiation (PAR) absorbed and the photosynthetic characteristics of the foliage. Predictions of growth and stand development are sensitive to values used for light interception parameters (Linder 1985; Mäkelä 1990b). Light interception varies by large amounts both spatially and temporally, as a result of shading by other elements in the canopy and changing irradiance above the canopy. For these reasons much effort has been applied to developing models of light interception for predicting photosynthetic production.

Ross (1981) and Oker-Blom (1986) have comprehensively reviewed and synthesized matters pertaining to the mathematical description of PAR absorption by canopies. The proportion of incoming PAR absorbed by foliage depends on numerous canopy attributes including the amount of foliage, its vertical and horizontal arrangement, the size and shape of leaves, twigs, and branches, and the inclinations of leaves with respect to the horizon and to the azimuth. These attributes vary with species, stage of development, and the arrangement of trees on the land. Some of these attributes also vary within crowns; for example, sun foliage is located near the top of crowns and shade foliage near the bottom of crowns. The degree to which interception of PAR depends on canopy structure varies between direct and diffuse radiation, with time of day and year, cloud cover and latitude.

The canopies of some stands, including those of certain agricultural crops and deciduous tree species have the following characteristics: (1) the canopies are closed and appear more or less flat-topped; (2) foliage is dispersed randomly within horizontal layers throughout the canopy; (3) every point within a horizontal layer has the same probability of having foliage; (4) the locations of foliage in each layer are independent of those in all other layers; and (5) all foliage elements in a layer have the same light intercepting properties. When the vertical and horizontal arrangement of foliage fits this description, the interception of PAR is estimated by using Beer's Law.

$$I = I_0 \times e^{-k \cdot L} \tag{3}$$

where I_0 is irradiance above the canopy, I is the mean irradiance at a horizontal level within the canopy, L is the leaf area index above the horizontal level, and k is the light extinction coefficient. The mean value of irradiance at each horizontal level is estimated with this equation. Only information about the vertical distribution of foliage and a single parameter, k, are required for predicting the interception of light. Because the arrangement of foliage in each horizontal layer is random, the same value of irradiance is expected at all points within any horizontal stratum. Beer's Law is usually used in combination with a model that describes the vertical distribution of foliage. Equations of the normal distribution (Stephens 1969, Beadle et al. 1982) and the β distribution (Kellomäki et al. 1981) have been used to describe the vertical distribution of foliage. This approach is applicable to situations where there is no difference between self-shading and shading by neighbors, and, therefore, is not directly applicable to estimating the photosynthetic production by trees in mixed species stands.

Clumped foliar distributions such as those of most conifers, complicates the prediction of light interception. Kellomäki et al. (1986) identified two forms of clumped foliar distributions for calculating light interception. Clumps of foliage are randomly distributed throughout the canopy volume in one type of clumped foliar distribution, but, because of the clumps, the foliage within a horizontal layer is not randomly distributed. The arrangement of clumps of foliage within each horizontal layer is independent of the locations of foliage in all other strata. This type of clumped foliage is similar to randomly distributed foliage described above except that it is clumps of foliage that are arranged in the canopy rather than leaves.

A second type of distribution describes how clumps of foliage are aggregated within crowns. In this type of distribution, clumps of foliage are not randomly located throughout each horizontal layer and the arrangement of clumps in one horizontal layer depends on that in other layers. Size and shape of the crown, and locations of neighboring crowns must be accounted for to describe foliar distribution for this type of clumped foliage. The method developed for predicting light interception in canopies with clumped foliar distribution, where foliage is aggregated into crowns, is the only one that is sufficient for estimating light interception for mixed stands, since it is the only one that estimates light interception for individual trees. This is necessary for mixed stands, since adjacent crowns can have different light intercepting and photosynthetic properties, when they are trees of different species.

Less light is intercepted by foliage in the upper horizontal layers of a canopy with clumped foliage than one with randomly distributed foliage (Oker-Blom and Kellomäki 1983; Kellomäki et al. 1986). This difference between foliar arrangements occurs because leaves within a clump overlap or shade one another. Because of overlapping foliage within a clump, the projected area of a clump of foliage is less than the sum of projected areas of leaves that comprise the clump. It follows that more light reaches the lower horizontal strata of foliage in a canopy with clumped foliage than one with randomly distributed foliage. Consequently the foliage in lower layers of a canopy with clumped foliage can do photosynthesis at higher rates than can the foliage in the lower layers of a canopy with randomly distributed foliage. This effect of clumped foliage on photosynthetic production counters to some extent the lower rate of interception per unit of foliage in comparison to a canopy with an equal amount of randomly distributed foliage (Grace et al. 1987b). The significance of this interplay between the effect of foliar arrangement on light interception and the nonlinear photosynthetic light response depends on whether photosynthesis saturates at a high light intensity or a low light intensity, and therefore it varies among species.

The expected irradiance is not the same at every point in a horizontal layer of a canopy with clumped foliage as it is for a canopy with randomly distributed light. For this reason predicting only the mean irradiance as is done with Beer's Law does not sufficiently characterize the light climate of a horizontal level of a canopy with clumped foliage. The irradiance expected to reach each point in a canopy with clumped foliage depends on the arrangement of foliage above the point and, because foliar distribution is non-random, this probability differs from point to point. Calculations must be repeated for many points before the light reaching a horizontal layer can be characterized.

More information is required to describe the arrangement of foliage in a canopy with clumped foliar distribution than in a canopy with randomly distributed foliage. First, the number and locations of trees in the stand must be known. Second, the shape of the crown, total tree height, height to the base of live crown, and the arrangement of foliage within the crown affect light interception (Oker-Blom and Kellomäki 1983; Grace et al. 1987a; Oker-Blom et al. 1989; Kuuluvainen and Pukkula 1989). The locations of foliage cannot be described with a single equation for the vertical distribution of foliage because the horizontal distribution is not random.

An approach to estimating the light interception of individual crowns in stands has emerged from the recent work of Oker-Blom and Kellomäki (1983), Grace et al. (1987a,b), Oker-Blom et al. (1989), and Kuuluvainen and Pukkula (1989). The interception of direct and diffuse radiation are predicted separately in these procedures. Each neighboring crown intercepts direct radiation according to the path length of the beam through the crown, and the probability of some foliage being on the path. The probability of light being intercepted by a neighboring tree is calculated by treating the foliage as randomly distributed or as randomly distributed clumps. Hence the average probability of intercepting light per unit of path length is a characteristic of the crown. Path length through a neighboring crown depends on shape of the crown, and its height and location relative to the sun and subject tree. Since it is the light reaching the outer surface of the crown that is predicted, the spaces between crowns which contributed to the non-random foliar distribution have been taken into account. The within-crown interception of direct radiation at some point in the crown of the subject tree, given that the light passed through the crown of the neighboring tree, depends on foliar distribution and path length through the crown of the subject tree. When it can be assumed that clumps of foliage are randomly distributed throughout the crown this calculation is the same as that for a canopy with randomly distributed clumps. The calculation is repeated for many points in each horizontal layer of a crown. Path lengths through the neighboring tree and through the subject tree differ for each point on a horizontal plane, and hence there is a different probability for each point of light being intercepted somewhere along the path. These procedures probably can be extended to mixed stands because the effects of neighboring trees is separated from that of the subject tree.

A large number of calculations are required when using these procedures for estimating light interception just to predict the rate of photosynthesis of a crown or canopy at a point in time. All of these calculations must be repeated many times to make estimates of daily and annual photosynthetic production. Hence, estimating photosynthetic production throughout the entire course of stand development requires a large input of weather data and extensive computer resources. In order to find less demanding ways of estimating photosynthetic production, Grace (1990) and Mäkelä (1990b) assessed relations between predictions of annual photosynthetic production made by detailed models of crown photosynthesis, and a number of easily measured crown and canopy attributes. Both studies found equations for predicting annual photosynthetic production of crowns using the quantity of foliage, shape of the crown, and number of trees per hectare as independent variables.

The work of Grace (1990), Mäkelä (1990b) and Ludlow et al. (1990) has much in common with competition indices developed in slightly more empirical approaches to predicting individual tree growth, such as that of Mitchell (1975), but these approaches also differ significantly. Mensurational competition indices cannot be measured and assessed independently of their predictions of tree growth, but the ecophysiological approaches can be verified independently of their predictions of tree growth by comparing measurements of irradiance with model predictions (Mäkelä and Hari 1986). Mensurational competition indices implicitly relate total light interception by the crown directly to stem production. In doing so, mensurational competition indices implicitly assign inflexible rates for other physiological processes, such as respiration, allocation and shedding that intervene between photosynthesis and growth. The ecophysiological approach does not attempt to predict production directly from estimates of light interception. Other physiological processes are dealt with explicitly after photosynthetic production is estimated in physiological models, and so they can be handled in a more flexible manner than in mensurational growth models.

Respiration

Respiration by foliage is included with photosynthesis, and therefore only the respiration by heterotrophic components is dealt with in this section. Respiration can be subdivided into two to four fractions based on how the products of the biochemical processes are used (Penning de Vries 1974, Amthor 1986, Lavigne 1988): (1) constructive respiration provides energy for the synthesis of new plant tissues; (2) maintenance respiration produces energy for the continuance of existing tissues; (3) respiration is required for the uptake of some nutrients; (4) it appears worthwhile to separate respiration used for translocation in large plants such as trees. This subdivision of respiration concerns only how the products are used, and the part of the plant that is respiring. It does not imply that the respiration fractions are biochemically different or distinct in any other way. The subdivision of respiration can be applied separately to each member of a stand. Hence, mathematical expressions used in models of pure stands can be applied directly to models of mixed stands.

Constructive respiration is easily incorporated in models. Approximately 0.2 grams of CO_2 are produced by respiration to produce 1 gram dry weight of stem tissue (Chung and Barnes 1977). These calculations were made using the procedures of Penning de Vries et al. (1974). The value of the constructive respiration parameter depends on the chemical composition of the tissues produced, and therefore can be expected to differ slightly between species and component of the tree. The constructive respiration of a component is proportional to the new production, i.e.,

$$r_c = c \times (NPP/S) \tag{4}$$

where r_c is annual constructive respiration per living unit of the component, c is rate of constructive respiration per unit of net primary production, S is some unspecified indirect measure of the quantity of living tissues in a component, and (NPP/S) is annual net primary production per living unit of the component.

A maintenance respiration parameter cannot be ascertained from theoretical, biochemical considerations as it is for constructive respiration (Penning de Vries 1975). McCree (1970) and Thornley (1970) proposed that maintenance respiration rate of a component was independent of growth rate. Subsequently it was found that maintenance requirements of fast-growing tissues of a component were greater than those of slow-growing tissues (Thornley 1977, Penning de Vries et al. 1979, McCree 1982, Lavigne 1988). The relationship between growth rate and maintenance respiration rate for stems of balsam fir (*Abies balsamea* (L.) Mill.) are illustrated in Figure 2, as an example. Many physiological models of tree or stand growth assume maintenance respiration of a component is independent of growth rate of the component. These models probably underestimate maintenance respiration when production rates are very high, and overestimate respiration when production rates are very slow. By assuming that maintenance respiration rate was independent of growth rate, and therefore using the same rate for thinned and unthinned stands, the maintenance respiration of the thinned stand would have been underestimated by 23%, and the maintenance respiration of the unthinned stand would have been overestimated by 37% (Lavigne 1988).

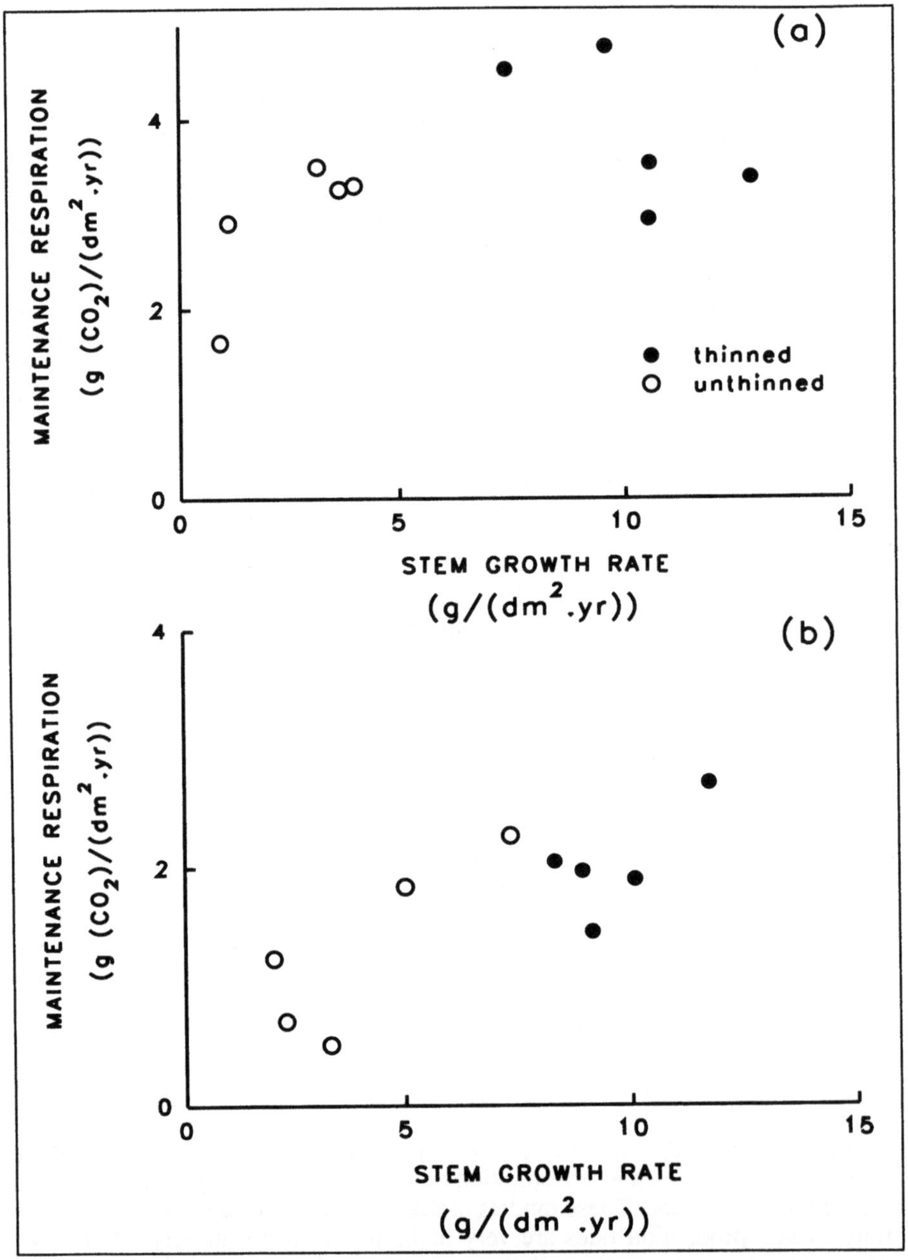

Fig. 2. Annual maintenance respiration rates of balsam fir trees growing in a thinned and an unthinned stand in Newfoundland, Canada. (a) Breast-height internodes, (b) mid-crown internodes. Redrawn from Lavigne (1988).

An indirect measure of the quantity of living tissues in long-lived components such as tree stems is required for assigning respiration to fractions because it is not feasible to measure it directly. Mäkelä and Hari (1986) and Mohren (1987) used sapwood volume to estimate the quantity of living tissues requiring maintenance in woody organs, rather than using stem surface area. Maintenance respiration rate is assumed to be independent of growth rate in these models. Ryan (1989) suggested that relating stem maintenance respiration to sapwood volume was theoretically superior to relating it to stem surface area, because sapwood volume gives a better estimate of the living tissues requiring support. However, Goodwin and Goddard (1940) measured lower rates of respiration for the inner sapwood than for the outer sapwood, so every unit of sapwood volume does not respire at the same rate, indicating that either units of sapwood volume contained different amounts of living tissue, or contained tissues with different metabolic rates. Moreover, Ryan (1990) found that the average rates of stem maintenance respiration per unit of sapwood volume were not the same for all trees or internodes of a tree, and therefore using sapwood volume was not shown to be clearly superior to using surface area.

Respiration is required for nutrient uptake (Amthor 1986; Lambers et al. 1983). Ledig et al. (1976) and Szaniawski (1981) estimated high rates of respiration by roots of seedlings in comparison to respiration by shoots. Respiration for nutrient uptake could be partially responsible for the higher respiration rates of roots, as this use of respiration products was responsible for a substantial proportion of root respiration by herbaceous species (Amthor 1986). When whole plant respiration is subdivided then energy consumption for nutrient uptake can be considered as part of constructive respiration (Penning de Vries et al. 1974). However, when respiration by components is the basis for subdividing respiration, the large distances between sites of nutrient uptake and sites of growth of above-ground components requires that respiration for nutrient uptake be separated from constructive respiration. Respiration for nutrient uptake has not been accounted for separately in physiological models. A method of describing respiration for nutrient uptake that can be incorporated into physiological models has not been proposed. Such a method would lead to improved paradigms for allocating production and photosynthate to root systems, in addition to improved accuracy in modelling total respiration.

Respiration for translocation has not been dealt with separately from respiration for other purposes in physiological models and no equations have been proposed for describing this fraction of respiration. However, in balsam fir stems the respiration rates of internodes near breast height were greater than those in the mid-crown with the same production rates (Figure 2). These differences might have been caused by higher rates of respiration for translocation below the crown than in the upper crown.

Allocation

The allocation of production to components is predicted with rules for maintaining functional relationships. Mathematical descriptions of allocation have improved recently. Paradigms and parameter values determined for use in models of pure stands can be applied directly in models of mixed stands. No additional knowledge of the phenomena are required to extend current practices to models of mixed stands. Two complementary approaches have been used in modelling allocation: the pipe model theory and the principle of a functional balance.

Pipe model theory. One rule for describing allocation in mathematical terms is based on the pipe model theory. Shinozaki et al. (1964a,b) theorized about a relationship between the cross-sectional area of sapwood at any point on a stem and the quantity of foliage above the point. The sapwood is modelled as an assemblage of pipes connecting foliage to feeder roots. Valentine (1985, 1988, 1990), Mäkelä (1986, 1988, 1990a) and Ludlow et al. (1990) used pipe-model theory to develop models relating foliar production to the production of conducting tissues in branches, stems, and roots. The correlation between production allocated to foliage and that allocated to stems, branches, and large roots changes as trees grow taller because the pipe length increases, and hence more sapwood volume is required for pipes of a given diameter. Mäkelä (1986) developed the following equation to estimate the proportion of total annual net primary production allocated to stem production:

$$ag_s = (\alpha_s \times ag_f + \beta_s)H \qquad (5)$$

where ag_s and ag_f are the proportions of total annual net primary production allocated to stems and foliage, α_s and β_s are parameters, and H is stem height. Similar equations were developed for branches (ag_b) and large roots (ag_k).

The correlation between foliage quantity and the sapwood cross-sectional areas differs between species (Whitehead et al. 1984, Pothier and Margolis 1990). Values of parameters of equations such as Equation 5 that relate growth of woody components to foliar growth could vary between species because of differences in stem permeability to water, abilities to acquire water, and water use efficiency. Differences in allocation between stems and branches could also arise between species because of dissimilar crown architecture and crown development patterns. For example, species having excurrent crown form might allocate more of total growth to foliar growth and stem growth, and less to branch growth, compared to species having decurrent crown form.

The values of coefficients fit to an equation relating foliar quantity and sapwood cross-sectional area depended on site (Albrekston 1984, Espinosa Bancalari et al. 1987), and therefore, the parameter values of relations between foliar growth and the growth of woody components, such as Equation 5, could also vary with site quality. Sapwood permeability increased as trees gained in stature (Pothier et al. 1989), which might offset the effect of increasing height on the additional energy required to move water from the soil to the foliage, and thereby help to keep the relationship between sapwood cross-sectional area and foliar quantity the same during stand development. Therefore, parameter values of allocation rules may not change simply because the stand ages.

Sapwood permeability of balsam fir and white birch (*Betula papyrifera* Marsh.) did not change during the first 2 years after thinning (Pothier and Margolis 1990). Brix and Mitchell (1983) did not find a clear effect of thinning on the relationship between foliar quantity and sapwood cross-sectional area of Douglas-fir (*Pseudotsuga menziesii* (Mirb.) Franco). These studies indicated that the relationship between allocation to woody components and allocation to foliage would not be affected by thinning. An example of the correlations between the production of water-conducting components and foliar production can be seen in thinned and unthinned balsam fir stands (Figure 3). The same proportional relationship between foliar production and the production of stems and branches existed in these thinned and unthinned stands (Lavigne 1991).

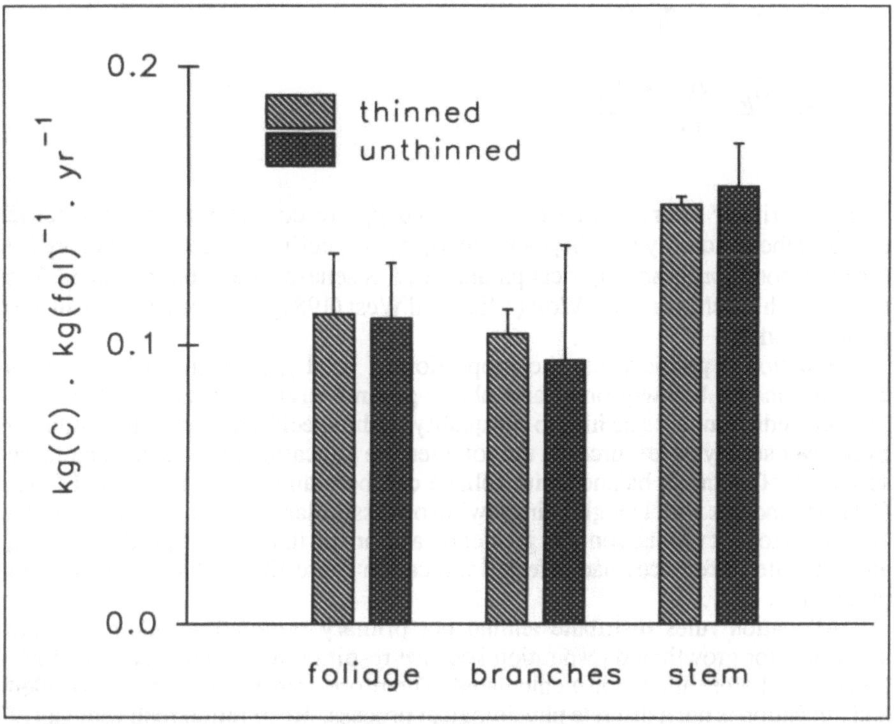

Fig. 3. Allocation of growth among above-ground components of balsam fir trees growing in a thinned and an unthinned stand in western Newfoundland. Error bars are +1 standard error. Redrawn from Lavigne (1991).

Principle of a functional balance. The second approach to allocation involves the recognition of a functional balance between foliage and roots; specifically the product of foliar specific activity and foliar weight is considered to be proportional to the product of root specific activity and root weight, and annual production by these components must maintain this relationship. This principle was used by Reynolds and Thornley (1982) to develop an equation relating foliar production to the production of fine roots. Mäkelä (1986) developed the following equation for a tree growth model:

$$ag_{fr} = \alpha_{fr} \times ag_f + \beta_{fr} \tag{6}$$

where

$$\alpha_{fr} = \pi_N \times \left(\frac{\sigma_c}{\sigma_N}\right)$$

$$\beta_{fr} = \frac{S_{fr} - \alpha_{fr} \times S_f}{G}$$

The subscript *fr* refers to fine roots, α_{fr} and β_{fr} are coefficients, σ_C is specific photosynthetic activity (kg C/(yr • kg(fol)), σ_N is specific root activity (i.e., kg N/(yr • kg(roots)), π_N is an empirical parameter, S is senescence of biomass, and G is total growth. McMurtrie and Wolf (1983) and West (1987) also used this rule in tree growth models.

Equation 6 predicts that the proportion of total growth allocated to roots depends on the below-ground and above-ground environments, and therefore growth predictions are sensitive to site quality. If the specific root activity is reduced by a lower supply of resources in the soil then the allocation to roots must increase so that the functional balance with foliage can be maintained. Keyes and Grier (1981) found this with Douglas-fir growing on dissimilar sites. They found that the allocation to root production was greater on a poor site than a more productive site, and that site differences had affected allocation more than it had affected total production.

Allocation rules distribute annual net primary production rather than the substrates for growth and respiration because respiration is subtracted from photosynthesis before allocation is calculated. Therefore, translocation is not modelled and allocation is not a discrete physiological process. Respiration requirements of heterotrophic components can change over time independent from any changes in their annual production because they increase in size relative to the quantity of foliage. Moreover, maintenance requirements per unit of the living heterotrophic component depend on production rates, as discussed previously. For these reasons, there is not a simple relation between allocation to production and allocation to respiration that can be used to predict one from the other during all of stand development, or after various silvicultural treatments have been applied. A comparison of a thinned stand and an unthinned stand of balsam fir illustrates some of the differences between allocation of annual production and allocation of photosynthates (Lavigne 1991). The allocation of production to stems in a thinned balsam fir stand was similar to that in an unthinned stand (Figure 4), but more total photosynthates were allocated to the stems in the unthinned stand. The additional allocation of photosynthates in the unthinned stand was used for stem respiration. However, the additional allocation was not proportional to the difference in stem surface area because the stem maintenance respiration per unit of stem surface area was lower in the unthinned stand.

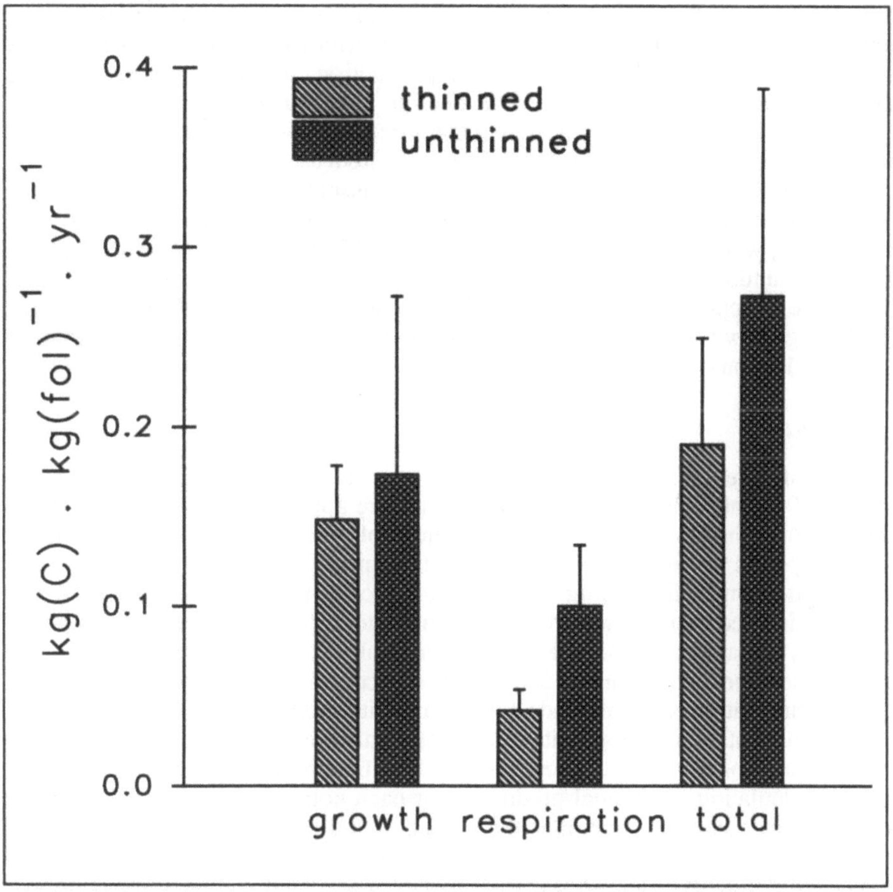

Fig. 4. Allocations to stem growth and stem respiration by balsam fir trees in a thinned and an unthinned stand in Newfoundland, Canada. Error bars are +1 standard error. Redrawn from Lavigne (1991).

Height growth

It is especially important to predict height growth accurately in models of mixed stands. The past sequence of annual height growth determines the current vertical distribution of foliage within a crown, and consequently it determines current annual photosynthetic production. In mixed stands the productivity of each species can change over time as the vertical arrangement of foliage of one species changes relative to another. Since the light intercepting properties and photosynthetic light responses differ between species, model predictions of productivity and stand development for mixed stands are sensitive to predictions of height growth.

Annual height increment has been estimated as part of predicting foliar production and crown development (i.e., Ludlow et al. 1990), or when predicting the distribution of stem growth over the bole (i.e., Valentine 1985, 1988, 1990). Ludlow et al. (1990) assumed rules about constant crown morphometry, and foliar area density, and about the impact of canopy closure on foliar production below the points of crown contact, to determine the dimensions of a shell of new foliage containing the previously estimated foliar production. The height increment was one of the calculated dimensions of the shell of new foliage.

Estimating height growth as part of the procedures for estimating stem growth is difficult, in part because extension of the main stem uses a small proportion of the biomass allocated to stem production. Arbitrary assumptions about lengthening of water conducting pipes are required when height growth is estimated in this way. In contrast, when height growth is treated as part of crown production then the biological understanding of shoot development can be used to make predictions. Therefore, it appears preferable to predict height growth as part of procedures for predicting crown development rather than as part of procedures for distributing stem production over the bole.

Computation

A simplified approach to estimating photosynthesis, such as those described by Grace (1990) and Mäkelä (1990b), appears more appropriate to models of stand development than making repeated predictions of instantaneous light interception at many points within the crowns. Taking the simplified approach probably does not obviate the need for dealing with more complicated models of light interception, but the detailed models would be used only to calibrate the simpler models. Problems remain that limit the capacity to make complicated models of light interception for mixed stands. This limits our confidence in predictions of photosynthesis, but does not preclude continuing with model development.

The allocation rules distribute annual net primary production to the components rather than photosynthates. Therefore, total annual production must be estimated before calculating the annual production of each component. This requirement dictates the order of calculation in the model. In most published models the order of calculations is as follows: (1) annual photosynthesis is estimated; (2) maintenance requirements of the whole plant are subtracted from annual photosynthesis; (3) the remaining photosynthate is converted into total annual net primary production by accounting for constructive respiration; and (4) the net primary production is distributed among components according to the allocation rules. It is necessary to assume that maintenance respiration rate is independent of growth rate with this order of calculations because maintenance expenses are paid before growth is estimated. As discussed in the respiration section, this assumption can lead to errors when the model is used to estimate stand development over many years. Also, this order of calculation makes it impossible to include the contribution of current-year foliage to photosynthetic production in a realistic manner because photosynthetic production is estimated before foliar growth.

Another order of calculations can be envisaged which permits maintenance respiration to depend on growth rate and incorporates a more realistic role for current-year foliage (Figure 5). This order of calculations has not been used in published models. This alternative order of calculations is as follows: (1) an initial estimate of foliar production is made; (2) annual photosynthesis is estimated, including the contribution made by the current year foliage; (3) production of heterotrophic components are estimated from the prediction of foliar production and the allocation rules; (4) these estimates of production are used to estimate constructive and maintenance respiration; (5) the total of estimated respiration and

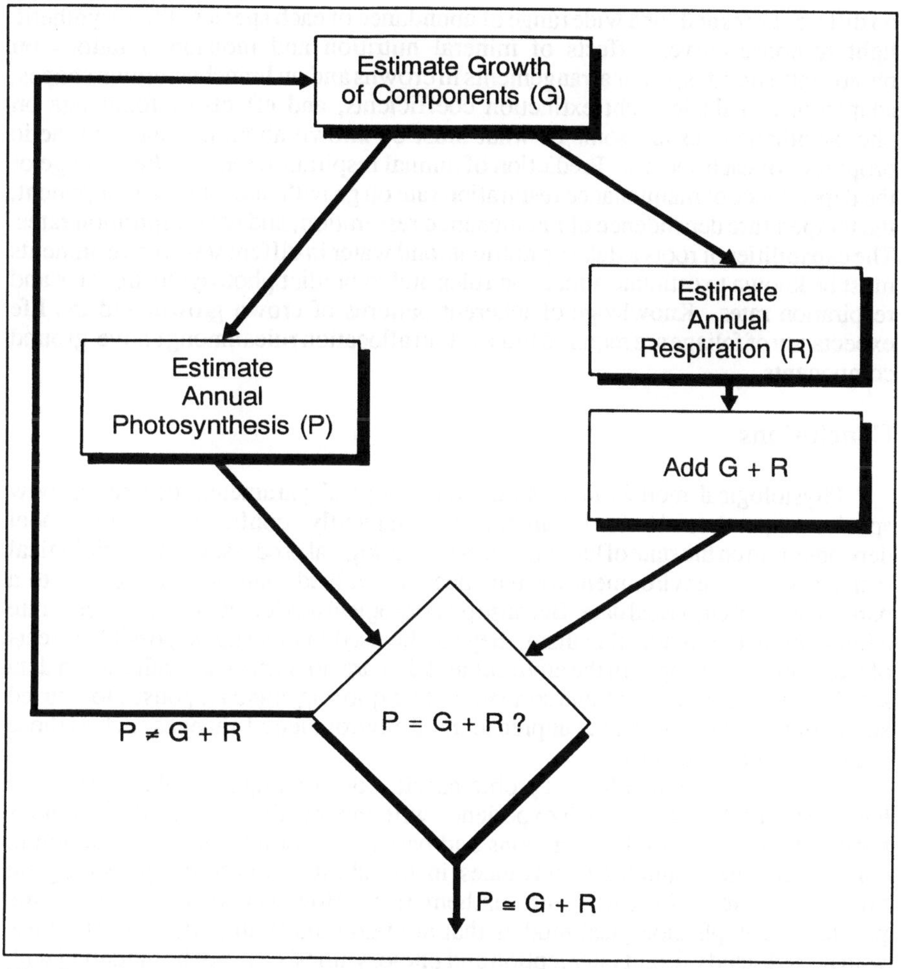

Fig. 5. Proposed order of calculations for estimating annual increments in stand development in a physiological model of mixed stands.

production is compared to estimated photosynthesis; (6) if the total of production and respiration is greater than photosynthesis then the estimate of foliar production is reduced and the procedures are repeated, or if the total of production and respiration is less than photosynthesis then the estimate of foliar production is increased before repeating the procedures. The calculations are repeated until estimated photosynthesis equals the estimated total of production and respiration.

Calibration

Ecophysiological and morphometric data are required for each species in order to calibrate a complete physiological model of stand development. A large amount of information is required if the model is constructed to predict stand development on different sites and for a wide range of abundance of each species. Photosynthetic light response curves, effects of mineral nutrition and moisture relations on photosynthetic rates, foliar arrangements in crowns and on branches, crown shapes, adaptations to shade, light extinction coefficients, and effects of foliar age on photosynthetic rates are some of what must be known about the photosynthetic properties of each species. Prediction of annual respiration requires knowledge of the dependence of maintenance respiration rate on growth rate for each component, the temperature dependence of maintenance respiration, and root respiration rates. The capabilities of roots to take up nutrients and water in different soil environments must be known to estimate allocation rules and to predict photosynthetic rates and respiration rates. Knowledge of inherent patterns of crown growth and the life expectancy of foliage are required to ascertain allocation rules among above-ground components.

Conclusions

Physiological models incorporate autecological parameters describing how species respond to the environment. Consequently, confidence in the model depends more on the state of knowledge of physiological processes and physiological responses to the environment for the species involved than on statistical fit to a particular set of empirical data. Because physiological models attempt to incorporate a fair amount of biological realism, they can be used to investigate possible effects of treatments or changes in the environment that are not part of any calibration data set. This makes models of this sort suited to exploring stand responses to untried silvicultural treatments and to unprecedented environmental changes such as those caused by climate change.

Many of the methodological problems with constructing physiological models have been resolved after much experience with models for pure stands. Methods for dividing respiration into fractions and paradigms for allocating production to components are examples of advances in the ability to describe physiological processes more realistically with mathematics. However, there are still some problems with physiological models that are particularly important for their use with mixed stands. Light interception and photosynthesis can not be modelled with a high level of realism. Also, prediction of changes in canopy structure over time, and hence changes in photosynthesis by each species is prevented by the absence of a realistic paradigm for height growth. It is hoped that these problems will be resolved in the not-too-distant future.

Large bodies of physiological data are required to calibrate physiological models. These data do not exist yet for most important tree species. The high potential utility of physiological models should be an incentive to do the necessary ecophysiological research.

Yahata et al. (1979) and Sievänen et al. (1988) have used ecophysiological knowledge to improve upon mensurational models of tree growth. Hybrid models of this sort can bridge the gap between mensurational and physiological models. These studies illustrated that more understanding of stand development can be gained from the analysis of traditional growth and yield data by considering physiological processes, and that it is feasible to use models incorporating more biological realism for a wider variety of purposes than is true for the standard yield models.

Literature cited

Albrektson, A. 1984. Sapwood basal area and needle mass of Scots pine (*Pinus sylvestris* L.) trees in central Sweden. Forestry 57: 35-43.

Amthor, J.S. 1986. Evolution and applicability of a whole plant respiration model. Journal of Theoretical Biology 122: 473-490.

Beadle, C.L., Talbot, H., and Jarvis, P.G. 1982. Canopy structure and leaf area index in a mature Scots pine forest. Forestry 55: 105-123.

Brix, H., and Mitchell, A.K. 1983. Thinning and nitrogen fertilization effects on sapwood development and relationships of foliage quantity to sapwood area and basal area in Douglas-fir. Canadian Journal of Forest Research 13: 384-389.

Chung, H.-H., and Barnes, R.L. 1977. Photosynthate allocation in *Pinus taeda*. I. Substrate requirements for synthesis of shoot biomass. Canadian Journal of Forest Research 7: 106-111.

Espinosa Bancalari, M.A., Perry, D.A., and Marshall, J.D. 1987. Leaf area - sapwood area relationships in adjacent young Douglas-fir stands with different early growth rates. Canadian Journal of Forest Research 17: 174-180.

Goodwin, R.H., and Goddard, D.R. 1940. The oxygen consumption of isolated woody tissues. American Journal of Botany 27:234-237.

Grace, J.C. 1990. Modeling the interception of solar radiant energy and net photosynthesis. In: Process Modeling of Forest Growth Responses to Environmental Stress. R.K. Dixon, R.S. Meldahl, G.A. Ruark, and W.G. Warren, eds. Timber Press, Portland, Oregon. pp. 152-158.

Grace, J.C., Jarvis, P.G., and Norman, J.M. 1987a. Modelling the interception of solar radiant energy in intensively managed stands. New Zealand Journal of Forestry Science 17: 193-209.

Grace, J.C., Rook, D.A., and Lane, P.M. 1987b. Modelling canopy photosynthesis in Pinus radiata stands. New Zealand Journal of Forestry Science 17: 210-228.

Jarvis, P.G., and Leverenz, J.W. 1983. Productivity of temperate, deciduous and evergreen forests. In: Physiological Plant Ecology IV. Ecosystem processes: mineral cycling, productivity and man's influence. Edited by P.S. Nobel, C.B. Osmond and H. Ziegler. Encyclopedia of Plant Physiology, New Series, Volume 12D. Springer-Verlag, New York. pp. 233-280.

Kellomäki, S., Hari, P., Kanninen, M., and Ilonen, P. 1980. Eco-physiological studies on young Scots pine stands: II. Distribution of needle biomass and its application in approximating light conditions inside the canopy. Silva Fennica 14: 243-257.

Kellomäki, S., Oker-Blom, P., and Kuuluvainen, T. 1986. The effect of crown and canopy structure on light interception and distribution in a tree stand. In: Crop physiology of forest trees. P.M.A. Tigerstedt, P. Puttonen and V. Koski, eds. Helsinki University Press, Helsinki. pp. 107-115.

Keyes, M.R., and Grier, C.C. 1981. Above- and below-ground net production in 40-year-old Douglas-fir stands on low productivity and high productivity sites. Canadian Journal of Forest Research 11: 599-605.

Kuuluvainen, T., and Pukkula, T. 1989. Simulation of within-tree and between-tree shading of direct radiation in a forest canopy: effect of crown shape and sun elevation. Ecological Modelling 49: 89-100.

Lambers, H., Szaniawski, R.K., and Visser, R. 1983. Respiration for growth, maintenance and ion uptake. An evaluation of concepts, methods, values and their significance. Physiologia Plantarum 58: 556-563.

Landsberg, J.J. 1986. Physiological ecology of forest production. Academic Press London. 198 p.

Landsberg, J.J., and McMurtrie, R. 1985. Models based on physiology as tools for research and forest management. In: Research for Forest Management. Edited by J.J. Landsberg and W. Parsons. CSIRO, Melbourne. pp. 214-228.

Lavigne, M.B. 1988. Stem growth and respiration of young balsam fir trees in thinned and unthinned stands. Canadian Journal of Forest Research 18: 483-489.

Lavigne, M.B. 1991. Effects of thinning on the allocation of growth and respiration in young stands of balsam fir. Canadian Journal of Forest Research 21: 186-192.

Ledig, F.T., Drew, A.P., and Clark, J.G. 1976. Maintenance and constructive respiration, photosynthesis and net assimilation rate in seedlings of pitch pine (*Pinus rigida* Mill.). Annals of Botany 40: 289-300.

Linder, S. 1985. Potential and actual production in Australian forest stands. In: Research for Forest Management. Edited by J.J. Landsberg and W. Parsons. CSIRO, Melbourne. pp. 11-35.

Ludlow, A.R., Randle, T.J., and Grace, J.C. 1990. Developing a process-based growth model for Sitka spruce. In: Process Modeling of Forest Growth Responses to Environmental Stress. R.K. Dixon, R.S. Meldahl, G.A. Ruark, and W.G. Warren, eds. Timber Press, Portland, Oregon. pp. 249-262.

Mäkelä, A. 1986. Implications of the pipe model theory on dry matter partitioning and height growth in trees. Journal of Theoretical Biology 123:103-120.

Mäkelä, A. 1988. Performance analysis of a process-based stand growth model using Monte Carlo techniques. Scandanavian Journal of Forest Research 3:315-331.

Mäkelä, A. 1990a. Modeling structural-functional relationships in whole-tree growth: resource allocation. In: Process Modeling of Forest Growth Responses to Environmental Stress. R.K. Dixon, R.S. Meldahl, G.A. Ruark, and W.G. Warren, eds. Timber Press, Portland, Oregon. pp. 81-95.

Mäkelä, A. 1990b. Adaptation of light interception computations to stand growth models. In: Modelling to understand forest functions. Edited by H. Jozefek. Silva Carelica 15, University of Joensuu. pp. 221-239.

Mäkelä, A., and Hari P. 1986. Stand growth model based on carbon uptake and allocation in individual trees. Ecological Modelling 33:205-229.

McCree, K.J. 1970. An equation for the rate of respiration of white clover plants grown under controlled conditions. In: Prediction and measurement of photosynthetic productivity. Proceedings of the IBP/PP Technical Meeting, Trebon, Czechoslovakia, 14-21 September 1969. PUDOC, Wageningen. pp. 221-229.

McCree, K.J. 1982. Maintenance requirements of white clover at high and low growth rates. Crop Science 22: 345-351.

McMurtrie, R., and Wolf. L. 1983. Above- and below-ground growth of forest stands: a carbon budget model. Annals of Botany 52:437-448.

Mitchell, K.J. 1975. Dynamics and simulated yield of Douglas-fir. Forest Science Monograph 17. 39 p.

Mohren, G.M.J., 1987. Simulation of Forest Growth, Applied to Douglas Fir Stands in The Netherlands. Ph.D. Thesis, Wageningen Agricultural University, The Netherlands. 184 p.

Mohren, G.M.J., Gerwen, C.P. Van, and Spitters, C.J.T. 1984. Simulation of primary production in even-aged stands of Douglas fir. Forest Ecology and Management 9:27-49.

Oker-Blom, P. 1986. Photosynthetic radiation regime and canopy structure in modeled forest stands. Acta Forestalia Fennica 197: 1-44.

Oker-Blom, P., and KellomÑki, S. 1983. Effect of grouping of foliage on the within-stand and within-crown light regime: comparison of random and grouping canopy models. Agricultural Meteorology 28: 143-155.

Oker-Blom, P., Pukkula, T., and Kuuluvainen, T. 1989. Relationship between radiation interception and photosynthesis in forest canopies: effect of stand structure and latitude. Ecological Modelling 49: 73-87.

Penning de Vries, F.W.T. 1974. Substrate utilization and respiration in relation to growth and maintenance in higher plants. Netherlands Journal of Agricultural Science 22: 40-44.

Penning de Vries, F.W.T. 1975. The cost of maintenance processes in plant cells. Annals of Botany 39: 77-92.

Penning de Vries, F.W.T., Brunsting, A.H.M., and van Laar, H.H. 1974. Products, requirements and efficiency of biosynthesis: a quantitative approach. Journal of Theoretical Biology 45: 339-377.

Penning de Vries, F.W.T., Witlage, J.M., and Kremer, D. 1979. Rates of respiration and of increase in structural dry matter in young wheat, ryegrass and maize plants in relation to temperature, to water stress and to their sugar content. Annals of Botany 44: 595-609.

Pothier, D., and Margolis, H.A. 1990. Changes in the water relations of balsam fir and white birch saplings after thinning. Tree Physiology 6: 371-380.

Pothier, D., Margolis, H.A., and Waring R.H., 1989. Patterns of change of saturated sapwood permeability and sapwood conductance with stand development. Canadian Journal of Forest Research 19: 432-439.

Reynolds, J.F., and Thornley, J.H.M. 1982. A shoot:root partitioning model. Annals of Botany 49: 585-597.

Ross, J. 1981. The radiation regime and architecture of plant stands. Dr W. Junk Publishers, The Hague. 391 p.

Ryan, M.G. 1989. Sapwood volume for three subalpine conifers: predictive equations and ecological implications. Canadian Journal of Forest Research 19: 13997-1401.

Ryan, M.G. 1990. Growth and maintenance respiration in stems of *Pinus contorta* and *Picea engelmannii*. Canadian Journal of Forest Research 20: 48-57.

Shinozaki, K., Yoda, K., Hozumi, K., and Kira, T. 1964a. A quantitative analysis of plant form - the pipe model theory. I. Basic analysis. Japanese Journal of Ecology 14: 97-105.

Shinozaki, K., Yoda, K., Hozumi, K., and Kira, T. 1964b. A quantitative analysis of plant form - the pipe model theory. II. Further evidence of the theory and its application in forest ecology. Japanese Journal of Ecology 14: 133-139.

Sievänen, R., Burk, T.E., and Ek, A.R. 1988. Construction of a stand growth model utilizing photosynthesis and respiration relationships in individual trees. Canadian Journal of Forest Research 18: 1027-1035.

Smith, D.M. 1986. The Practice of Silviculture, Eighth Edition. John Wiley & Sons, New York. 527p.

Stephens, G.R. 1969. Productivity of red pine. 1. Foliage distribution in tree crown and stand canopy. Agricultural Meteorology 6: 275-282.

Szaniawski, R.K. 1981. Growth and maintenance respiration of shoots and roots in Scots pine seedlings. Pflanzenphysiol. 101: 391-398.

Thornley, J.H.M. 1970. Respiration, growth and maintenance in plants. Nature 227: 304-305.

Thornley, J.H.M. 1977. Growth, maintenance and respiration: a re-interpretation. Annals of Botany 49: 257-259.

Valentine, H.T. 1985. Tree-growth models: derivations employing the pipe-model theory. Journal of Theoretical Biology. 117:579-585.

Valentine, H.T. 1988. A carbon-balance model of stand growth: a derivation employing pipe-model theory and the self-thinning rule. Annals of Botany 62: 389-396.

Valentine, H.T. 1990. A carbon-balance model of tree growth with a pipe-model framework. In: Process Modeling of Forest Growth Responses to Environmental Stress. R.K. Dixon, R.S. Meldahl, G.A. Ruark, and W.G. Warren, eds. Timber Press, Portland, Oregon. pp. 33-40.

West, P.W. 1987. A model of biomass growth of individual trees in forest monoculture. Annals of Botany 60: 571-577.

Whitehead, D., Edwards, W.R.N., and Jarvis, P.G. 1984. Conducting sapwood area, foliage area and permeability in mature trees of *Picea sitchensis* and *Pinus contorta*. Canadian Journal of Forest Research 14: 940-947.

Yahata, H., Suzaki, T., and Miyajima, H. 1979. Some methods to estimate parameters in relating photosynthesis to tree growth and application to *Cryptomeria japonica*. Journal of the Japanese Forestry Society 61: 151-162.

Silviculture and Management of Mixed-Species Stands

IV

Stand development patterns in Allegheny hardwood forests, and their influence on silviculture and management practices

10

DAVID A. MARQUIS

Introduction

Management of Allegheny hardwood forests is complicated by difficulties in securing adequate regeneration and problems in regulating current stands to maximize benefits. An understanding of the history and origin of present stands, plus knowledge of typical stand development patterns can provide the basis for appropriate silvicultural practices.

History and origin of present stands

The original forests of the Allegheny Plateau in northern Pennsylvania and southern New York were primarily hemlock-beech and beech-maple associations, with smaller amounts of pine-oak-chestnut in the larger stream valleys (Illick and Frontz 1928; Hough and Forbes 1943). (Scientific names of tree species are listed at end of paper).

The original stands were uncut until about 1800, when some lumbering began for white pine adjacent to streams. From about 1850 to 1890, white pine, hemlock, and selected hardwood sawlogs were cut throughout the region; these partial cuts were restricted to the larger and better trees because markets were limited to these products, and transporting large volumes of timber from the upland areas was too expensive and difficult. After about 1890, logging railroads and band sawmills became practical. These technological advances, coupled with the development of markets for small products such as pulpwood and chemical wood, provided the incentive for clearcutting over extensive areas. Between 1890 and 1930, nearly the entire Allegheny Plateau of northwestern Pennsylvania was cut over (Marquis 1975a).

Because the chemical wood plants prevalent on the Plateau at that time used trees of all species and sizes down to 5 to 8 cm (2 to 3 inches) in diameter, the railroad-era clearcuts were very complete. This gave rise to second- and third-growth hardwood stands that are as truly even-aged as any hardwood forests resulting from commercial logging (Elliot 1927).

Within this general pattern of cutting, there was great variation from place to place. The number of partial cuts made in any individual stand during the 1850-1890 period ranged from none to several; their severity ranged from light to heavy. Many stands were clearcut between 1890 and 1930, but some never received a chemical wood clearcut, while others were clearcut twice.

Market conditions and transportation largely determined the extent of cutting at a specific time and place. Some of the partial cuts were made for a single product: first for white pine, later for hemlock, still later for high-quality hardwoods. In many stands there were two separate cuts for hemlock 10 to 20 years apart. In other stands, some combination of these products were taken in a single entry. After logging

M. J. Kelty (ed.), The Ecology and Silviculture of Mixed-Species Forests, 165–181.

railroads were built, it was common to cut for multiple products. Sawlog cuts after this time generally removed all species of sawtimber-size trees, and chemical wood or pulpwood was often removed at the same time, or shortly after.

Each of the partial cuts was followed by a surge of regeneration, with species composition and density determined by the seed sources present, residual overstory density, and other factors. Often, these younger stems were too small to be merchantable when the stands were re-entered during the railroad logging era, and some were left as scattered residuals, even after the relatively complete chemical wood clearcuts. On the basis of a dozen stands on which there are records, these residuals typically ranged from 5 to 20 cm (2 to 8 inches) dbh and numbered 125 to 1000 per hectare (50 to 400 per acre), representing .5 to 2.3 sq. m/ha (2 to 10 sq. ft./acre) of basal area. These residuals had an important impact on the new stand that developed.

Stands that were clearcut for sawtimber, but not chemical wood or pulpwood, are even more heavily influenced by the residuals. In these instances, the residuals may have amounted to 4.6 to 11.5 sq. m/ha (20 to 50 sq. ft./acre) of basal area after cutting. Such stands are really multi-aged, although they are commonly lumped with, and treated like, even-aged stands.

Silvical characteristics of major species

To understand stand development processes in Allegheny hardwood stands, it is necessary to understand the silvical characteristics of the major species. Black cherry, sugar maple, and red maple are the predominant species, and usually represent 65 to 95 percent of stand basal area. American beech, eastern hemlock, yellow birch, sweet birch, white ash, yellow-poplar, cucumbertree, and other species are common associates (Marquis 1980).

Black cherry is a shade intolerant species. It is the fastest growing of the major species in the type, especially during juvenile years. Black cherry often outgrows and overtops the other species during the first 10 years of stand life. However, those individual cherry trees that are unable to maintain a dominant or codominant crown position generally do not survive in lower crown strata.

Black cherry produces some seed every year, and abundant seed about every other year. The seed is capable of lying dormant in the forest floor for 3 to 5 years before germinating, so cherry seed is usually abundant in stands where there is a seed source. In addition, birds and mammals sometimes introduce significant quantities of cherry seed into stands that do not have a seed source.

Although intolerant of shade, small black cherry seedlings are common beneath even fairly dense canopies if there is an adequate seed source. The seedlings survive only a few years at these low light levels, but are constantly replaced by new germinants, and are capable of responding if released. Cherry regeneration may arise from these advance seedlings, from new seedlings that develop after cutting, or from stump sprouts (Marquis In press).

Sugar maple and American beech are at the other end of the spectrum. Both are quite tolerant of shade. While they are capable of making fairly good growth when exposed to adequate sunlight, both species will survive for many decades after being overtopped. Under such conditions, they make imperceptible growth. But if released, they are capable of responding and moving up into the main crown canopy.

Sugar maple produces moderate quantities of seed at 3 to 6 year intervals. Beech produces small quantities of nuts at intervals of 5 or 6 years. However, overtopped and suppressed individuals do not produce significant quantities of seed.

Advance regeneration of both sugar maple and beech is common if seed sources are adequate and deer browsing not severe. These species are capable of surviving for many years in the understory, and will make moderate growth at fairly low light levels. Large seedlings and saplings of these species often dominate the understory in stands maintained at densities suitable for timber production.

Sugar maple sprouts readily from cut stumps. American beech may also sprout from stumps, but it more commonly produces root suckers, which are a major source of reproduction. Beech root suckers can be very dense beneath even a heavy overstory canopy.

Most sugar maple and beech that become part of a new stand originate well before the final harvest – as large advance regeneration seedlings, or root suckers. Seedlings of these species that originate at the same time as faster-growing associates are usually left quickly behind in the race for a main canopy position (Gabriel and Walters In press; Tubbs and Houston In press).

Red maple is intermediate between the cherry and the sugar maple-beech groups in both shade tolerance and juvenile growth rate. Red maple produces an abundance of seeds at frequent intervals, and sprouts prolifically from cut stumps. Red maple advance seedlings survive longer than cherry under low light, and grow faster than sugar maple and beech at higher light levels. Successful regeneration may arise from new seedlings, from advance seedlings, or from sprouts (Walters In press).

Most of the other common associates, such as white ash, yellow birch, and cucumbertree are intermediate in tolerance and growth rate, and occupy a position similar to red maple. Of these species, white ash produces seed that remain viable in the forest floor for 3 to 5 years; birch produces tremendous quantities of seed that may be dispersed long distances by wind. Birch is the only major Allegheny hardwood species likely to reproduce in abundance from seed dispersed to the area after harvest cutting. All three species sprout from cut stumps (USDA Forest Serv. In press).

One other factor of importance in Allegheny hardwood stands is susceptibility to deer browsing (Marquis 1981a; Marquis and Brenneman 1981). Deer populations were extremely low in the early 1900's when present stands originated, but are currently high in many parts of Pennsylvania (Marquis 1975a). Eastern hemlock, sugar maple, red maple, white ash, cucumbertree, and yellow-poplar seedlings are highly preferred by deer, and are often eliminated by browsing. Black cherry is utilized, but is less preferred. American beech and striped maple are also utilized, but are able to recover from and thereby withstand browsing better than other species. The ability of beech to recover from browsing is due largely to the root sucker origin of many stems. As a result, cherry and beech are often the only commercial species that regenerate in current Allegheny hardwood stands in Pennsylvania.

Stand development patterns in Allegheny hardwoods

Continuous 50-year records of stand development in several stands on the Kane Experimental Forest in northwestern Pennsylvania provide data to illustrate stand development patterns in Allegheny hardwoods (Marquis 1981b). In most cases, stand data were collected immediately before and immediately after harvest cutting, and at 5 to 10 year intervals thereafter.

Truly even-aged stands

Although chemical wood clearcuts of the early 1900's removed nearly all trees over 5 to 8 cm (2 to 3 inches) dbh, there were almost always a few residuals left uncut. One such stand that was clearcut in 1937 was incorporated into an experiment in which all trees over 1.3 cm (0.5 inches) dbh were mowed down after commercial logging. Thus, the resulting second-growth stand provides an ideal case history of stand development in a truly even-aged situation.

The original old-growth stand was a beech-sugar maple-hemlock stand that had been partially cut in the 1880's, removing nearly all of the hemlock for sawlogs and bark (bark was used for tanning leather). In 1937, most of the hardwood was overmature and decadent, with trees up to 94 cm (37 inches) dbh. Total basal area was 24 sq. m/ha (106 sq. ft./acre) in trees 1.3 cm (0.5 inches) dbh and larger. Beech represented 74 percent of the basal area, and sugar maple 17 percent. The remaining 9 percent was divided about equally among black cherry, red maple, yellow and sweet birch, and miscellaneous other species.

There was a dense, but variable, understory as a result of the earlier partial cutting and decadence. This consisted of about 12,355/ha (5,000/acre) stems of advance regeneration between .6 m (2 ft.) in height and 1.3 cm (0.5 inches) dbh, and an additional 3,210/ha (1,300/acre) saplings in the 2.5 to 7.6 cm (1 to 3 inch) classes. Nearly all of this advance regeneration was sugar maple and beech.

The overstory was clearcut in 1937 for sawlogs and chemical wood. Only about 250/ha (100/acre) widely scattered saplings over 3.8 cm (1.5 inches) dbh were left standing--a total of 1.2 sq. m/ha (5 sq. ft./acre) of basal area. Immediately after cutting, all residuals over 1.3 cm (0.5 inches) dbh were mowed down on two 0.1 hectare (1/4 acre) experimental plots. This left 11,930/ha (4,830/acre) advance seedlings, distributed as follows: 54 percent sugar maple, 44 percent beech, 1 percent birch, and 1 percent striped maple. There were an average of only 10/ha (4/acre) black cherry seedlings (less than .001 percent of the advance growth), and no red maple.

Although the new second-growth stand consisted entirely of shade tolerant maple and beech in year 1, black cherry and other less tolerant species that regenerated at the time of harvest cutting quickly outgrew the tolerants. By age 10, black cherry represented 30 percent of the basal area, and that proportion increased to 75 percent by age 50 (Fig. 1).

This systematic increase in the proportion of basal area represented by cherry is a result of its faster growth in both height and diameter. Starting as new seedlings in competition with well-established maple and beech advance regeneration, the cherry caught up with the maple and beech and already dominated the stand by age 10. By age 35, the dominant and codominant crown strata was 100% black cherry (Fig. 2).

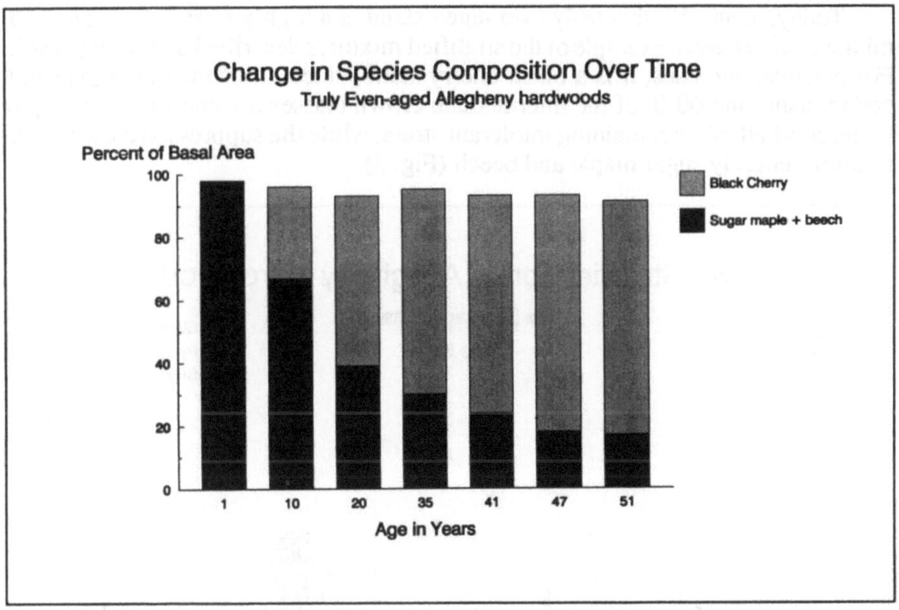

Fig. 1. Proportion of basal area by species for a truly even-aged Allegheny hardwood stand from age 1 to 50 years.

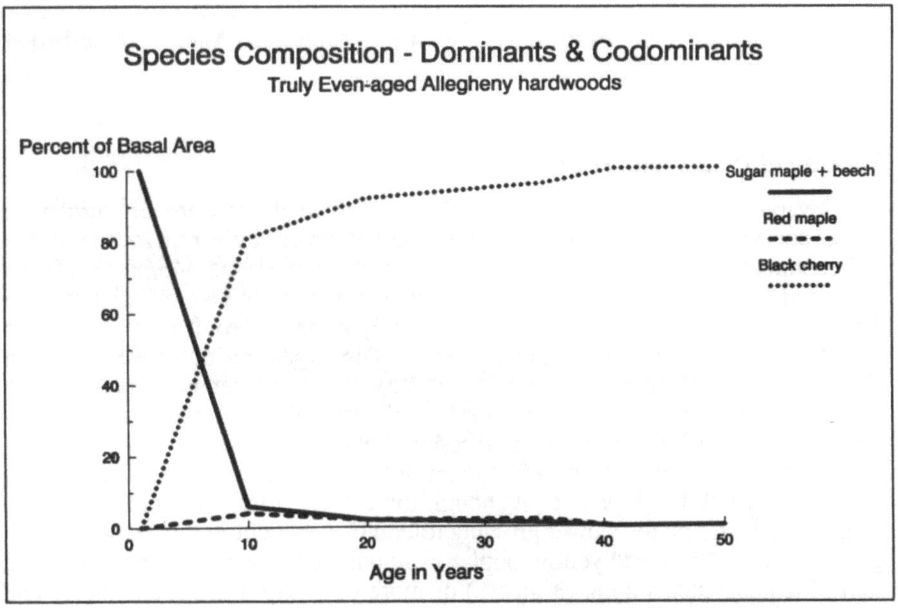

Fig. 2. Proportion of basal area by species in the dominant and codominant crown strata as a function of age for a truly even-aged Allegheny hardwood stand.

Today, at age 50, this truly even-aged stand is a highly stratified, even-aged mixture; it is a classic example of the stratified mixtures described by Smith (1986). For practical purposes, it is a black cherry stand, since all of the dominants and codominants and 60 % of the intermediate crown classes are cherry. Red maple represents half of the remaining intolerant strata, while the suppressed crown layer is almost entirely sugar maple and beech (Fig. 3).

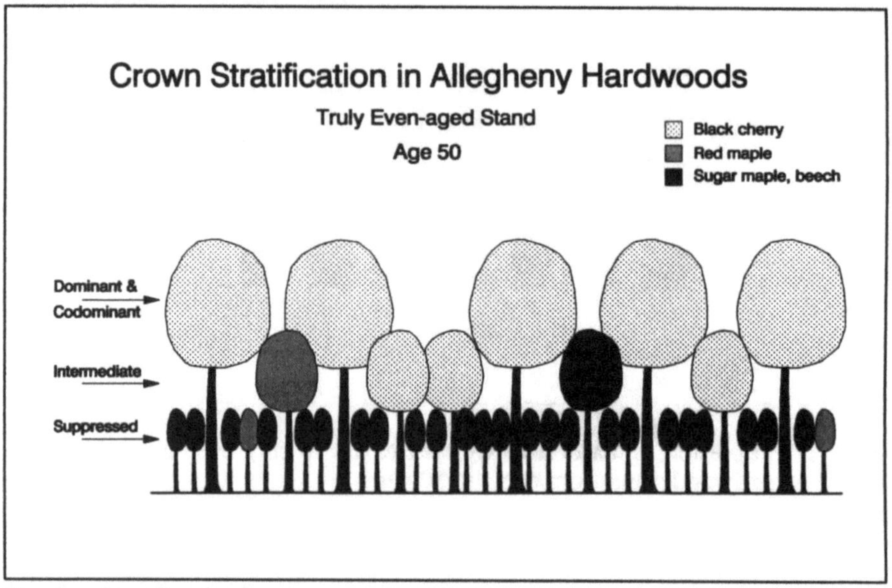

Fig. 3. Stratification into crown strata for a truly even-aged Allegheny hardwood stand at age 50.

Even-aged stands with residuals

Several companions to the mowed plots described above were left unmowed. These plots were in the same stand, and were part of the same clearcut for sawtimber and chemical wood. The residuals left after cutting averaged 445/ha (180/acre) stems and 2 sq.m/ha (8.7 sq. ft./acre) basal area in trees over 3.8 cm (1.5 inches) in diameter, and 930/ha (376/acre) stems and 2.2 sq. m/ha (9.7 sq. ft./acre) basal area in trees over 1.3 cm (0.5 inches) in diameter. The largest residuals were 12.7 cm (5 inches) dbh, and 99 percent of their basal area was either sugar maple or American beech. These numbers, sizes, and species of residuals are typical of the fairly complete clearcuttings for chemical wood in that era.

These small residual stems left after cutting have had a considerable impact on subsequent stand development. Although the tendency for intolerant species to outgrow and overtop the slower growing tolerants is evident in this stand too, fast growing black cherry and yellow-poplar were unable to overcome the large head start of residual stems 8 cm (3 inches) or more dbh. By age 50, the cherry and yellow-poplar had caught up to the maple and beech residuals in both height and diameter, but did not overtop them. Having achieved a position in the upper canopy, these shade tolerants have been able to grow fast enough to maintain their position.

Thus, the dominant and codominant canopy in this even-aged stand with residuals includes representation of all species groups (Fig. 4). The shade tolerant stems in dominant and codominant positions are residuals left after cutting. The shade tolerants in suppressed crown positions are stems that originated as seedling-sized advance regeneration, and were quickly overtopped.

Stratification in Allegheny hardwood stands that are essentially even-aged, but include some shade tolerant residuals, is more complex than in truly even-aged stands. It includes stratification based both on species and – in the case of the tolerants – on age (or initial size).

Although the two stands illustrated here are single case histories, all stands in which I have been able to document similar cutting histories have followed similar patterns. Truly even-aged stands tend to have (Marquis 1981c):

1) higher proportions of intolerant species than even-aged stands with residuals. The more complete the final cutting, the higher the percentage of cherry and yellow-poplar.
2) more complete crown stratification by species. Shade tolerant species are almost never found in the dominant and codominant crown strata.
3) lower stand diameters, and slower development. Part of this is simply the lack of residual trees that had a head start, so average diameter is lower. But even the cherry seem to grow slower in truly even-aged stands where they are in severe competition with other cherry. In stands with residuals, the residuals create an uneven canopy that seems to result in earlier differentiation into crown classes, and faster growth.

From the standpoint of timber production, these differences can have important impacts on early thinning possibilities. In the examples above, the even-aged stand with residuals contained 41 percent more pulpwood volume, 48 percent more sawtimber volume, and 90 percent more value than the truly even-aged stand. This more rapid development into merchantable sizes makes commercial thinning possible 10 to 15 years earlier in stands with residuals.

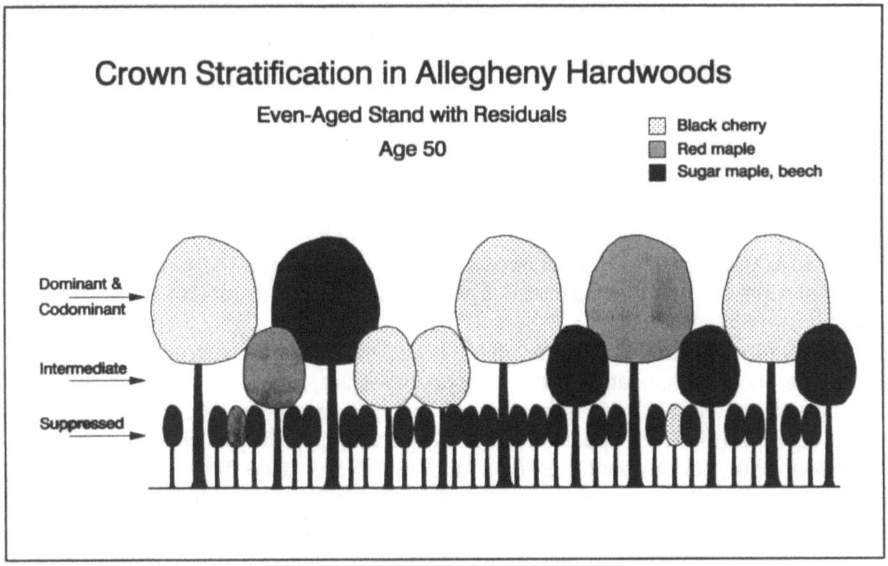

Fig. 4. Sratification into crown strata for even-aged stand with residuals at age 50.

Multi-aged stands

Multi-aged stands in the Allegheny hardwood region are the result of commercial clearcutting for sawtimber only. The youngest age class consists of stems that became established or were released by this heavy cutting. The older age class consists of residuals, much as described for even-aged stands with residuals. But in the multi-aged stands, there were more residuals, and some of them were larger trees up to 30 cm (12 inches) or more dbh at the time of cutting (because no pulpwood or chemical wood products were removed). These residuals were often stems that regenerated as a result of some earlier partial cutting, so they were often no more than 20 to 40 years older than the youngest age class.

To illustrate stand development of a multi-age stand, and compare it to the truly even-aged and even-aged with residuals stands, data are included from a multi-aged stand on the Kane Experimental Forest. The original stand on this area was an old-growth hemlock-sugar maple-beech stand similar to the one clearcut to produce the two even-aged stands above. There was a heavy cutting for hemlock and hardwood sawlogs in about 1895, but there were no further cuttings after that time.

Although no data are available immediately after the 1895 cutting in this stand, an attempt was made to reconstruct it in 1932 from tallies and age counts of trees cut for this purpose in adjacent plots. The stand at that time was found to be a mixture of second-growth 35 to 40 years old that originated from the 1895 cut, plus scattered residuals from the old growth. Most of these residuals were 50 to 80 years old, but a few stems up to 200 years were found. The annual rings of many of these older stems showed a distinct release 37 years earlier.

The maximum diameters of trees in the older age class at the time of release (1895) were 30 to 36 cm (12 to 14 inches). Trees larger than 20 cm (8 inches) were almost entirely hemlock and beech, while sugar maple was also prevalent below 20 cm (8 inches). This evidence suggests the stand was cut to a diameter limit, with a lower minimum diameter for sugar maple. Precise data on numbers of trees or basal area left after the 1895 cut are not available, but projections of the older age class backward from the 1932 inventory suggests that there was at least 4.6 sq. m/ha (20 sq. ft/acre) of basal area in residuals.

Data presented on this multi-aged stand are for 1943, when the new age class was 48 years old. This makes it comparable to the 50-year data presented for the two even-aged stands (Fig. 5).

In this multi-aged stand, black cherry represents only 12 percent of the basal area, as opposed to 75 and 46 percent for the truly even-aged and even-aged with residuals stands. More shade tolerant residuals being left at the time of final harvest results in a higher proportion of those species in the next stand, with the less intolerant reproduction that survives. Thus, multi-age stands tend to be dominated by sugar maple and beech.

The extreme initial size advantage allowed many shade tolerant residuals to make it into the dominant and codominant crown strata, where they were eventually joined by the new age class of intolerants. Thus, the crown stratification so evident in even-aged stands is not apparent in multi-age stands. All species are intermixed in the dominant, codominant, and intermediate layers. However, the suppressed layer is still primarily sugar maple and beech, because the intolerants cannot survive there.

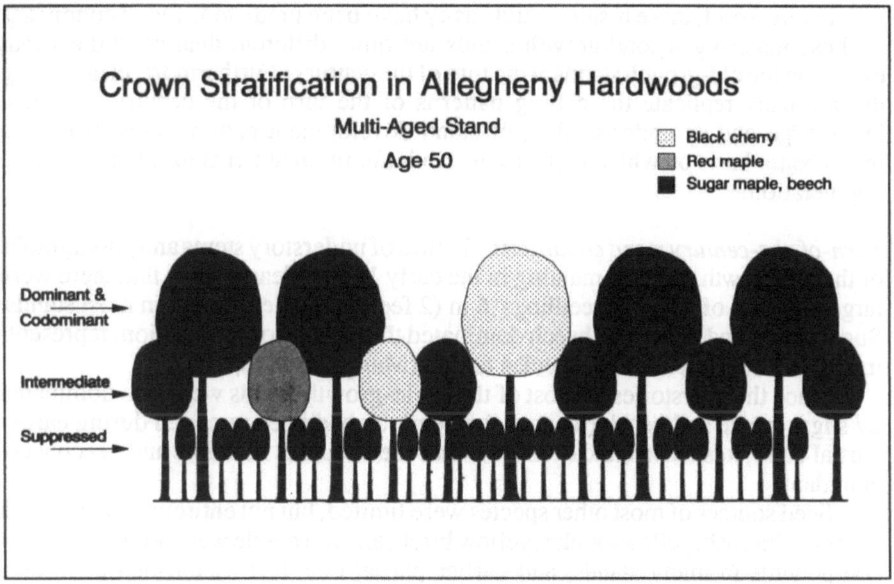

Fig. 5. Stratification into crown strata for multi-aged stand at age 50.

Multi-aged stands are almost always older (that is, the youngest age class is older) than even-aged stands on the Allegheny Plateau. In our examples, the multi-aged stand reached 50 years in 1945, but the two even-aged stands did not reach 50 years until 1988. This is not a chance occurrence of these particular examples. The incomplete clearcutting for sawlogs occurred early, before logging railroads and chemical wood plants created markets for small trees. After these markets developed, most stands were clearcut more completely. So, even-aged Allegheny hardwoods are usually younger than their multi-aged counterparts (Marquis 1981b).

Management implications

Regeneration

When even-aged management was instituted in Allegheny hardwoods in the mid-1960's, it was assumed that clearcutting would produce third-growth stands much like current second-growth. Clearcutting of the old-growth had created the second-growth, so it seemed logical that similar procedures would work again.

That has not happened. Nearly half of the early clearcuts on the Allegheny Plateau of Pennsylvania failed to develop regeneration. Many of these cutover areas became herbaceous openings. Most of the remainder have regenerated to either low-value beech and striped maple thickets, or to nearly pure stands of black cherry (Marquis and Brenneman 1981; Marquis 1983; Tilghman 1989). Even though black cherry is the most valuable hardwood in the region, monocultures of cherry are highly undesirable because of their susceptibility to insect and disease catastrophes, and for the extreme lack of diversity for all resource values, but especially wildlife.

In retrospect, these results could easily have been predicted. Stand conditions in these maturing second-growth stands are quite different than conditions that existed in the old-growth stands at the turn of the century. Furthermore, clearcutting did not truly replicate the cutting patterns of the turn of the century. Silvical knowledge and an understanding of stand development patterns now provide a much better basis on which to prescribe silvicultural treatments to achieve desired regeneration.

Turn-of-the-century stand conditions. Tallies of understory stems and photographs of the old growth stands remaining in the early 1930's clearly show that there were large numbers of advance seedlings .6 m (2 feet) and taller present in most stands. Sugar maple and American beech dominated this understory vegetation, representing 90 to 100 percent of the species tallied (Marquis 1981a, 1981c).

Since the overstories of most of these old-growth stands were also dominated by sugar maple and beech (most of the hemlock had been removed during earlier partial cuts), it is reasonable to assume that seed sources of maple and beech were abundant.

Seed sources of most other species were limited, but not entirely absent. Black cherry, white ash, yellow-poplar, yellow birch, and red maple were present as minor components in many stands, and earlier partial cuts had established additional stems. Some of these younger stems reached seed producing age by the time of final harvest cutting. Furthermore, seed dispersal by birds (cherry) and wind (birch) brought in additional quantities to many stands, and the ability of cherry, ash, and yellow-poplar seed to lie dormant in the forest floor helped to build up seed supplies of these species.

Present stand conditions. Present stands have understory conditions markedly different than those that existed in the 1890 to 1930 era. In most second-growth stands, sugar maple, beech, and hemlock are rarely found in the upper crown layers where seed production is concentrated. Even when they occur in the upper canopy, these species produce relatively few seed, at infrequent intervals. That seed is not dispersed over great distances, and it does not remain viable in the forest floor for more than a year. As a result, seed supply for these species is extremely limited in many present-day stands.

In contrast, black cherry seed is now extremely abundant. One study estimated nearly 1.2 million/ha (.5 million/acre) viable black cherry seeds in the forest floor of typical second-growth cherry-maple stands (Marquis 1975b).

Understories of present stands are dominated by black cherry advance regeneration, beech root suckers, striped maple stems, or ferns. Seedlings of sugar maple, hemlock, red maple, white ash, and birch are scarce due to lack of seed, and excessive deer browsing eliminates those few seedlings of these species that do become established. Cherry is browsed somewhat less than many other species, and new cherry seedlings germinate almost every year. Thus, cherry seedlings are often the only seedling-origin advance regeneration present.

Deer browsing and shade keep advance regeneration from growing very large; when present, it tends to be less than .3 m (1 foot) tall. In contrast, beech root suckers, striped maple, and fern survive the shade and browsing very well, and often expand gradually to the point where they interfere with establishment and survival of desirable advance regeneration.

The more nearly even-aged the second-growth stand, the higher the proportion of cherry it will have in the overstory, and the greater the chances of finding abundant cherry advance regeneration in the understory. Sources of beech root suckers and striped maple are scarce in these stands; if interfering understories exist, they are most likely to be ferns.

Multi-aged stands have less cherry seed source, and fewer cherry advance regeneration seedlings. Residual beech provide a source of beech root suckers, and striped maple stems are often associated with the beech. Without cherry, advance regeneration of any kind is generally scarce.

Why turn-of-the-century practices produce different results now. Since cherry was able to become established after harvest cutting at the turn of the century, and quickly outgrow the already established maple and beech advance regeneration, it should be no surprise that clearcutting of a stand with dense cherry advance regeneration and no maple or beech advance regeneration would produce nearly pure stands of cherry.

For example, the old-growth stand that produced the truly even-aged second-growth described above contained about 5000 stems of maple and beech advanced regeneration over .6 m (2 feet) tall at the time of cutting, and almost no cherry advance regeneration. In the second-growth stands of today that reproduce to third-growth cherry monocultures, cherry advance regeneration averages in excess of 99,000/ha (40,000/acre) stems (sometimes exceeding 250,000/ha (100,000/acre) stems), and there is almost no advance regeneration of sugar maple.

In stands where the advance growth consists of undesirable plants (ferns, beech root suckers, striped maple), the next stand will usually be dominated by these species. Regeneration of desirable woody species rarely develops unless it was abundant as advance seedlings prior to overstory removal (Grisez and Peace 1973).

The failure of black cherry, red maple, and birch to develop as new seedlings germinating after cutting – as they did at the turn of the century – could not easily have been predicted from past history. This is caused by excessive deer browsing, which did not exist then.

Browsing since the 1930's has resulted in increases in fern, beech, and striped maple in the understory, loss of advance regeneration of most desirable species other than cherry, and continued pressure against desirable species in the developing stand after final harvest (Horsley 1982; Marquis and Grisez 1978; Tilghman 1989). When present in very large numbers, advance regeneration is able to respond quickly to release, and the large numbers provide more than deer can eat in the few years it takes for some of them to grow out of reach. But when numbers of seedlings are small, or time required to grow out of reach is long, deer can totally prevent regeneration.

Although the 1890 to 1930 era cuttings were very complete clearcuttings, they were often preceded by one or more partial cuts. Stands present at that time, with their over-mature trees, were also well into the understory reinitiation stage or old growth stage of stand development, whereas second-growth stands of today are often still in the stem exclusion stage (Oliver 1981). Thus, conditions at the turn of the century were much more like those expected in a shelterwood sequence than in a clearcut.

Thus, stand conditions and cutting practices at the turn-of-the-century were markedly different than those of today. And these differences have produced different results.

Abundant (and large-sized) advance regeneration and residual saplings of tolerant species, ample seed and vigorous growth habit of the intolerant species, and lack of deer browsing combined to produce fully-stocked second-growth stands containing a rich mixture of species. But clearcutting today generally produces different results: 1) lack of advance growth of tolerants, plus dense advance growth of cherry and severe deer browsing now produce cherry monocultures; or 2) lack of advance growth of any desirable species, interfering plants, and severe deer browsing now produce regeneration failures.

Recommended regeneration practices. Given the current situation and a knowledge of stand development patterns, what silvicultural practices might be used to ensure adequate regeneration containing a mixture of desired species? There are several techniques now in use that have helped to overcome regeneration problems.

Adequate quantities of advance regeneration of desired species are clearly needed to ensure regeneration success after final harvest. Therefore, guidelines have been developed to indicate how many advance seedlings are needed under a wide variety of conditions (Marquis and Bjorkbom 1982). Shelterwood cuttings provide a way to increase the amount and size of advance seedlings where they are inadequate (Marquis 1978, 1979a, 1979b; Horsley 1982).

Guidelines on interfering plants also indicate situations under which fast-growing cherry can outgrow them, and situations in which an herbicide is needed to control the interfering plants (Marquis and others 1984).

Advance regeneration of shade tolerant species must be large sapling or even pole size to ensure they make it into the main crown canopy of the next stand. Hence, a practice of retaining carefully selected residuals of sugar maple, beech, or hemlock at the time of final harvest provides a way to ensure representation of those species in the next stand. These residuals are larger and older than what is normally considered advance regeneration. However, they are able to respond to overstory release along with the smaller seedlings of less tolerant species, and they clearly function as a form of advance regeneration. In a sense, this practice creates a two-aged, even-sized stand that can more easily be managed by traditional even-aged procedures than a truly even-aged stand with its distinct species stratification (Marquis 1981d).

Advance regeneration of other less tolerant, highly-preferred browse species is more difficult to obtain. The shade of a fairly heavy shelterwood overstory helps to encourage intermediate species like red maple to increase in size and abundance relative to cherry in some stands. Fairly dense canopies following the seed cut, plus retention of that canopy for an extended period, will favor a shift from pure intolerant reproduction to a mixture of intolerant and less tolerant species (Marquis 1988; Horsley 1982).

Deer fencing also shifts the balance toward highly preferred browse species (Marquis and Brenneman 1981; Marquis 1974). For maximum benefit, the fencing should be erected at the time of the shelterwood seed cut rather than at the time of final harvest. In areas of heavy deer browsing and limited seed production of desired species, direct protection is often the only way to secure reproduction of desired species.

Intermediate treatments

The highly stratified nature of most even-aged Allegheny hardwood stands makes even-aged culture difficult. Both the application of intermediate thinnings, and the determination of stand maturity are much more complex than in single-species or non-stratified stands.

In theory, even-aged stands have a bell-shaped diameter distribution, and intermediate thinnings are designed to cut more heavily from the smaller sizes. This increases average stand diameter, and gives the largest and best trees more room to grow, so that rotations are reduced.

However, stratified, even-aged Allegheny hardwood stands have an inverse-J diameter distribution, which is usually associated with all-aged stands. This is true for all three types of Allegheny hardwood stands discussed here, including truly even-aged ones (Fig. 6). Early thinnings that cut more heavily from below remove mostly unmerchantable tolerant saplings. On the other hand, if the saplings are ignored and the cut taken entirely from the merchantable sizes, the thinning tends to be a high-grading that removes primarily the faster-growing intolerant species. Thus, the sizes of trees removed may have much more impact on ultimate stand yield than more traditional density or spacing considerations (Marquis 1986).

Another problem with even-age culture is that the species in the several crown strata mature at markedly different times. The dominant cherry and yellow-poplar reach financial maturity in about 80 years, whereas the lower strata sugar maple and beech may require 120 to 150 years to reach maturity. This makes it difficult to determine when the stand as a whole is mature. Numerous rotation strategies must also be evaluated, from harvesting and regenerating when the dominant strata reaches maturity, to managing each strata to its individual maturity.

Recommended even-age cultural practices. No single thinning or rotation strategy is appropriate to all Allegheny hardwood stands. The degree of species and age stratification and the overall species composition determine which strategies will yield maximum volumes or values.

In stands with a high degree of stratification and a high proportion of intolerant cherry, the best strategy is to manage for the cherry alone. Thinnings should attempt to bring the cherry to maturity as quickly as possible, and the entire stand should be harvested and a new one regenerated when the cherry reaches maturity. In these stands, most of the value is in the cherry component, and the heavily suppressed tolerant strata will not reach maturity for 30 to 60 years after the cherry has matured. A second crop of cherry can be grown in not much longer than is required to bring the tolerant strata to maturity, and the cherry is much more valuable.

However, in stands where stratification is less pronounced, and where there is less of a cherry component, it is usually more profitable to manage for an extended rotation, removing each strata as it achieves its individual maturity (Smith 1986). Some, but not all, of the cherry is harvested when it reaches financial maturity, but stand management is continued for up to an additional 30 years to bring more tolerant species from the lower strata to their own maturity. Stands of this type have enough of the more tolerant species to provide a fully-stocked stand after the cherry is removed, and the more tolerant species are not as far behind the cherry since they have been represented in the dominant crown strata. This practice requires care to maintain cherry seed sources, and/or cherry advance regeneration to the end of the extended rotation, so that cherry can be regenerated at that time.

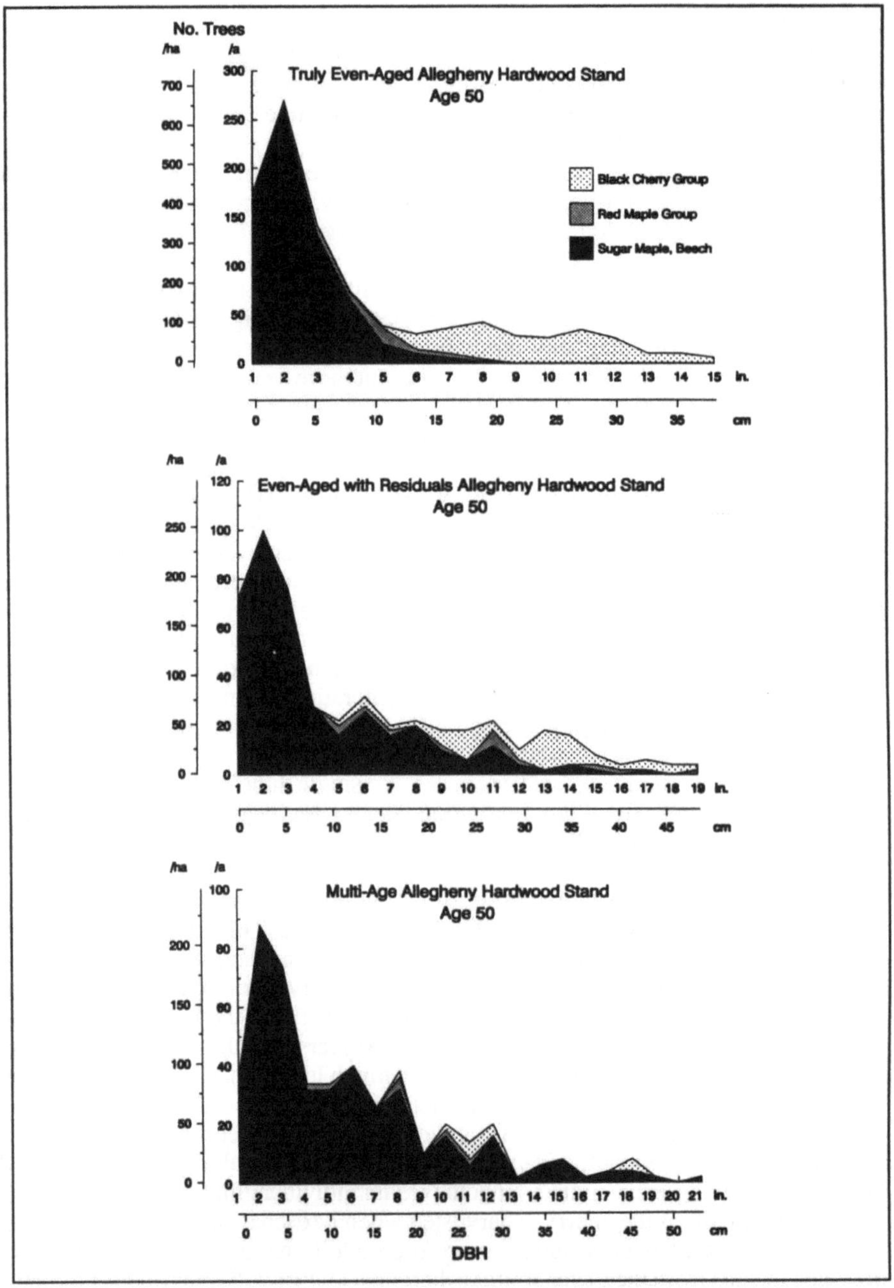

Fig. 6. Diameter distributions for three 50-year-old Allegheny hardwood stands, as a function of age arrangement.

All thinnings in stratified stands must pay special attention to stand structure, as well as to density, quality, and spacing. Again, there are several variations of thinning appropriate to stands of different types (Marquis 1986).

In young stands with extreme stratification (with more than about 4.6 sq. m/ha (20 sq. ft./acre) of basal area in the suppressed sapling class), it is rarely possible to make a commercial thinning without high-grading the stand. It is generally recommended that thinnings be deferred in such stands until natural mortality has reduced the suppressed strata to a more appropriate level. If early thinnings are considered necessary to accelerate growth of the best stems, some non-commercial reduction in the sapling class is also needed.

In intermediate aged stands, commercial thinnings will achieve all the usual objectives of thinning if the cut is concentrated in the smaller diameters of the merchantable size classes. Stand specific guidelines are available to regulate density, structure, and quality in such thinnings (Marquis and others 1984).

Toward the end of the rotation, most stands with moderate stratification and moderate to low proportions of cherry benefit from a combined thin-harvest cutting. This cutting harvests some of the already mature upper strata, while continuing to thin from below in the more tolerant species of the lower strata. Again, specific guidelines on structure, density, and quality are available (Marquis and others 1984).

Summary

Knowledge of the history and origin of present stands, and of the silvical characteristics of the species are important to the understanding of stand development patterns in Allegheny hardwood stands. Such knowledge and understanding can provide a sound basis for the development of appropriate silvicultural practices in this forest type.

Literature cited

Elliot, H. E. 1927. What follows pulp and chemical wood cuttings in northern Pennsylvania? Pa. Dept. For. and Waters Bull. 43. 7 p

Gabriel, W. J. and Walters, R. S. In press. Striped maple. pp. 44-49. In: Silvics of forest tree species in the United States. USDA Forest Serv., Washington, DC. Agric. Handb. 654.

Grisez, T. J. and Peace, M. R. 1973. Requirements for advance reproduction in Allegheny hardwoods--an interim guide. USDA Forest Serv., Northeast. Forest Exp. Sta., Upper Darby, PA. Res. Note NE-180. 5 p.

Horsley, S. B. 1982. Development of reproduction in Allegheny hardwood stands after herbicide-clearcuts and herbicide-shelterwood cuts. USDA Forest Serv., Northeast. Forest Exp. Sta., Broomall, PA. Res. Note NE-308. 4 p.

Hough, A. F. and Forbes, R. D. 1943. The ecology and silvics of forests in the high plateau of Pennsylvania. Ecol. Monogr. 13:299-320.

Illick, J. S. and Frontz, L. 1928. The beech-birch-maple forest type in Pennsylvania. Pa. Dept. For. and Waters Bull. 30. 14 p.

Marquis, D. A. 1974. The impact of deer browsing in the Allegheny hardwood regeneration. USDA Forest Serv., Northeast. Forest Exp. Sta., Upper Darby, PA. Res. Pap. NE-308. 8 p.

Marquis, D. A. 1975a. The Allegheny hardwood forests of Pennsylvania. USDA Forest Serv., Northeast. Forest Exp. Sta., Upper Darby, PA. Gen. Tech. Rep. NE-15. 32 p.

Marquis, D. A. 1975b. Seed storage and germination under northern hardwood forests. Can. J. For. Res. 5:478-484.

Marquis, D. A. 1978. The effect of environmental factors on the natural regeneration of cherry-ash-maple forests in the Allegheny Plateau region of the eastern United States. pp. 90-99. In: Proc., Establishment and treatment of high-quality hardwood forests in the temperate climatic region held September 11-15, 1978 at Nancy-Champenoux, France. IUFRO, Div. 1, Fevillus Precieux.

Marquis, D. A. and Grisez, T. J. 1978. The effect of deer exclosures on the recovery of vegetation in failed clearcuts on the Allegheny Plateau. USDA Forest Serv., Northeast. Forest Exp. Sta., Broomall, PA. Res. Note NE-270. 5 p.

Marquis, D. A. 1979a. Ecological aspects of shelterwood cutting. pp. 40-56. In: Proc., National Silviculture Workshop. Workshop held September 17-21, 1979 at Charleston, SC.

Marquis, D. A. 1979b. Shelterwood cutting in Allegheny hardwoods. J. For. 77:140-144.

Marquis, D. A. 1980. Black cherry-maple. pp. 32-33. In: Eyre, F. H. (ed). Forest cover types of United States and Canada. Society of American Foresters, Washington, DC.

Marquis, D.A. 1981a. Effect of deer browsing on timber production in Allegheny hardwood forests of northwestern Pennsylvania. USDA Forest Serv., Northeast. Forest Exp. Sta., Broomall, PA. Res. Pap. NE-475. 10 p.

Marquis, D. A. 1981b. Even-age development and management of mixed hardwood stands: Allegheny hardwoods. pp. 213-226. In: Proc., National Silviculture Workshop on Hardwood Management held Roanoke, VA. USDA For. Serv., Washington, DC.

Marquis, D. A. 1981c. Removal or retention of unmerchantable saplings in Allegheny hardwoods: Effect on regeneration after clearcutting. J. For. 79:280-283.

Marquis, D. A. 1981d. Survival, growth, and quality of residual trees following clearcutting in Allegheny hardwood forests. USDA Forest Serv., Northeast. Forest Exp. Sta., Broomall, PA. Res. Pap. NE-477. 7 p.

Marquis, D. a. and Brenneman, R. 1981. The impact of deer on forest vegetation in Pennsylvania. USDA Forest Serv., Northeast. For. Exp. Sta., Broomall, PA. Gen. Tech. Rep. NE-65. 7 p.

Marquis, D. A. and Bjorkbom, J. C. 1982. Guidelines for evaluating regeneration before and after clearcutting Allegheny hardwoods. USDA Forest Serv., Northeast. Forest Exp. Sta., Broomall, PA. Res. Note NE-307. 4 p.

Marquis, D. A. 1983. Regeneration of black cherry in the Alleghenies. pp. 106-119. In: Proc. 11th annual hardwood symp., Hardwood Research Council: The hardwood resource and its utilization: where are we going? Symposium held May 10-13, 1983 at Cashiers, NC. USDA Forest Serv., South. Forest Exp. Sta., New Orleans, LA.

Marquis, D. A., Ernst, R. L. and Stout, S. L. 1984. Prescribing silvicultural techniques in hardwood stands in the Alleghenies. USDA Forest Serv., Northeast. Forest Exp. Sta., Broomall, PA. Gen. Tech. Rep. NE-96. 90 p.

Marquis, D. A. 1986. Thinning Allegheny hardwood pole and small sawtimber stands. pp. 68-84. In: Smith, H. C. and Eye, M. C. (eds). Guidelines for managing immature Appalachian hardwoods. Proceedings held on May 18-30, 1986 in Morgantown, WV. West Virginia University, Morgantown, WV.

Marquis, D.A. 1988. Guidelines for regenerating cherry-maple stands. In: Smith, H. C., Perkey, A. W. and Kidd, W. E., Jr. (eds). Guidelines for regenerating Appalachian hardwood stands. Workshop held May 24-26, 1988 in Morgantown, WV. West Virginia Univ. and USDA Forest Serv., Northeast. For. Exp. Sta., Morgantown, WV. 293 p.

Marquis, D. A. In press. Black cherry. pp. 551-560. In: Silvics of forest tree species in the United States. USDA Forest Serv., Washington, DC. Agric. Handb. 654.

Oliver, C. D. 1981. Forest development in North America following major disturbances. For. Ecol. Manage. 3:169-172.

Smith, D. M. 1986. The practice of silviculture. John Wiley and Sons, Eighth Edition, New York, NY. 527 p.

Tilghman, N. G. 1989. Impacts of white-tailed deer on forest regeneration in northwestern Pennsylvania. J. Wildl. Manage. 53(3):524-532.

Tubbs, C. H. and Houston, D. B. In press. American beech. pp. 299-305. In: Silvics of forest trees in the United States. USDA Forest Serv., Washington, DC. Agric. Handb. 654.

USDA Forest Serv. In press. Silvics of forest trees in the United States. USDA Forest Serv., Washington, DC. Agric. Handb. 654.

Walters, R. S. and Yawney, H. W. In press. Red maple. pp. 56-59. In: Silvics of forest trees in the United States. USDA Forest Serv., Washington, DC. Agric. Handb. 654.

Scientific names of trees mentioned

Common Name	Scientific Name
white pine	*Pinus strobus* L.
eastern hemlock	*Tsuga canadensis* (L.) Carr.
black cherry	*Prunus serotina* Ehrh.
sugar maple	*Acer saccharum* Marsh
red maple	*Acer rubrum* L.
striped maple	*Acer pensylvanicum* L.
American beech	*Fagus grandifolia* Ehrh.
yellow birch	*Betula allegheniensis* Britton
sweet birch	*Betula lenta* L.
white ash	*Fraxinus americana* L.
yellow-poplar	*Liriodendron tulipifera* L.
cucumbertree	*Magnolia acuminata* L.
oak	*Quercus* spp.
chestnut	*Castanea dentata* (Marsh) Borkh.

Experiments in mixed mountain forests in Bavaria

11

PETER BURSCHEL, HANY EL KATEB,
and REINHARD MOSANDL

Introduction

The maintenance of the mixed mountain forests of the Bavarian Alps has been the goal of forestry in this region for the past 150 years. The main tree species are spruce (*Picea abies*), fir (*Abies alba*), beech (*Fagus sylvatica*), and maple (*Acer pseudoplatanus*). *Taxus baccata*, *Sorbus* spp., *Ulmus glabra*, *Fraxinus excelsior* and some others are of minor importance.

This type of forest:

- in many respects resembles the natural situation;
- is considered the best for resisting such hazards of mountainous areas as storms, snow, and insects;
- represents the most effective type of forest for protection of steep slopes against landslides, avalanches, and erosion;
- meets all aesthetic expectations of tourists as well as of the local population; and
- is of high productivity for timber on many sites.

However, the findings of several large- and medium-scale inventories conducted in the 1970's clearly indicate that these mountain forests are rapidly losing their species mixture (Burschel et al. 1977, Schreyer and Rausch 1978). Fir, maple, and all minor species—and to a lesser extent beech—are already disappearing in younger stands. Consequently, the Chair of Silviculture and Forest Management of the University of Munich initiated comprehensive studies in order to investigate the dynamics of this type of forest as a means to detect the causes of the undesirable reduction of tree species diversity. Some of the more important findings of these investigations are presented here.

The experiment

A permanent research project was initiated in fall 1976. The results discussed here are mainly from the first decade of observation (1976 to 1986). The experiment consists of 25 main plots, each of 0.5 ha (71 m x 71 m). Its geographical location is shown in Figure 1. Different silvicultural regeneration treatments were applied, and subplots were fenced to prevent deer and cattle from entering. Table 1 contains some basic information of site characteristics, stand properties, and silvicultural measures applied. Within each main plot most measurements were taken in a central area of 0.1 ha (33 m x 33 m). A brief outline of measurements and observations taken is presented in Table 2.

M. J. Kelty (ed.), The Ecology and Silviculture of Mixed-Species Forests, 183–215.

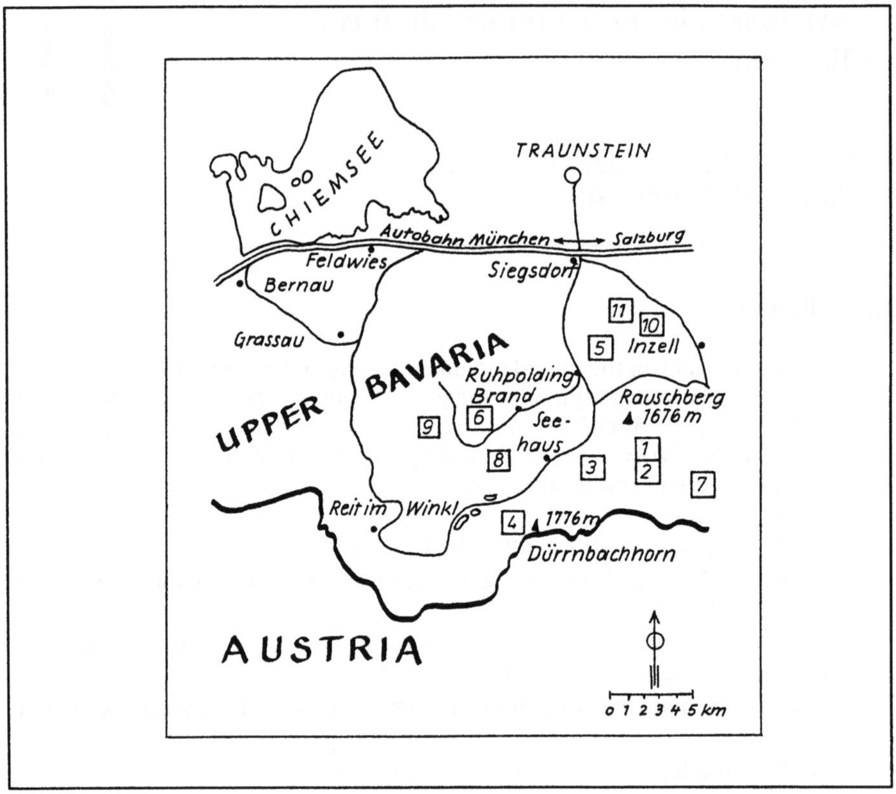

Fig. 1. Location of research area in the eastern Bavarian Alps.

Results

Ecological consequences of silvicultural measures

The opening up of the canopy of an old stand by silvicultural means has several effects:

- it changes the bioclimatological conditions near the ground;
- it accelerates the mineralization of litter and influences the quality of humus;
- it improves the photosynthetic potential of the remaining old trees, regeneration and ground vegetation;
- it increases the stand's aerodynamic roughness, which augments storm risk.

The results concerning the first two aspects will be presented in the following subchapter, while the last two have been used for the interpretation of the findings referring to the regeneration process.

Table 1. Some characteristics of experimental plots (the figures indicate the great natural lack of homogeneity of what is called mixed mountain forest. Control: no intervention; light shelterwood cutting: reduction of b.a. by 30%; heavy shelterwood cutting: reduction of b.a. by 50%; gap cutting: circular opening of 30 m diameter; clear cutting: elimination of whole stand by one cutting operation; pasture forest: opened by grazing long ago, leaving a dense cover of grassy ground vegetation.
S = spruce; F = fir; B = beech; M = maple; O = other species.

Plot No.	Site	Soil type	Elevation a.m.s.l.	Slope(°), direction	Silviculture	Age years	Canopy density %	Basal area, m^2	Species, % b.a. S	F	B	M	O
1.0	Hauptdolomit (Nor period of alpine Trias and corresponding slope deposites)	rendzina, loam-rendzina, terra-fusca, brown soil-terra-fusca; humus : mull; pH : 4.6-7.6	900 (lower slope)	22	control	106	68	45	33	32	19	8	8
1.1				22	light shelterwood	106	56	33	44	34	14	5	3
1.2				26 NW	heavy shelterwood	144	49	27	31	35	24	10	0
1.3				28	clear cutting	110	0	0	0	0	0	0	0
1.4				23	gap cutting	114	-	-	31	23	42	3	1
2.0			950 (upper slope)	30	control	96	76	40	43	16	40	0	1
2.1				27	light shelterwood	96	60	30	48	29	14	2	7
2.2				21 NW	heavy shelterwood	113	39	20	14	66	15	5	0
2.3				27	clear cutting	103	0	0	0	0	0	0	0
2.4				31	gap cutting	107	-	-	49	16	35	0	0
3.0			800	12 NW	control	104	80	50	80	11	9	0	0
3.2				18	heavy shelterwood	104	42	21	55	30	11	4	0
4.0			1230	28 NW	control	120	80	45	24	15	34	1	26
4.2				29	heavy shelterwood	270	43	20	54	19	19	2	6
5.0			900	32 N	control	103	80	56	79	7	14	0	0
5.2				31	heavy shelterwood	103	45	27	58	11	13	8	10
6.0			900	21 S	control	170	86	49	23	30	34	12	1
6.2				25	heavy shelterwood	170	51	23	42	32	17	9	0
7.2			900	18 NW	pasture forest	135	51	28	58	37	2	0	3
8.4			1250	24 W	gap cutting	125	-	41	98	1	1	0	0
9.4			1200	23 S	gap cutting	120	-	32	90	5	5	0	0
10.0	Flysch	brown soils from Flysch-sand-stone	810	18 N	control	140	83	50	77	17	6	0	0
10.2				23	heavy shelterwood	140	54	29	78	20	2	0	0
11.0			980	13 N	control	150	87	65	56	44	0	0	0
11.2				18	heavy shelterwood	150	58	33	45	54	1	0	0

Table 2. Measurements made on experimental plots.

Soil:	soil type, pH, type of humus, plant associations.
Stand characteristics:	diameter, height, crown height, crown projection, canopy density (periscope), age (from stumps after cutting, increment bores on strips surrounding the central plot).
Bioclimate:	radiation, relative illumination (RI), air temperature, air humidity, precipitation, behavior of snow (some plots, some periods).
Seed and litter production:	seed (litter) traps (0.25 m^2), 30 per plot, for the first two years on all plots, for the whole period of 10 years on 3 and for varying periods on several other plots.
Ground vegetation:	estimation of area coverage, determination of biomass and species composition.
Natural regeneration:	determination of seedling number and heights on 96 sampling units of one square meter each on every plot (a total of 2400), marking of every seedling according to species and year of germination. Determination of biomass for a represen tative number of individuals.
Artificial regeneration:	On every plot seedlings of all important species were planted. Results of this part of the experiment are not presented here.
Browsing, grazing:	On each plot 2/3 of area was fenced while 1/3 remained unfenced. In addition a special study was carried out to distinguish between browsing and trampling by deer and cattle.
Rodents:	Trapping, feeding trials, exclusion trials.
Fungi:	Determination of damage and species.

Change of bioclimatological conditions by different silvicultural measures. A change in stand density leads to change in light conditions near the ground. The relative illumination (RI) is a good measure of many ecological effects caused by silvicultural treatments in old stands. As expected, an acceptable relationship was found between the conventional indicators of stand density like basal areal or volume and the relative illumination. By applying a stand density measure (BD) which includes both stem number and basal area, we were able to compensate for the large structural variation in stands of mixed mountain forests, and were able to quantify this light-density relationship statistically to a very reliable level (Figure 2). For practical reasons, however, canopy density (CD) was used frequently as a substitute indicator of the ecological situation on a given plot.

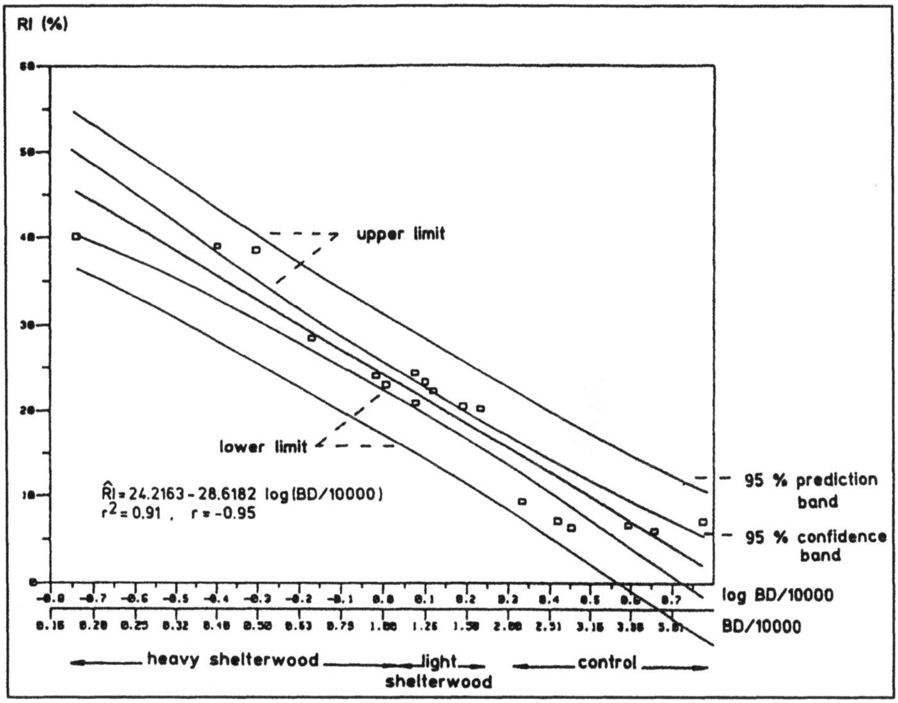

Fig. 2. Relationship between relative illumination (RI) and stand density (BD = n^2/b.a.) where n = no. of trees per hectare and b.a. = basal area per hectare (El Kateb 1990).

The silvicultural measures applied to old stands by more or less drastically reducing their leaf area exert a strong influence on the interception of precipitation. The corresponding reaction of the stands studied can be derived from Figure 3. While light interventions only have small effects, heavyshelterwood cutting may already increase throughfall by about 100 mm, or 14%. The maximum increase was found on the clear cut, amounting to about 25%.

Since the average precipitation in the research area is fairly high, and dry periods are not at all frequent, such gains of water are not of too much ecological importance. However, it is important to mention that the considerably larger amount of additional rain reaching the ground on the relatively small clear cut areas (0.5 ha) did not visibly lead to run off and erosion.

Findings of the kind presented here must be interpreted very differently in areas where precipitation is less abundant and soil conditions might be not as favourable.

Biological processes in mountainous areas are strongly influenced by snow conditions during winter. They are determined by two factors:

- snow interception of stands decreases with increasing openness;
- wind pressure as well as thawing and refreezing under open conditions increase the density of the snow.

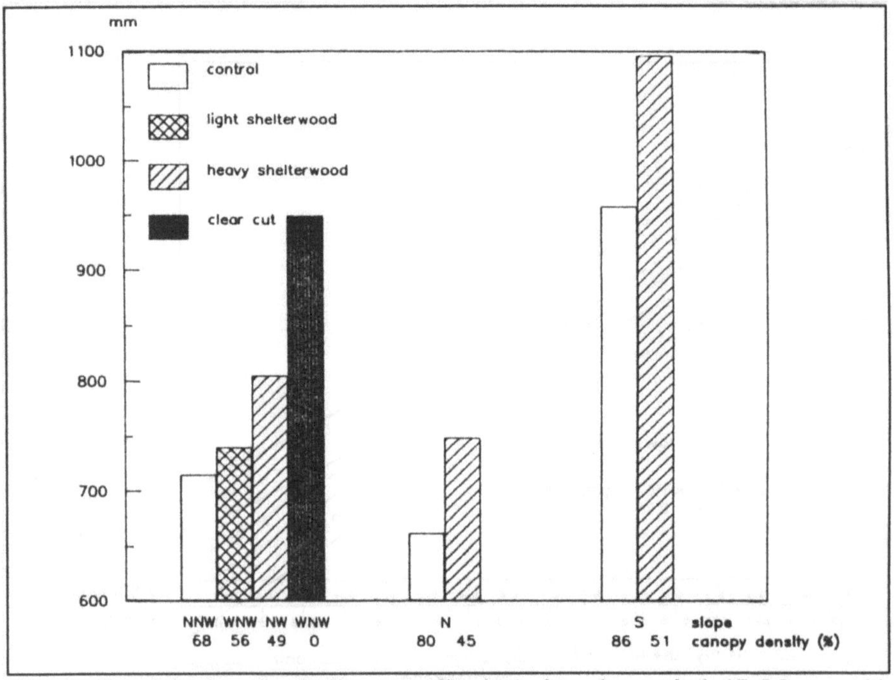

Fig. 3. Precipitation (excluding stemflow) during the period 17 May to 11 October 1977 after different silvicultural treatments in 3 stands of mixed mountain forests (Mayer 1979).

As a consequence of both effects, the snow cover on gaps lasts almost a month longer than under dense forest (Table 3). This effect is even more pronounced on clear cut areas. Such prolongation of snow cover can have two detrimental consequences: the danger of snow movements which can damage young plants increases with weight and duration of snow layer and the attack of *Herpotrichia nigra*, a fungus which develops under snow, may severely damage young spruce plants. Also, the time period decreases in which young plants may photosynthesize. The higher the snow cover and the longer it lasts, the more pronounced are the two risks mentioned. In the altitudinal range of our research area, considerable snow movement on gaps and clear cuts could be proved by corresponding measurements but detrimental effects on young plants were insignificant. The fungus mentioned did not play an important role during the whole observation period. It is, however, a very important factor where high altitude forests or reforestation above actual timber line are concerned.

Table 3. Duration of snow cover and average snow density as influenced by silvicultural treatments during winter 1979/80 (Bertold 1980).

Silviculture	Control	Shelterwood Cutting		Gap Cutting
		light	heavy	
Snow cover > 3 cm, days	143	149	158	172
Snow density, g/cm^3	0.13	0.14	0.15	0.18

The quality of humus. The condition of the humus layer was studied four years after initiation of the experiment (Schörry 1980). It could be proved that with decreasing density of canopy, the frequency of humus types "mull" and "F-mull" increased while "moder" and "mulltype moder" decreased correspondingly (Table 4).

Table 4. Area percentage of different humus types found four years after initiation of the experiment under different silvicultural conditions (Schörry 1980).

Silviculture	Type of humus	
	mull, F-mull	moder mulltype moder
control	59	41
light shelterwood	75	25
heavy shelterwood	81	19

More light, higher temperatures, and greater humidity in the light and heavy shelterwoods apparently have increased the rate of decomposition of the accumulated dead organic matter, which was present on about 41% of the surface area on control plots. As a consequence, more than 50% of the less favourable types of humus disappeared, lifting the proportion of mull to 81%. Though this effect is not of too much ecological importance on sites as good as the ones of the research area, the type of silvicultural system may become of great significance when applied on less favorable soils.

The response of old stands to different silvicultural measures

The response of old stands to regeneration cuttings of different intensities will be shown for the basal area increment of individual trees, volume increment of stands, and mortality rates. As can be seen in Figure 4, the basal area increment of individual spruce, fir, and beech is largely diameter dependent; the bigger the trees, the more increment. This fundamental relationship holds true under very different density conditions, but attains considerably higher increment levels with increasing openness of the stand. The most spectacular effect is achieved with beech, a tree species well known for flexibility to react to any improvement of living conditions.

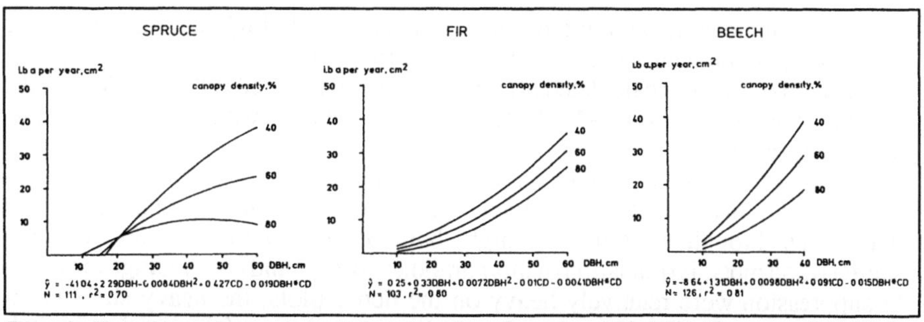

Fig. 4. Basal area increment (i.b.a./year) as function of initial diameter (DBH) for spruce, fir, and beech for three different canopy densities: 40%-heavy shelterwood cutting, 60%-light shelterwood cutting, 80%-control (Mosandl 1990).

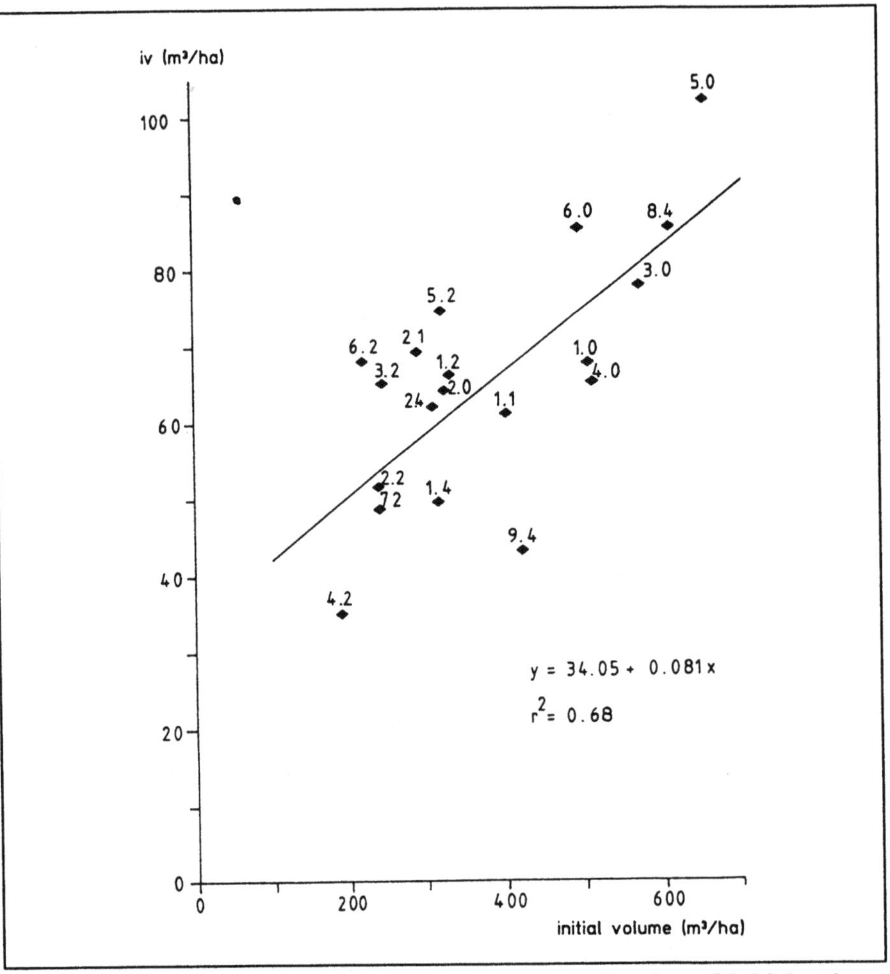

Fig. 5. Stand volume increment over ten years as function of initial volume (Mosandl 1990).

The volume increment of the stands is unexpectedly high and reflects closely the initial stocking density, as can be derived from Figure 5. The increment for a period of 10 years was calculated after the Swiss "check method": iV = V at end of period - V at start of period + V of dead trees (i.e., all trees which died accidentally or by suppression).

To determine the dynamics on the plots representing different situations, the "natural mortality" has to be examined separately as given in Figure 6. It indicates that losses through mortality are high, both on the control and on the heavy shelterwood plots. Explanation of this paradox is simple and convincing: losses due to suppression were relatively heavy on the dense plots; the heavy shelterwood cutting, on the other hand, decreased the collective stability of the stand and left the remaining trees individually not resistant enough against windstorms. In addition, such silvicultural measures augment the aerodynamical roughness of the canopy which may lead to increase of wind velocity locally. Both effects led to similar results. On most control plots, as well as on some heavy shelterwood plots, these types of losses compensated the increment almost completely. Moderate opening up of stands had the advantage of stopping losses by suppression without increasing storm risk too much.

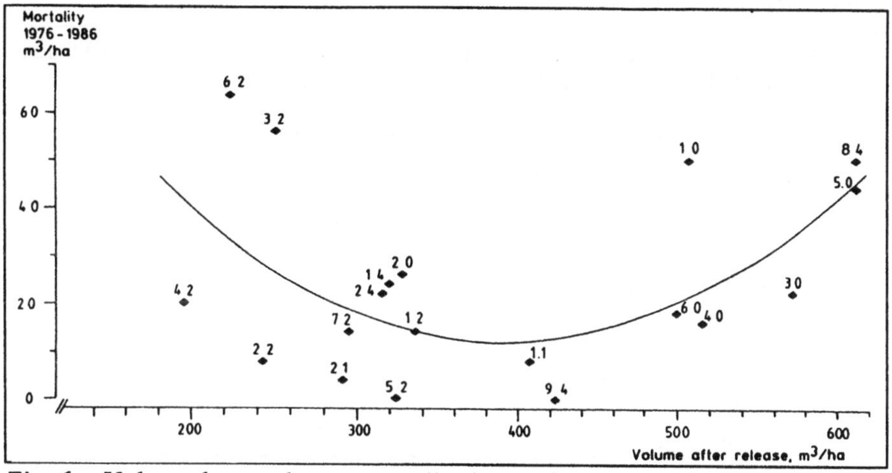

Fig. 6. Volume losses due to mortality in relationship to initial stand density (Mosandl 1990). Numbers = plot numbers (details given in Table 1).

Litter and seed production

Litter. The amount of tree litter produced in a given stand of the forest type investigated can be derived from two independent variables: the maximum production of the control stand (untreated) and the degree of opening produced by silvicultural treatments. The corresponding function is linear and statistically significant, as seen in Figure 7.

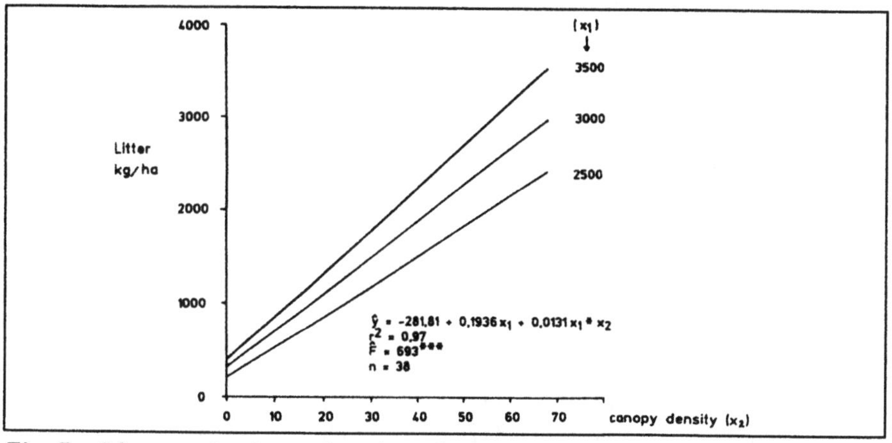

Fig. 7. Litter production as function of canopy density (x_2) and amount produced in a completely dense stand (x_1) (Mosandl 1990).

Seeds. The main purpose of litter trapping in this research project was to quantify the process of fructification. Although very time consuming and expensive, the results were worthwhile. Table 5 shows that all tree species produced large numbers of seeds, with spruce, fir, and maple shedding at least some every year. Beech on the other hand was the least effective seed producer and only fructified periodically at long intervals.

Table 5. Seed production (thousand per ha) on a control plot (68% canopy density) during the observation period (Veltsistas 1980, Mosandl 1990).

Year	Spruce	Fir	Beech	Maple
77	251	681	8	37
78	2600	51	529	471
79	49	8	1	20
80	17	173	0	69
81	588	1029	0	63
82	20	1	0	419
83	557	113	4	29
84	33	52	3	168
85	19	51	60	393
86	12	19	0	120

The production of about 1285 viable seeds per square meter for spruce, 697 for fir, 555 for maple, and 254 even for beech over a 10-year period should not cause any regeneration problem due to lack of seeds. In fact, the reverse is true--the mountain forests of the Bavarian Alps are sufficiently fructifying to ensure dense regeneration.

The influence of the canopy density on seed production can be derived from Table 6. There is a decrease in seed production with a decrease in canopy density of the residual stand, of course reaching a minimum on the clear cut. This relationship is least pronounced with maple, a species which is able to cover large areas almost uniformly with their aerodynamically efficient seeds, even if there are very few mother trees. The reverse is true for beech because their nuts can only be found below or near the crowns of the mother trees.

Table 6. Seeds (thousand per ha) on differently treated plots, observation period: 1976/77 to 1980/81 (Veltsistas 1980, Mosandl 1990).

Species	canopy density			
	68% control	56% light shelterwood cut	39% heavy shelterwood cut	0% clear cut
spruce	3505	3883	2145	356
fir	1942	747	970	91
beech	538	312	116	3
maple	660	492	547	107

The data presented in Tables 5 and 6 give an impression of the seed supply for a given stand. In order to evaluate the possible influence of the silvicultural measures applied on the fructification potential of a given species, the seed density was related to a unifying factor. This has been done for spruce in Figure 8. In good seed years (e.g., 1977/78), it is fairly evident that the improvement of the ecological situation of the crowns of the trees left on the shelterwood plots led to an increase of the seed production of these individuals. In less fertile years this relationship is less evident.

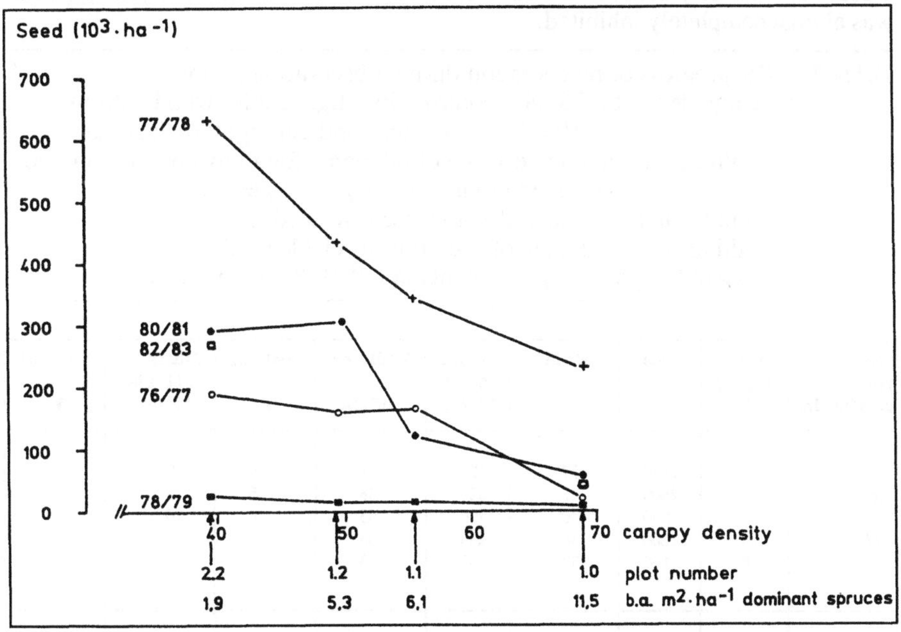

Fig. 8. Density of spruce seeds (thousand/ha) per square meter basal area of mother trees as a function of canopy density (Mosandl 1990).

Establishment of regeneration

From these studies, it was evident that seeds of all major species were present and plentiful. The next steps in the process of regeneration to be investigated were germination and establishment of new plants.

First of all, it can be stated that germination as well as establishment of regeneration took place on all 25 research plots in a very abundant manner. Plant densities ranged from a minimum of 5000 plants per ha on a very dark control plot (canopy density 80%), to a maximum of nearly 500,000 per ha on the heavy shelterwood with 39% canopy density. On all the plots and under all silvicultural situations mixed species regeneration developed, with the mixture normally more pronounced than in the old stand. The process of regeneration establishment will be discussed here in a more detailed manner for some typical situations.

Table 7 presents something like a "regeneration balance sheet" for some important silvicultural situations. For this balance sheet, the fate of all seedlings and young plants from one especially good seed year were followed. It is evident that the number of seedlings found in the first year after seed fall depends more on the silvicultural circumstances under which they germinated than on the number of viable seeds. Looking at the remaining plants at the end of the observation period, this important finding becomes even more evident. For spruce, fir, and maple, the survival rates generally increase with decreasing crown density, with the heavy shelterwood cutting offering the best conditions. On the other hand, where crown closure is extremely dense in the control stand, the establishment of young plants was almost completely inhibited.

Table 7. The progress of regeneration during observation period.
canopy densities: 76,68 = control; 56 = light shelterwood cutting; 39 = heavy shelterwood cutting; 0 = clear cut
seedlings: accumulated value of all plants found by counting in two-week intervals during year of germination
viability %: percentage of viable seeds to seeds
seedling %: percentage of seedlings to viable seeds
survival %: percentage of plants in fall 1986 to seedlings

species year of germination	canopy density %	seeds 10^3/ha	Viability %	viable seeds 10^3/ha	Seedlings 10^3/ha	seedling %	Plants fall 1986 10^3/ha	Survival %
	76	1670	37	625	-	-	0	-
spruce	68	2600	32	825	38	5	1	2
	56	2080	36	756	43	6	10	23
1978	39	1200	36	433	33	8	16	49
	0	160	37	60	3	4	<1	12
	76	80	26	21	-	-	1	-
fir	68	681	35	236	22	9	5	24
	56	179	31	55	18	34	8	41
1977	39	339	27	92	25	28	17	67
	0	29	36	11	2	15	1	56
	76	209	46	96	-	-	0	-
beech	68	529	42	224	135	60	23	17
	56	287	48	137	18	13	2	9
1978	39	109	52	57	29	50	1	5
	0	no seed supply on clear cuts						
	76	16	31	5	-	-	1	-
maple	68	471	35	163	90	55	15	17
	56	172	40	68	42	62	7	17
1978	39	231	44	101	68	68	24	35
	0	19	43	8	7	83	2	33

The situation on clear cut areas was different because the seed supply was scarce and competition of ground vegetation was severe. The data indicate another important peculiarity of the regeneration process in mixed stands: some species, in this case spruce, require a much higher number of seeds in order to achieve a density of plants comparable to other ones, such as fir, beech, and maple, as shown here.

The balance sheet of Table 7 compares only the situation at the beginning and end of the process. The course of events is presented for spruce and fir in Figure 9. In both cases, the main lossed occurred in the very first years, followed by a long period of merely small changes. Again, this figure depicts the role of the silvicultural measure applied for the survival of the new plants.

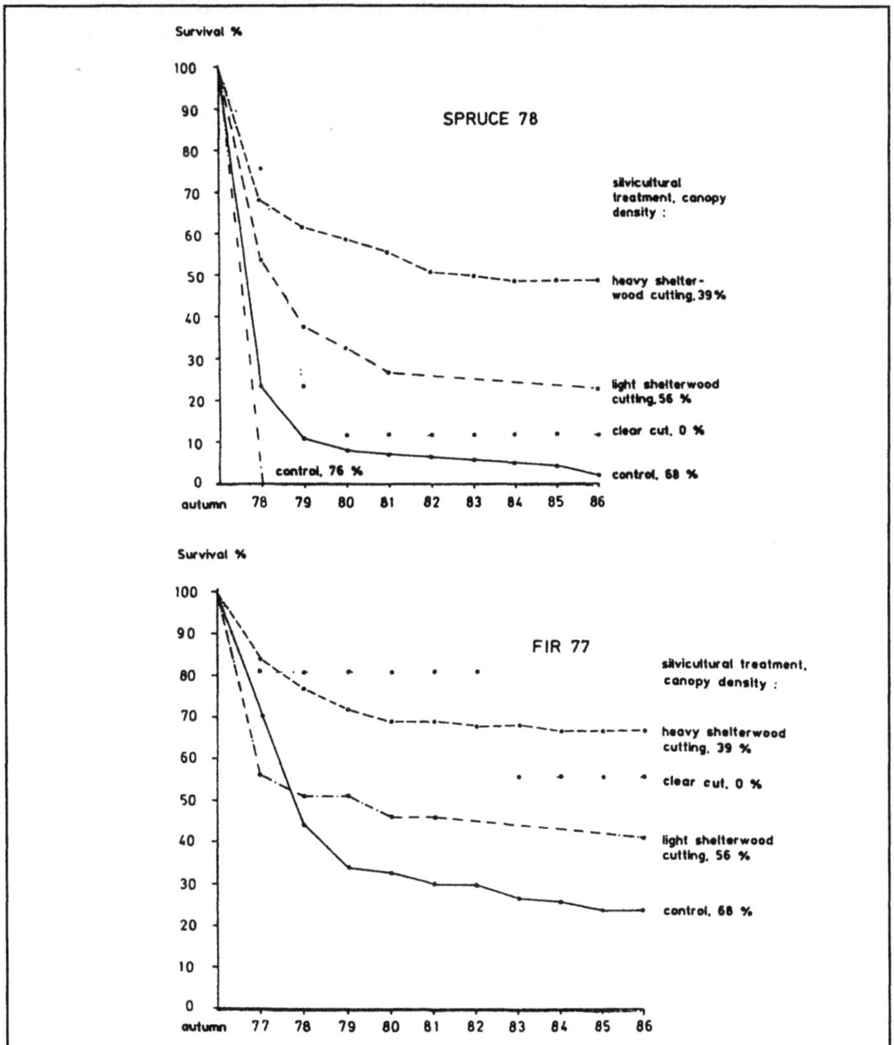

Fig. 9. Survival rates for given age classes of spruce and fir. Survival of 100% = total number of seedlings counted in year of germination (Mosandl and El Kateb 1988).

The different behavior of spruce and fir on the clear cut plots is mainly due to the fact that young spruces suffer more from competition for light by ground vegetation. The generally greater hardiness of fir is also reflected by the position of the survival lines within the coordinates. It is important to note, however, that for both species the ranking of the silvicultural measures remains the same, except in the case of clear cuts for the reasons already explained.

To this point, only individual seed years have been evaluated with respect to the regeneration process. In Table 8 the sum of events over the whole observation period is presented for two pairs of plots: a control and a heavy shelterwood plot on "Hauptdolomit" (plots 1.0 and 2.2), an important geological formation of the Bavarian Alps, and another pair on "Flysch" (plots 10.0 and 10.2), a basic rock weathering to very productive soils. The tremendous regenerative activity becomes evident from these data, which is typical for the life cycle of the mixed mountain forests investigated. For example, a total of 540×10^3 plants per hectare were counted over a 10-year period on the heavy shelterwood plot (2.2), of which 461×10^3 were still there at the end of the observation. The corresponding plot (10.2) on the Flysch site produced even more seedlings (898×10^3), of which considerably less (287×10^3) survived until the last counts were done. Again, the strong influence of canopy closure on seedling number becomes very evident (except for beech because of the scarcity of seed trees), and a generalization can be made: the same ecological conditions which seem to be favourable for the germination of seedlings also increased their survival chances. This in quantitative terms demonstrates the effectiveness of silvicultural measures applied in practical forestry.

Table 8. Sum of plants (10^3/ha) counted in fall of year of germination during observation period (brackets = plants as % of viable seeds) and number of plants found at the end of observation period, fall 1986 (brackets = survival rate in %). The observation period for plots 1.0 and 2.2 is 1977-1986, and for plots 10.0 and 10.2 is 1980-1986.

species	sum of plants in	HAUPTDOLOMIT		FLYSCH	
		control (1.0)	heavy shelter-wood (2.2)	control (10.0)	heavy shelter-wood (10.2)
spruce	year of germination	40 (3)	78 (9)	509 (10)	741 (14)
	fall 1986	8 (20)	53 (68)	130 (26)	202 (27)
fir	year of germination	62 (9)	100 (28)	220 (19)	155 (32)
	fall 1986	36 (58)	86 (86)	100 (45)	84 (54)
beech	year of germination	59 (23)	23 (10)	16* (26)	0* -
	fall 1986	26 (45)	21 (93)	12 (75)	0 -
maple	year of germination	119 (21)	339 (58)	4* (52)	2* (136)
	fall 1986	76 (64)	301 (89)	3 (83)	1 (67)

* numbers are low because of lack of seed trees and therefore statistically little meaningful.

From seed to seedling: the role of rodents and fungi

Rodents. It is well known that only a small proportion of seeds produced has a chance to develop into plants, and this could be shown for these plots (Tables 7 and 8). Large numbers of seedlings are destroyed by various agents. In some years and under certain ecological conditions, fungi may play the major part (Burschel et al. 1964); in others, rodents become the main factor. The only mast year during the observation period for all tree species, including beech, was 1977 (Table 5). This year was used to make a comprehensive study on the population development of mice, among which *Clethrionomys glareolus* and *Apodemus flavicollis* were the most frequent species. From systematic trapping experiments the relationship shown in Figure 10 could be derived.

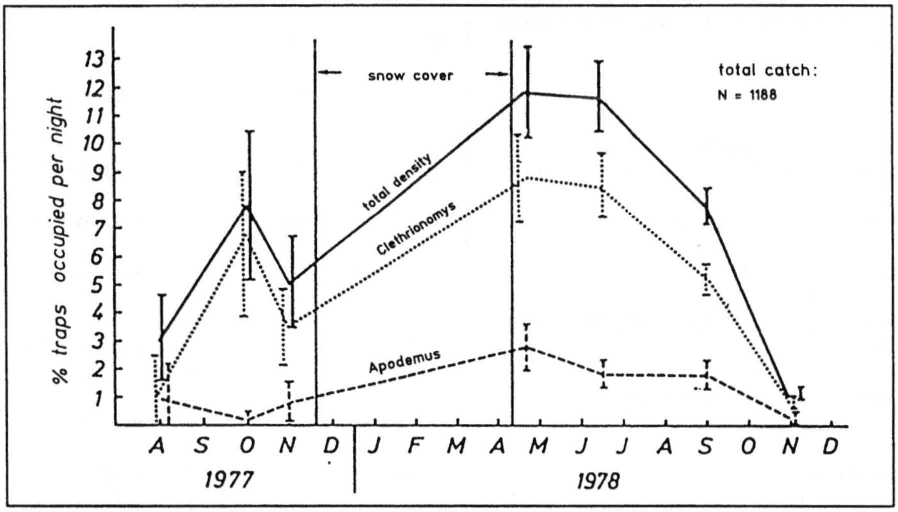

Fig. 10. Development of density of mice after seedfall 1977 (Bäumler and Hohenadl 1980).

Figure 10 indicates clearly that the mice population reacted strongly to the improvement of the food supply by falling seeds, among which beech nuts played the dominant role as far as weight and nutritive value are concerned. The living conditions were so favorable that both species were able to increase their population density to a maximum in winter time under the 5 months of snow cover. This density was maintained until summer 1978. It began to decrease when the seedlings became uneatable because of lignification. Already in late fall of 1978, the mice population had returned to the low initial density.

The role of mice in the regeneration process can be seen from Table 9. Seeds and seedlings in less abundant mast years can be almost completely consumed by rodents. On the other hand, it can be deduced that prolific fructifications with more than 20 to 30 viable beechnuts per m² will hardly be endangered by rodents. Such densities will be achieved easily in good seed years and in pure beech or beech-dominated mixed stands (Burschel 1966). However, it might become critical where beech trees are not frequent or the intervals between seed years are very large, as is the case near the upper limit of the species.

Table 9. Losses of beech nuts and of seedlings caused by *Clethrionomys glareolus* (Cl.) and *Apodemus flavicollis* (Ap.) after seed year 1977 (Bäumler and Hohenadl 1980). The data are based on field catches, feeding trials, and exclusion experiments.

Item	Cl.	Ap.	unit
average density of mice in research area	19	6	animals/ha
daily consumption of beech nuts (feeding trials)	30	24	beech nuts/mouse
proportion of beech nuts in stomach of catches	80	90	%
daily consumption	456	130	beech nuts/ha
additional consumption of lactating animals + nestlings (feeding trials)	96	44	beech nuts/ha
total daily consumption	552	174	beech nuts/ha
total population	726		beech nuts/ha
Approximation of total loss, Oct. 1977-Oct. 1978	182000		beech nuts/ha
Loss of seedlings from spring 1978 (exclusion trials)	13000		seedlings/ha
sum of losses	≈200000		beech nuts + seedlings/ha

For other tree species of the mixed mountain forests, strong evidence of mice eating both seeds and seedlings was found, but it was not possible to quantify observations, as in the case of beech.

Fungi. Fungi are another important agent causing losses of young plants. In order to distinguish attack by fungi as cause for the loss of seedlings as distinct from other mortality reasons (such as lack of illumination because of canopy density), a special trial was carried out with seedlings of spruce, fir, beech, and maple on the control and heavy shelterwood plots. Half of the seed beds were treated with fungicides while the other half remained untreated. The experiment was carried out during the 1982 growing season. The seedlings suffered heavy losses during the first months after germination. As can be seen from Figure 11, the losses are much more severe on the control plot with its dense crown cover than on the heavy shelterwood plot. The reduction of losses through application of fungicides indicates the degree to which fungi were interfering in the process. (The matrix for the ANOVA is shown in Table 10).

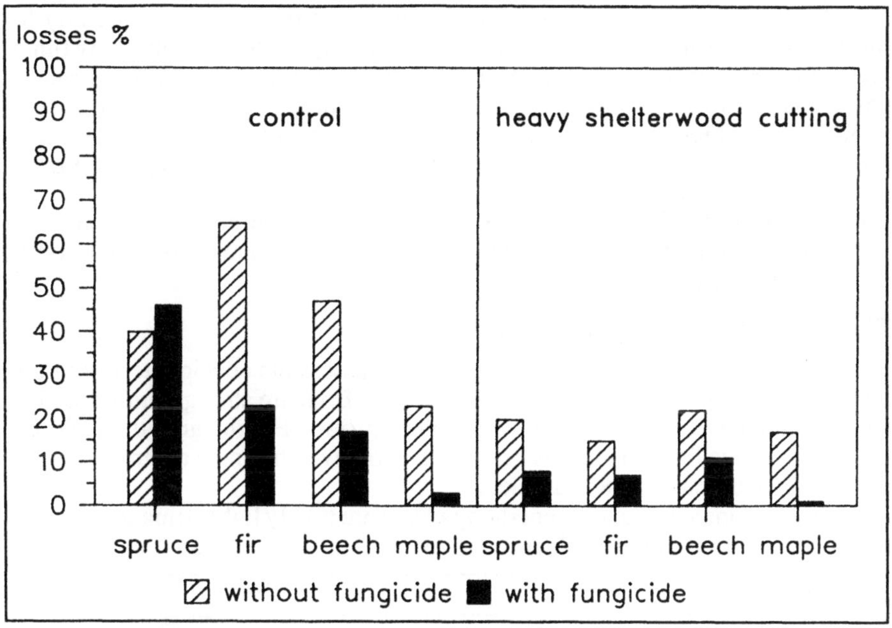

Fig. 11. Seedling losses (Mosandl and Aas, 1986).

Table 10. Result of ANOVA to test the influence of different factors on the rate of seedling losses (Mosandl and Aas 1986).

	crown density A	fungi-cide B	inter-action AxB	tree species C	inter action AxC
total analysis	**	**	n.s.	**	*
species-wise:					
spruce	**	n.s.	*		
fir	**	**	*		
beech	n.s.	n.s.	n.s.		
maple	n.s.	*	n.s.		

Crown density was the main factor for the losses of seedlings, especially for spruce and fir. Plants weakened by lack of light easily succumb to attack by fungi. These findings concerning the impact of mice and fungi on the survival rate of young plants explain at least in part the factors which determine the process of establishment of regeneration in forests. It is important, however, to mention that:

- there was no case in which rodents or fungi were able to interrupt the process of regeneration completely.
- the degree of illumination, as a consequence of the silvicultural measure applied, was more important for the fate of the plants than the fungal or rodent damaging agents.

The development of regeneration

Ecologically, the most important criteria to assess the development of young trees are height growth and increase of biomass. Both are strongly influenced by the silvicultural condition under which they grow. Height growth is closely dependent on the light conditions near the ground, which can be controlled by the type of silvicultural treatment applied.

Figure 12 indicates that regenerating plants are highly light-demanding. Even heavy shelterwood cutting, which removed 50% of the old stand, only slightly

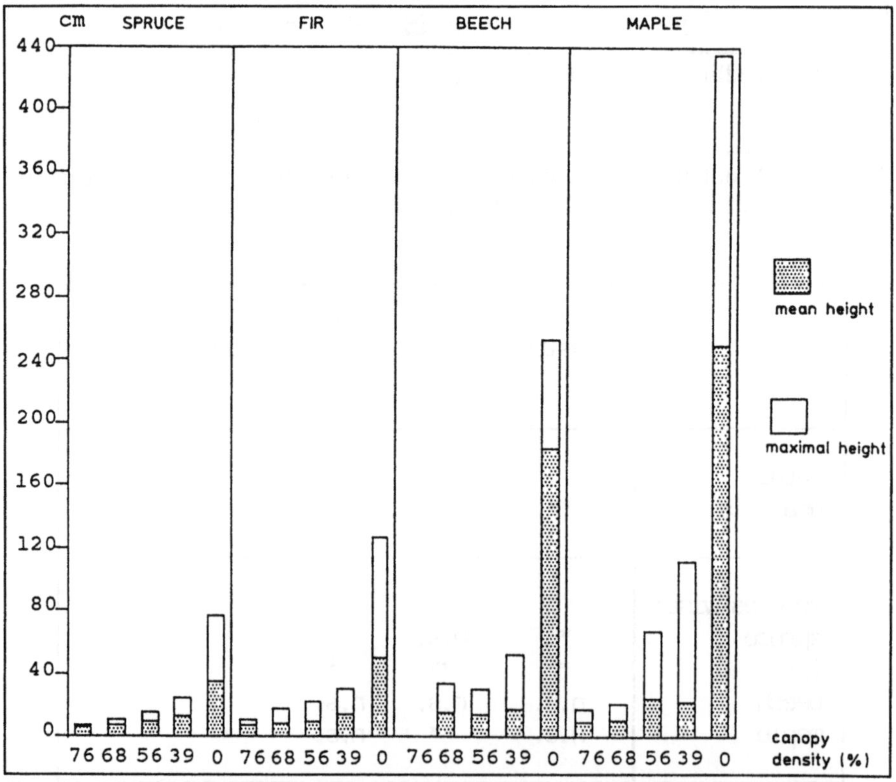

Fig. 12. Height of natural regeneration in stands of different canopy density in autumn 1986 (Mosandl and El Kateb 1988). Mean: average height of tallest plants of 32 sampling units; maximal height: tallest plant on 32 sampling units.

accelerated the height development of young plants. The situation changes completely on clear cut areas where full light was available. Here growth was strongly accelerated. This was most explicit for maple, but clearly visible for all species. The profuse development with full illumination could be expected, but the reasons for the clear differentiation among species requires explanation: on the clear cut plots the small young plants which were already established when felling operation took place, reacted immediately to the improvement of their living conditions. Each species grew as fast as its genetic potential allowed. As a consequence, the faster a species grew, the sooner it outcompeted slower ones. Since maple grew considerably faster, all conifers had to grow in an ecological situation in which a slight but evident light reduction was caused by the higher layer of maple. This situation represents a typical positive feed back: one species is growing slightly slower than another one, at least at the beginning. It will be overtopped by the faster one and as a consequence be relegated to the understory, if not eliminated eventually. In the case of the relatively light-permeable canopy of maple, however, it might well be that shade-tolerant fir may remain for decades in a suppressed stage but eventually become dominant when the accelerated early growth of young maples is followed by a rapid decrease in height growth, which is typical for the species. The statistically strong correlation that exists between availability of light and growth of young trees can be demonstrated very clearly even under field conditions (Figure 13); the

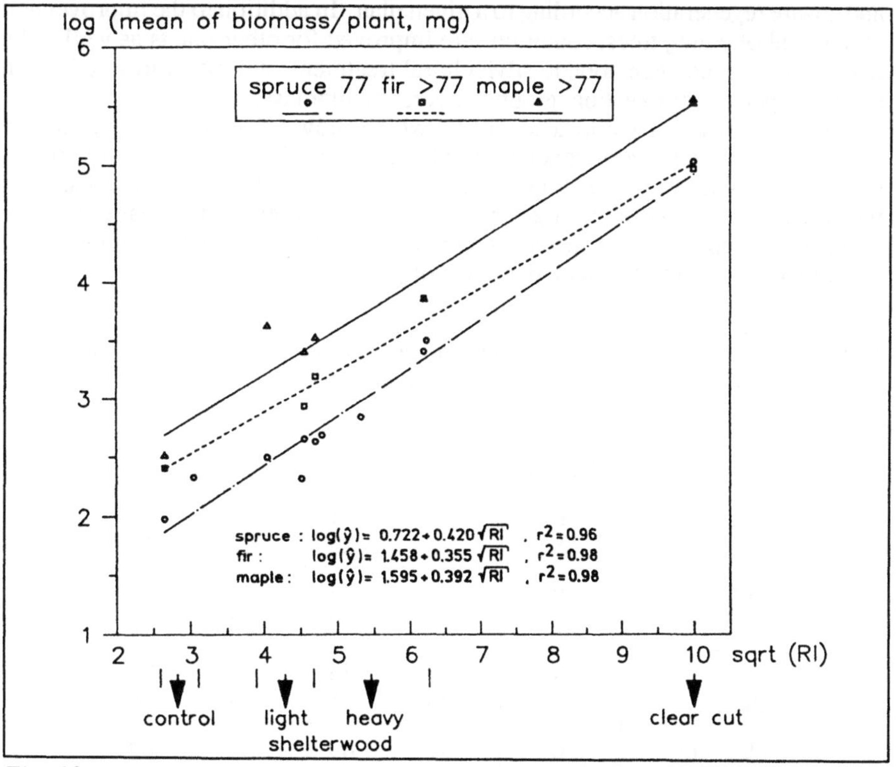

Fig. 13. Average biomass in fall 1986 (dry matter, mg) of regenerating plants under different silvicultural conditions for fir and maple (year of germination: before 1977) and spruce (year of germination: 1977).

mean biomass of regenerating seedlings, comprising shoot and root, was the criterion studied. The different silvicultural systems are indicated within the range of relative illumination (Figure 13) and their effect on the growth of regeneration is clearly visible. In this context, it is again evident that any kind of opening of the canopy of an old stand improves the living conditions of the regeneration on the ground. The improvement is very gradual up to situations which correspond to heavy shelterwood cuttings, but will increase drastically with further opening, and culminate under completely open conditions. This pattern offers a wide range of silvicultural options. The options include low cutting intensities, if gradual establishment and slow growth of regeneration is acceptable and the improved increment of the remaining old trees for a couple of decades is the principal goal. They also include a quick removal of the overstory if shelter for the regeneration as a means of protection against frost and weeds is the principal purpose of maintaining the canopy; in this case, the whole process of conversion to a new stand can be completed within a few years.

Development of ground vegetation as a consequence of silvicultural measures

The main purpose of applying silvicultural measures to old stands is to enforce and ensure regeneration according to a given plan. In addition to the improvement of survival of young trees, conditions are improved for other plants as well. The reaction of ground vegetation to silvicultural treatments will be expressed here in terms of species composition, ground cover, and biomass.

The research area was located on sites which allow a rich ground vegetation to develop. Even under very dark conditions 20 to 40 different species are present (Table 11). With the opening up of the crown layer, the number of plant species which can survive becomes larger, reaching a maximum on the clear cuts. The process of reappearance of species is a gradual one, leading finally to about twice as many in clearcuts as under a dense canopy.

Table 11. Number of species of the ground vegetation, excluding trees and shrubs (Uebelhoer 1978, Kotru 1985, Mosandl 1990).

silviculture	crown density	1977	1978	1984	1986
control	76	31	31	31	42
	68	21	23	27	36
light shelterwood	60	33	35	42	53
	56	17	22	33	45
heavy shelterwood	49	35	38	48	57
	39	23	35	43	56
clear cut	0	37	49	54	69
	0	37	44	54	76

The considerable increase of species numbers even on the control plots during the period 1984-86 might be explained as a consequence of the phenomenon "new forest decline", which by reducing the needle and leaf density increases the amount of light reaching the ground. As the ground vegetation can not react immediately to the better light conditions, our observations correspond well to the symptoms of forest decline which appeared in 1982 and 1983. Decline occurred on the shelterwood plots also, but for methodological reasons its possible additional effects cannot be separated from the ones caused by the silvicultural measure applied.

The development of the layer of ground vegetation is presented in Figure 14, based on estimating the proportion of the total surface area covered by ground vegetation. As crown density decreases, more light reaches the ground, which leads to an increase in the plant coverage. Almost complete coverage is attained on the clear cut areas outside the fencewhere browsing reduced tree regeneration. Inside the fence the coverage reached a maximum in 1983, but then decreased as a consequence of suppression from tree seedling regeneration.

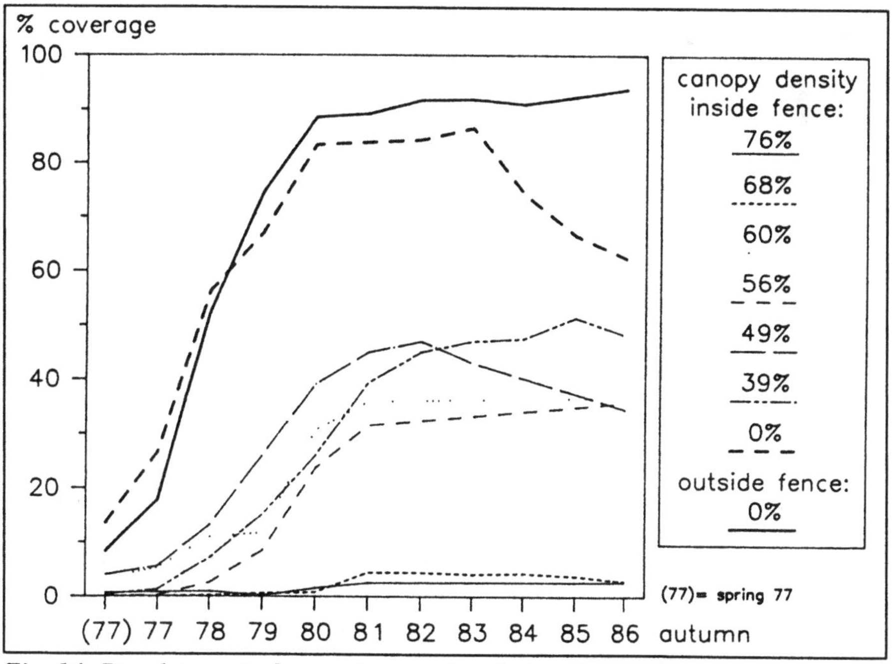

Fig. 14. Development of ground vegetation during observation period; % coverage: percentage of area covered by ground vegetation (excluding regeneration); each value is an average of 32 sampling units.

The response of the ground vegetation to changing light conditions was not immediate. It was only in the fourth or fifth growing season after the cutting operation that maximum expansion was attained, and this level was maintained later on. The most illustrative criterion for the development of ground vegetation is biomass (Table 12). The range of values vary from a few kg dry matter per ha to more than 2 tons on the unfenced parts of the clear cuts. Every silvicultural treatment represents a profound increase in productivity of ground vegetation with respect to the next less open treatment.

Table 12. Biomass (dry matter, kg/ha) of ground vegetation (without regeneration), fall 1986; each value is an average of 8 sampling units. (ANOVA: differences between silvicultural treatments significant). (Mosandl 1990).

species	silviculture, canopy density %							
	control		light shelterwood		heavy shelterwood		clear cut	
	76	68	60	56	49	39	0	0
fenced								
rubus idaeus	0	2	19	286	314	263	361	303
grasses	5	1	165	117	141	412	555	326
other	14	17	137	83	77	103	347	205
Sum	19	20	321	486	532	778	1263	834
unfenced								
rubus idaeus	0	0	2	201	68	192	374	624
grasses	27	0	147	209	143	485	1101	1141
others	31	7	120	55	94	89	464	449
Sum	58	7	269	465	305	766	1909	2214

Finally it is necessary to determine the competitive interplay between ground vegetation and natural regeneration, in order to determine when competition begins. Figure 15a permits a comparison between the biomass of natural regeneration and competing ground vegetation produced under different canopy densities. In very dark situations, the energy available for any type of plant production near the ground is so limited that competition between young trees and other ground vegetation is negligible.

With more light penetrating through the crowns the productivity increases and the ground vegetation achieves higher values than the regeneration. With this increase of production a mutual influence between the two groups of plants becomes probable. The height growth of maple regeneration of maple under different light conditions clearly reflects such a situation (Figure 15a): under a dense canopy (68%) maples which germinated four or more years later than the ones which

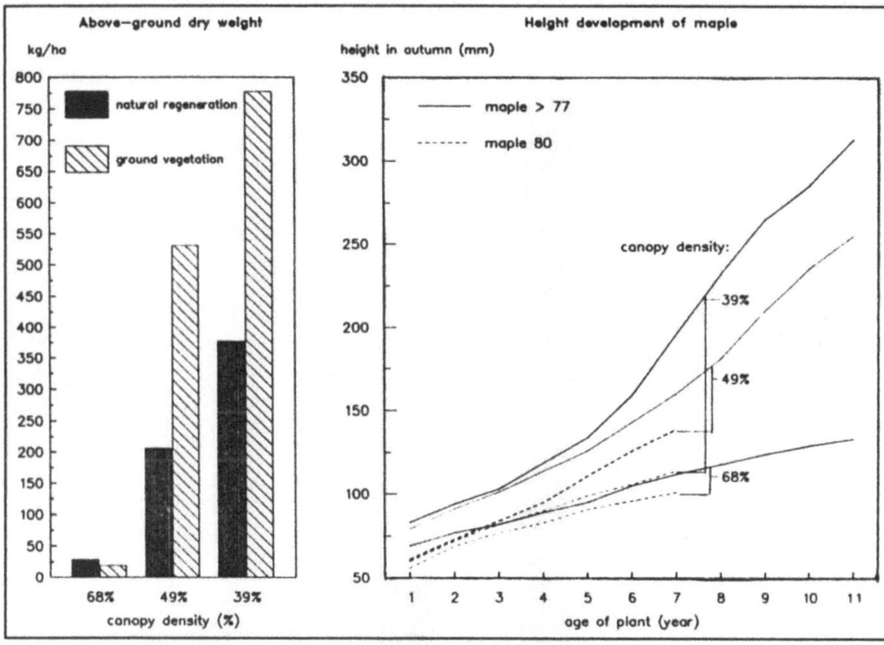

Fig. 15a. Above ground dry weight (kg/ha) of ground vegetation and of natural regeneration, fall 1986 (left). Height development of maple germinated before 1977 and in 1980 (right). All values inside fence.

already were established at the initial stage of the experiment, developed exactly in the same way as the older ones. The same holds true for 49% canopy density. Further opening up of the old stand to 39% canopy density changes the situation. Here the young maple plants (germinated in 1980) were suppressed by ground vegetation and show a height growth which corresponds to that under a dense canopy. The ground vegetation on this plot evidently has compensated for the reduction of leaf area in the tree canopy by silvicultural treatment. In Figure 15b another form of competitive interaction between regeneration and ground vegetation becomes appreciable. On the clear cuts young maple plants were able to outgrow the rapidly expanding ground vegetation from the beginning, so that after 10 years their biomass was twice as great as that of the ground vegetation which hardly maintained itself. The trees won the race. With the access of browsing animals on the unfenced areas the situation changes completely: the development of seedling regeneration, mainly of fast growing maple, was so much retarded that ground vegetation developed freely and, as far as competitive power is concerned, dominated the situation.

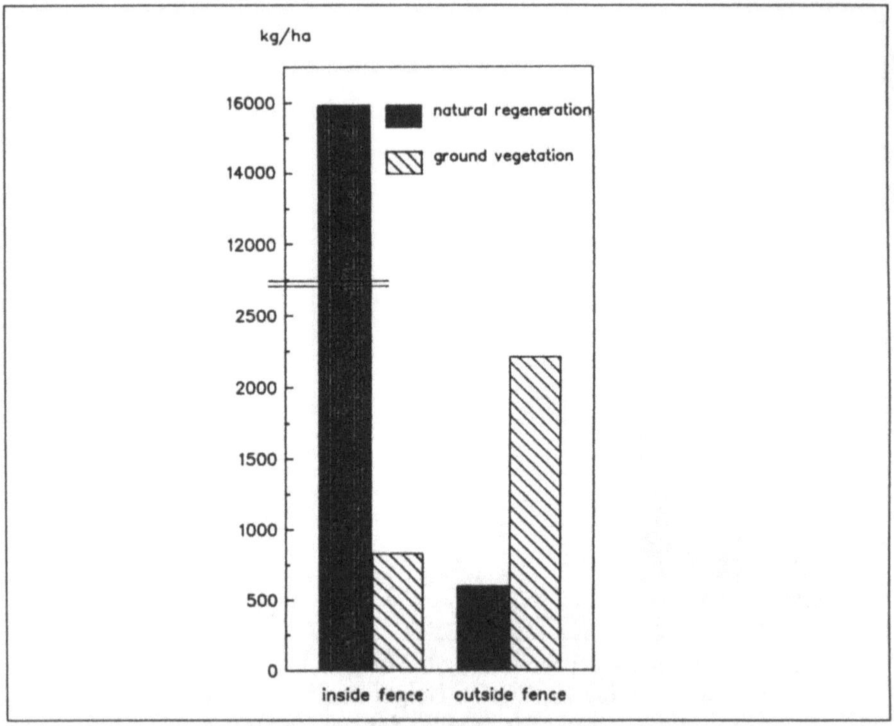

Fig. 15b. Above ground dry weight (kg/ha) of competing ground vegetation and of natural regeneration on the clear cut plot.

Browsing

The impact. The impact of browsing is evident on all plots outside fences regardless of the density of canopy (Table 13). The browsing effect on the individual plants (e.g., height and biomass per plant) and on plant density is reponsible for the decrease in biomass production. Figure 16 presents the situation for fir and maple, both species heavily affected by browsing. In both cases the effect of fencing is statistically significant.

The browsing effect occurs over the whole range of the canopy densities and badly hampers the development of regeneration. Its severity, however, increases drastically with increasing openess of the old stand. The maximum effect was found on the completely open clear cuts. Under such conditions the competitive potential of the regenerating trees is severely inhibited so that ground vegetation becomes able to outgrow the young trees (see Figure 15b).

Deer or cattle? In some parts of the Bavarian Alps—as a matter of fact on large areas of the whole Alps—cattle are kept in forests during summer, due to old grazing rights. Additional experiments were carried out to distinguish browsing damage caused by cattle from that of wild deer (*Capreolus capreolus*, *Cervus elaphus*, *Rupicapra rupicapra*). These were designed as exclosure trials, with some plots

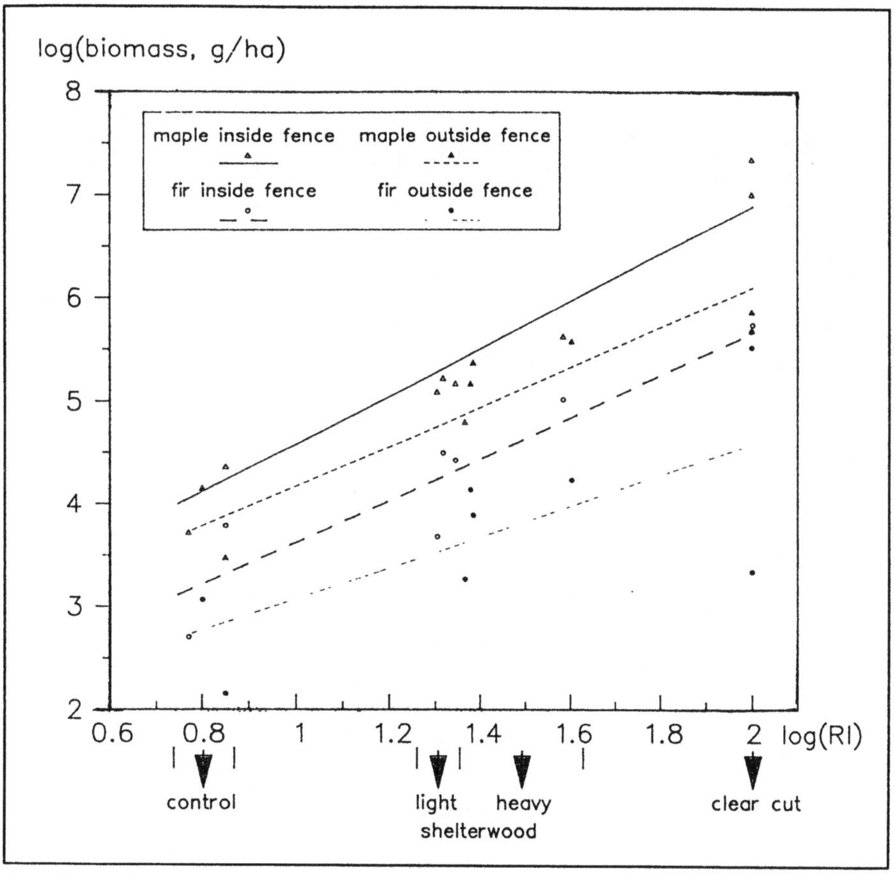

Fig. 16. The influence of fencing on the biomass production of fir and maple under different light conditions.

Maple: adj. $r^2 = 0.90$

inside fence: $\log(y) = 2.1816 + 2.3509 \log(x)$

outside fence: $\log(y) = 2.1816 + 1.9544 \log(x)$

Fir: adj. $r^2 = 0.69$

inside fence: $\log(y) = 1.5147 + 2.0698 \log(x)$

outside fence: $\log(y) = 1.5147 + 1.5354 \log(x)$

site	HAUPTDOLOMIT																FLYSCH			
species and year of germination	FIR 77								MAPLE 77								FIR 80			
canopy density	68 % control		56 % light shelter-wood cutting		39 % heavy shelter-wood cutting		0 % clear cut		68 % control		56 % light shelter-wood cutting		39 % heavy shelter-wood cutting		0 % clear cut		83 % control		54 % heavy shelter-wood cutting	
fence	+	-	+	-	+	-	+	-	+	-	+	-	+	-	+	-	+	-	+	-
density in spring 77; 1000/ha	7.2	10.6	2.5	3.1	5.9	7.2	4.1	6.9	19.7	20.0	18.1	29.7	31.9	46.6	62.8	36.3	241.9*	242.2*	163.4*	113.1*
mortality until autumn 86;	48	71	37	100	21	83	23	64	62	78	43	55	31	38	13	43	66	92	54	57
density in autumn 86; 1000/ha	3.8	3.1	1.6	0	4.7	1.3	3.1	2.5	7.5	4.4	10.3	13.4	21.9	28.8	54.7	20.6	81.3	20.0	75.0	49.1
% of plants damaged (autumn 86)	17	30	0	-	0	75	10	100	8	21	57	65	17	44	27	73	1	20	1	63
maximal height in autumn 86; cm	18	10	11	-	30	12	127	95	22	21	67	27	73	69	435	136	19	17	48	30
mean of bio-mass/plant in autumn 86; g	0.3	0.2	0.8	-	7.0	0.5	88.7	64.3	0.3	0.4	2.4	3.3	6.9	2.6	343.2	25.8	-	-	-	-

+ = inside fence, - = outside fence; * = density in spring 80

Table 13. The influence of browsing on regeneration of fir and maple.

inaccessible to both cattle and deer, other plots where only deer could enter, and other plots that were completely open to deer as well as to cattle. The findings are presented below.

Relatively little browsing damage occurred during summer. Beech, spruce, and maple were somewhat more heavily affected by additional cattle browsing than by deer browsing alone. Fir was apparently not browsed at all by cattle. Considering winter browsing, which can only be caused by deer, the situation changes drastically (Figure 17). In spring 1988, at the end of four years of observation, almost all of the leading shoots of fir, beech, and maple plants were damaged or destroyed by browsing, regardless of whether deer alone or deer as well as cattle had access to them. This clearly indicates that by far the most damaging browsing is done by deer. Even the relatively less attractive spruce had lost the main shoots on almost half of the plants. This means that the damage to the young plants would hardly decrease, even if grazing by cattle could be stopped.

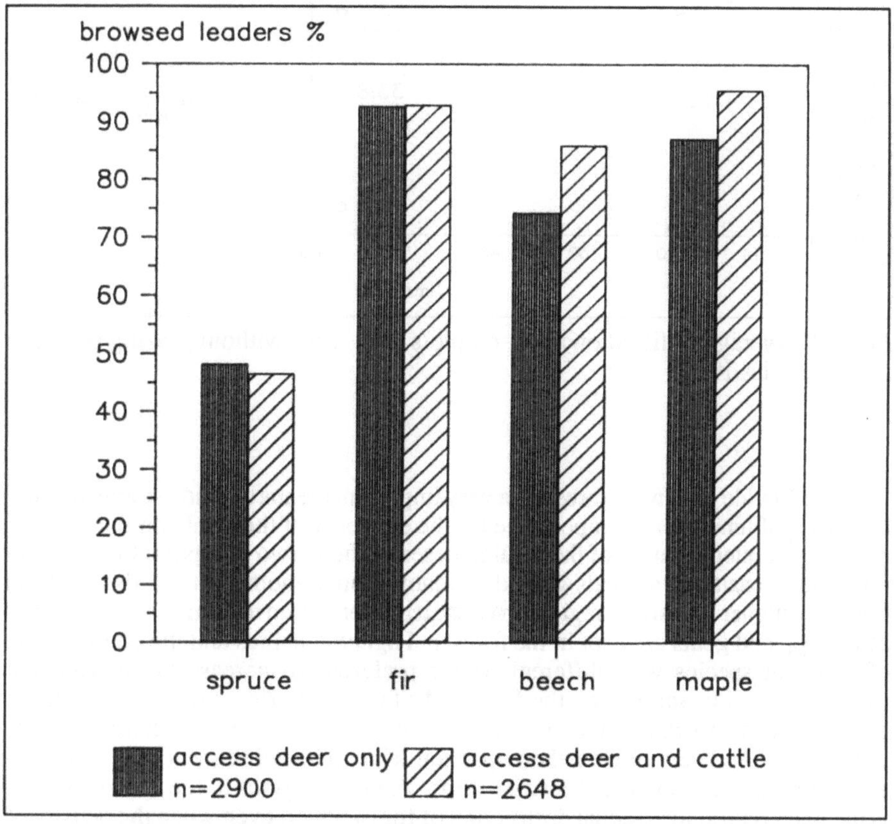

Fig. 17. Frequency (%) of main shoots browsed on 5548 marked regeneration plants in spring 1988 (Liss 1989).

This finding should in no way give the impression that cattle grazing in forests is a harmless activity. Large grazing animals do considerable damage to young plants by trampling. Perhaps even more detrimental is the compaction of the soil caused by continuous grazing, which greatly reduces its infiltration capacity (Figure 18), increases erosion, and may hinder the process of germination and establishment of young plants. The problem of soil compaction becomes greater on more susceptible soils—ones with higher clay content, shallower depths, and steeper slopes.

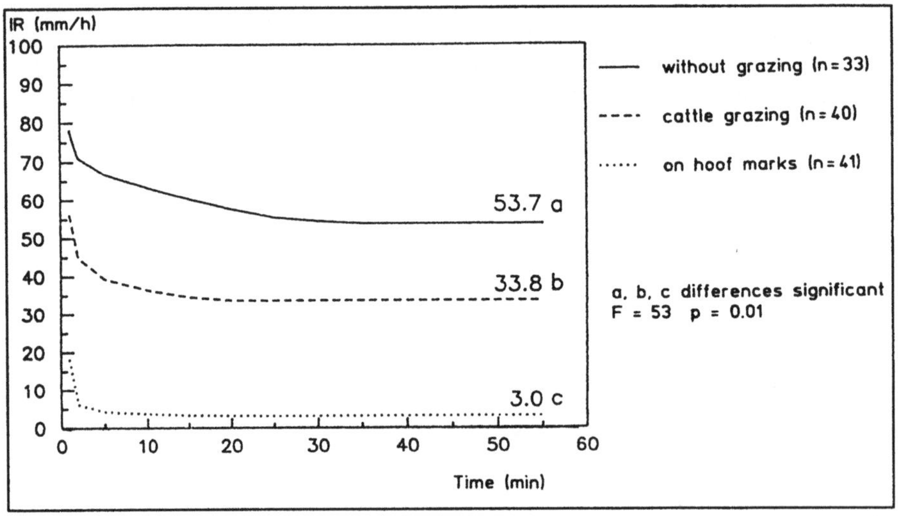

Fig. 18. Average infiltration rates on plots with and without grazing by cattle (Liss 1989).

Worldwide grazing in forests is a very important means of land use and in many regions is of fundamental importance for the economy of the rural population. For meaningful interpretation of the results presented here, two things are to be kept in mind: the density of animals grazed in many mountainous parts of the world is normally higher than in the region investigated. Very often not only cattle alone but also sheep and goats are kept in the forests. Higher densities and the combination of different species with different forage preferences increase the detrimental impact of grazing animals on the forest. As far as the effect on the regeneration process is concerned, it will come close to the combined impact of cattle and deer in our case. And very often soil compaction with all its consequences is considerably more severe than in the Bavarian Alps. In many regions of the world, grazing is the main reason for gradual destruction of forest cover, even when the density of browsing wild animals has become–or is–very low.

The ecosystem approach

The comprehensive set of data available after 10 years of investigation permits a rough and tentative approach to classifying the different silvicultural methods according to ecological principles. For this purpose a first approach towards calculating an ecological balance of living biomass was established (Table 14).

Table 14. Above ground biomass balance in a mixed mountain forest (+ = input; – = output). Approximate data, kg/ha/a. (Methodological details: Mosandl 1990).

silviculture	control	shelterwood cutting light	shelterwood cutting heavy		clear cut fenced	clear cut unfenced
	(1.0)	(1.1)	(1.2)	(2.2)	(1.3)	(1.3)
increment:						
wood	+2449	+2202	+2364	+1899	0	0
leafes	+2659	+2168	+1817	+1569	+191	+191
seeds	+16	+10	+14	+10	+1	+1
nat. reg.	+3	+13	+21	+38	+1594	+60
ground veget.	+20	+486	+532	+778	+834	+2214
sum input	+5147	+4879	+4748	+4294	+2620	+2466
mortality, trees	-1881	-294	-483	-324	0	0
litter	-2659	-2168	-1817	-1569	-191	-191
ground veget.	-20	-486	-532	-778	-834	-2214
sum output	-4560	-2948	-2832	-2671	-1025	-2405
balance	+587	+1931	+1916	+1623	+1595	+61

The forest investigated represents an aggrading ecosystem: living biomass is permanently accumulated. According to the silvicultural method applied, this process of accumulation occurs in a characteristic manner. Because of high mortality by suppression on the control plots, the proportion of total production which remains for accumulation is surprisingly low. This forest appears to be gradually approaching a steady state situation, in spite of its relatively low average age of about 100 years. All types of shelterwood cuttings brusquely interrupt this development towards more mature successional stages. Although one-third to one-half of the initial volume had been cut, the increment of wood was only slightly reduced, and the decrease in leaf production became almost fully compensated by ground vegetation. The proportion of natural regeneration within the whole production process was insignificant under the shelterwood canopy. Nevertheless it is worthwhile to mention that there is a more than 10-fold rise in regeneration from the control to the most open shelterwood plot. Since the production of ground vegetation in the same sequence of plots increases by a factor of more than 20, it becomes apparent that this vigorous layer of plants may dominate the understory if young trees are not able to outgrow them from the very beginning.

The clear cut plots were resettled by low vegetation in an almost explosive manner. Since a large number of tree seedlings were already present when the felling operation was carried out (although almost too small to be visible), they dominated the open area before the ground vegetation was able to react fully. This allowed the forest to completely determine the very early successional phase of colonization of the clear cut. A new forest ecosytem of surprisingly high productivity from the beginning replaced the harvested old one; but (and here is the restriction) this process occurs only if the impact of browsing is excluded. With free access of browsing animals, in this case wild deer, the situation changes completely. There, the forest plants (most importantly maple, which dominates the fenced area) are so heavily damaged that they become unable to compete with the ground vegetation, which in turn develops vigorously, forming a typical grazing ecosystem. Its productivity is about the same as found in the forest ecosystem on the clear cut. The main difference is that in one case—the forest—accumulation of biomass takes place, while in the other one—the pasture—the above ground production is lost every fall. But even with the browsing pressure, the grazing system will not last. In the long run (another one or two decades) spruce and some beech, as less heavily browsed species, will finally overcome the competition of the ground vegetation and reestablish a new forest. As a matter of fact, the absolute dominance of spruce in the young stands of the region investigated is the consequence of the interference of wild deer, maintained for hunting reasons in densities that are by far too great.

So far, the comprehensive study proved the tremendous productive potential of the forest investigated, including its vigorous capacity to regenerate. One aim for the continuation of the experiment is to develop a systematic process of regeneration to ensure that the new stand will not be a casual type of mixture with maple often dominating, but rather contain an economically necessary proportion of spruce and fir and an ecologically desirable participation of beech and all minor species of the Bavarian mountain forests. Since this type of desirable mixed forest—still present in the old stands—was maintained in spite of centuries of heavy but controlled grazing pressure, it must be possible to conserve and reestablish it in our ecology-conscious time. To reach this goal, two prerequisites have to be fulfilled:

- ecologically (silviculturally) founded densities of browsing wild deer must be maintained;
- refinement of the applied silvicultural measures must be made in order to ensure a planned type of mixture according to economical as well as ecological requirements.

Summary

Results of a comprehensive silvicultural experiment in mountain forests of spruce, fir, beech, maple, and minor species are presented. On 25 research plots, different regeneration procedures were applied and their consequences studied from several ecological and silvicultural aspects. The finding of the first decade of investigations are presented here in a condensed way. They cover the following aspects:

Ecological factors: light, precipitation, snow, humus quality

Old stands: increment, mortality, stability, litter production, seed production

Regeneration: establishment of seedlings, development of plants

Ground vegetation: development, competition factor for regeneration

Browsing: effect on regeneration and soil, differentiation between cattle and deer

Ecosystem approach: balance of biomass increment and mortality

The forests studied proved to be highly productive. All tree species are able to regenerate readily, with maple being most successful in this respect. Ground vegetation under certain silvicultural practices may hinder the development of the young trees but not prevent it. The establishment of the ground vegetation and its competitive interplay with forest regeneration could be demonstrated. The regeneration process can be severely hampered and altered by browsing of wild deer. Overbrowsing by deer proved to be the only way by which the existence of the mixed mountain forests studied can be seriously endangered as far as promptness of the regeneration process and (even more) maintenance of the species mixture are concerned.

The results in all are a contribution to an improved understanding of the dynamics of mixed forests.

References

Aas, G. 1984. Vorkommen und Bedeutung von Keimlingspilzen in der natürlichen Verjüngung des Bergmischwaldes. Diplomarbeit, Forstwiss. Fak. Univ. München. 78 p.

Bäumler, W. 1981. Zur Verbreitung, Ernährung und Populationsdynamik der Rötelmaus (*Clethrionomys glareolus*) und der Gelbhalsmaus (*Apodemus flavicollis*) in einem Waldgebiet der Bayerischen Alpen. Anz. Schädlingskde., Pflanzenschutz, Umweltschutz, 54:49-53.

Bäumler, W., and W. Hohenadl. 1980. Über den Einfluß alpiner Kleinsäuger auf die Verjüngung in einem Bergmischwald der Chiemgauer Alpen. Forstwiss. Cbl., 99:207-221.

Benra, G. 1989. Der Einfluß der Waldweide auf die Bodenvegetation im Bergmischwald. Diplomarbeit, Forstwiss. Fak. Univ. München. 56 p.

Berthold, J. 1980. Schnee im Bergmischwald. Diplomarbeit. Forstwiss. Fak. Univ. München. 83 p.

Binder, F. 1982. Das Ankommen und die Entwicklung der Naturverjüngung im Bergmischwald bei dichter Bodenvegetation. Diplomarbeit, Forstwiss. Fak. Univ. München. 77 p.

Bormann, F.H., and G.E. Likens. 1979. Pattern and process in a forested ecosystem. New York, Heidelberg, Berlin: Springer. 253 p.

Burschel, P. 1966. Untersuchungen in Buchenmastjahren. Forstwiss. Cbl., 85: 204-219.

Burschel, P., H. El Kateb, J. Huss, and R. Mosandl. 1985. Die Verjüngung im Bergmischwald. Forstwiss. Cbl., 104:65-100.

Burschel, P., J. Huss, and G. Kalbhenn. 1964. Die natürliche Verjüngung der Buche. Schriftenr. Forstl. Fak. Univ. Göttingen, 34:186 p.

Burschel, P., H. Löw, C.H. Mettin. 1977. Waldbauliche Untersuchungen in Hochlagen des Werdenfelser Landes. Forschungsberichte d. Forstl. Forschungsanst. München, 37: 193 p.

Burschel, P., R. Mosandl. 1981. Nachwuchsprobleme im Bergwald (Regeneration of mountain forests poses problems). Mitteilungen der Deutschen Forschungsgemeinschaft, 3/81:6-9. (Reports of the DFG, 3/81, 6-9).

El Kateb, H. 1990. Der Einfluß waldbaulicher Maßnahmen auf die Sproßgewichte von Naturverjüngungspflanzen im Bergmischwald. Biometrische Auswertung eines waldbaulichen Versuches. M. Sc thesis, Univ. of Stellenbosch. 193 p.

Feulner, T. 1979. Der Einfluß alpiner Kleinsäuger auf die Verjüngung in einem Bergmischwald bei Ruhpolding. Diplomarbeit, Forstwiss. Fak. Univ. München.

Grosse, H.-U. 1983. Untersuchungen zur künstlichen Verjüngung des Bergmischwaldes. Forschungsberichte d. Forstl. Forschungsanstalt München, 55: 215 p.

Hillenbrand, V. 1986. Wirkung unterschiedlicher Überschirmung auf Einzelbaumparameter von fünf Baumarten im Bergmischwald unter besonderer Berücksichtigung der Biomassenproduktion. Diplomarbeit. Forstwiss. Fak. Univ. München. 198 p.

Hohenadl, W. 1981. Untersuchungen zur natürlichen Verjüngung des Bergmischwaldes – Erste Ergebnisse eines Forschungsprojektes in den ostbayerischen Kalkalpen. Diss. Forstwiss. Fak. Univ. München. 197 p.

Huhn, S. 1979. Wachstumsreaktionen gepflanzter Fichten, Tannen, Buchen, Ahorne und Lärchen bei unterschiedlich starker Überschirmung und Höhenlage im Bergmischwald. Diplomarbeit, Forstwiss. Fak. Univ. München. 96 p.

Kirchen, M. 1989. Die Wirkung unterschiedlicher Überschirmungsgrade auf die Blattmorphologie bei vier Baumarten im Bergmischwald der Chiemgauer Alpen. Diplomarbeit, Forstwiss. Fak. Univ. Freiburg. 51 p.

Kirches, E. 1987. Genökologische Untersuchung an der Naturverjüngung eines Bergmischwaldes der ostbayerischen Kalkalpen. Diplomarbeit, Mathematisch-Naturwiss. Fak. Univ. Bonn. 100 p.

Kotru, R. 1985. Die Entwicklung der Bodenvegetation unter verschiedenen Überschirmungsvarianten im Bergmischwald. Diplomarbeit. Forstwiss. Fak. Univ. München. 73 p.

Laar, A. van. 1980. Quantitative studies of natural regeneration in the mountain forests of Bavaria. Lehrstuhl für Waldbau und Forsteinrichtung d. Univ. München (ed.). 110 p.

Liss, B.-M. 1988. Versuche zur Waldweide—der Einfluß von Weidevieh und Wild auf Verjüngung Bodenvegetation und Boden im Bergmischwald der ostbayerischen Alpen. Forschungsberichte d. Forstl. Forschungsanstalt München, 87: 221 p.

Liss, B.-M. 1988. Verjüngungsprobleme im Bergmischwald unter dem Einfluß von Weidevieh und Wild. Mitteilungen aus der Wildforschung. Hrsg: Wildbiologische Gesellschaft München, 91:209 p.

Liss, B.-M. 1988. Der Einfluß von Weidevieh und Wild auf die natürliche und künstliche Verjüngung im Bergmischwald der ostbayerischen Alpen. Forstwiss. Cbl., 107:14-25.

Liss, B.-M. 1989. Die Wirkung der Weide auf den Bergwald. Forschungsberichte d. Forstl. Forschungsanstalt München, 99:106 p.

Loy, S. 1989. Das Ankommen und die Entwicklung von Keimlingen der Baumarten Fichte, Tanne und Bergahorn auf durch Waldweide verdichteten Böden. Diplomarbeit, Forstwiss. Fak. Univ. München. 78 p.

Mayer, H. 1978. Mikroklimatische Verhältnisse im Bergmischwald bei verschiedenen Schlagverfahren. Schweiz. Meteor. Zentralanstalt, 40:113-116.

Mayer, H. 1979. Mikroklimatische Untersuchungen im ostbayerischen Bergmischwald. Arch. Met. Geoph. Biokl., Ser. B. 26 p.

Mayer, H. 1980. Schnee im ostbayerischen Bergmischwald unter verschiedenen Überschirmungen. Proc. XVI. Kongr. Alpine Meteorologie. 249-254.

Mayer, H. 1981. Globalstrahlung im ostbayerischen Bergmischwald unter verschiedenen Überschirmungen. Arch. Met. Geoph. Biokl., Ser. B, 2:283-292.

Mishra, V.K. 1982. Genesis and classifications of soils derived from Hauptdolomit (Dolomite) in Kalkalpen and effects of soil type and humus form on some features of forest natural regeneration. Diss. Forstwiss. Fak. Univ. München. 165 p.

Mosandl, R. 1984. Löcherhiebe im Bergmischwald. Ein waldbauökologischer Beitrag zur Femelschlagverjüngung in den Chiemgauer Alpen. Forschungsberichte d. Forstl. Forschungsanstalt München, 61: 317 p.

Mosandl, R. 1990. Die Steuerung von Waldökosystemen mit waldbaulichen Mitteln—dargestellt am Beispiel des Bergmischwaldes. Habil. Forstwiss. Fak. Univ. München. 246 p.

Mosandl, R., and G. Aas. 1986. Vorkommen und Bedeutung von Keimlingspilzen im Bergmischwald der ostbayerischen Kalkalpen. Der Forst- und Holzwirt, 41: 471-475.

Mosandl, R., H. El Kateb. 1988. Die Verjüngung gemischter Bergwälder—Praktische Konsequenzen aus 10-jähriger Untersuchungsarbeit. Forstwiss. Cbl., 107: 2-13.

Schörry, R. 1980. Bodenformen und Ansamungserfolg im Bergmischwaldprojekt Ruhpolding. Diplomarbeit, Forstwiss. Fak. Univ. München. 82 p.

Schreyer, G., and V. Rausch. 1978. Der Schutzwald in der Alpenregion des Landkreises Miesbach. Bayer. Staatsmin. Ern. Landw. Forsten. 116 p.

Smith, D.M. 1986. The practice of silviculture. 8th ed. New York: J. Wiley & Sons. 527 p.

Stölb, W. 1978. Das Vorkommen von Mäusen in einem Bergmischwald und deren Einfluß auf die Verjüngung der Hauptholzarten. Diplomarbeit, Forstwiss. Fak. Univ. München. 73 p.

Straka, G. 1989. Entwicklung von Pflanzungen im Bergmischwald unter dem Einfluß von Wild und Weidevieh. Diplomarbeit, Forstwiss. Fak. Univ. München. 114 p.

Uebelhör, K. 1979. Die Reaktionen der Bodenvegetation auf unterschiedlich starke Überschirmung im Bergmischwald bei Ruhpolding. Diplomarbeit, Forstwiss. Fak. Univ. München. 92 p.

Veltsistas, T. 1980. Untersuchungen über die natürliche Verjüngung im Bergmischwald—Die Fruktifikation in den Jahren 1976/77 und 1977/78 auf Versuchsflächen im Forstamt Ruhpolding. Diss. Forstwiss. Fak. Univ. München. 130 p.

Wilhelms, J.M. 1988. Die Wirkung unterschiedlicher Überschirmungsgrade auf die Blattinhaltsstoffe bei mehreren Baumarten im Bergmischwald der Chiemgauer Berge. Diplomarbeit Forstwiss. Fak. Univ. Freiburg. 69 p.

Wilke, B.M., V.K. Mishra, and K.E. Rehfuess. 1984. Clay mineralogy of a soil sequence in slope deposits derived from Hauptdolomit (Dolomite) in the Bavarian Alps. Geoderma 32:103-116.

Zwirglmaier, G. 1977. Waldbauliche Charakterisierung von Eingriffen in Altbestände des Bergmischwaldes. Diplomarbeit, Forstwiss. Fak. Univ. München. 115 p.

The red spruce-balsam fir forest of Maine: Evolution of silvicultural practice in response to stand development patterns and disturbances

12

ROBERT S. SEYMOUR

Silvical properties

Red spruce (*Picea rubens*) and balsam fir (*Abies balsamea*) are so similar that "spruce-fir" is often used as if it were a single species. The early monographs of Zon (1914) and Murphy (1917) accurately characterize both species as occupying a similar ecological niche: late-successional, very tolerant of shade, shallow rooted, and widely adapted to a variety of site and stand conditions. These species differ in important ways that influence silvicultural treatment (Fowells 1965). Fir produces abundant seeds, but is so susceptible to various heart-rot fungi that its potential life span is limited by the high risk of wind breakage or uprooting. Balsam fir is often cited as the classic example of a species ruled by a pathological rotation, effectively limited to ages 40-70, depending on site quality. Fir is also the preferred host and suffers extensive mortality from defoliation by the spruce budworm (*Choristoneura fumiferana*). The introduced balsam wooly adelgid (*Adelges piceae*) also is a serious pest of fir in coastal regions but does not cause serious damage inland.

In contrast, red spruce produces seeds infrequently, but is quite resistant to decay and tends to survive budworm defoliation. As a result, red spruce is inherently long-lived, and 300+ year-old trees were not uncommon in virgin forests (Cary 1894a; Oosting and Billings 1951; Leak 1975). Perhaps its most important silvical properties are the abilities to persist as advance regeneration and to respond well to release after many decades of suppression in very low-light conditions in the understory. Cary (1896) was the first to document the capability of red spruce to respond to release at advanced ages, but this remarkable quality was also highlighted by other early foresters (Graves 1899; Hosmer 1902). Subsequent studies (Westveld 1931; Davis 1989) have found that the ability of spruce to develop slowly in the understory gives this species an initial height advantage that allows it to compete successfully with faster growing fir and hardwoods after release from overstory cover.

Forest types and abundance

The natural range of both red spruce and balsam fir encompasses virtually all of Maine; however, stands where spruce and fir dominate occur primarily in the northern and eastern parts of the state. Within this zone, red spruce is ubiquitous on soils developed from glacial till, but stand species composition varies greatly with soil drainage and topography. In the most recent compilation of forest types for North America (Eyre 1980), red spruce is named as a major component of six associations and occurs in nine others. Spruce and fir dominate Maine's forests in terms of area and volume; 46% of the State's forest area falls into the spruce-fir forest type. Red spruce and balsam fir also rank first and second, respectively, in growing

M. J. Kelty (ed.), The Ecology and Silviculture of Mixed-Species Forests, 217–244.

stock volume. Together they comprise 60% of the softwood volume and 39% of all volume in the State (Powell and Dickson 1984). Interestingly, this dominance of Maine's forests by spruce may be relatively recent. Although spruce-fir forests were once regarded as stable features of the Maine landscape since deglaciation (Westveld 1953), recent analysis of pollen records suggests that today's red spruce-fir forest emerged only ca. 1000 years ago, corresponding to a decline in hemlock and beech abundance (Jacobson et al. 1987).

Large pulp and paper companies have long dominated ownership of the Maine's spruce-fir forest, and Maine currently has much more industrially owned timberland (over 3 million ha) than any other state. Logging has been so widespread (potentially altering natural species composition extensively) that the best ecological descriptions come from observations of virgin forests by early foresters (e.g., Graves 1899; Hosmer 1902). Based on relationships of vegetation with soils and landforms along with observed successional patterns, Westveld (1931, 1951, 1953) described what would now be called a series of habitat types similar to those formulated by Leak (1982) for the White Mountains of New Hampshire. "Dominant softwood" sites are those where a combination of soil drainage, nutrient status, and topographic position tends to exclude the more demanding northern hardwood species (e.g., sugar maple, *Acer saccharum*). These can be further subdivided into four sub-types. "Spruce swamps" support nearly pure stands of black spruce (*P. mariana*) in mixture with tamarack (*Larix laricina*) and northern white-cedar (*Thuja occidentalis*) on organic or very poorly drained mineral soils. "Spruce flats" occur on shallow glacial tills with impeded drainage at low elevations; red spruce and balsam fir, in various mixtures, dominate these sites with minor components of paper birch (*Betula papyrifera*) and red maple (*Acer rubrum*). "Spruce slopes" occur on mountainsides above an elevation of ca. 800m on shallow, very rocky soils. Fir and paper birch represented a minor component of the spruce slope type prior to human disturbance. A fourth variant of the dominant softwood type occurs along the Maine coast as a result of maritime influences (Davis 1966).

So-called "secondary softwood" sites occur on mid-slopes supporting well drained soils. Deeper rooting zones and improved nutrient status allow various hardwood species to form an important stand component. Yellow birch (*Betula alleghaniensis*), red maple, American beech (*Fagus grandifolia*) and sugar maple are the principal associates of red spruce and fir in what is commonly called the "mixedwood" type. Pure spruce stands of old-field origin also fall naturally into this type, since these agricultural soils originally had a strong hardwood component and revert to hardwoods after disturbances. In Maine, old-field spruce stands tend to be dominated by the more aggressive white spruce (*P. glauca*); old-field red spruce stands were more common in Vermont (Westveld 1931).

The natural ranges of red spruce and eastern hemlock (*Tsuga canadensis*) overlap in eastern and central Maine, where significant areas of mixed-conifer stands occur. Eastern white pine (*Pinus strobus*) frequently forms an important component of these spruce-hemlock stands, and also occurs as a scattered, valuable trees on spruce-flat sites. The boreal white and black spruces are uncommon associates of red spruce-dominated stands in Maine. In 1982 white and black spruce comprised only 10% and 8%, respectively, of all spruce volume statewide (Powell and Dickson 1984). White spruce often forms a minor component of pure spruce-fir stands, especially those dominated by fir. Black spruce is limited to very poorly drained swamps, and is rarely found on upland sites. It has recently been recognized that red and black spruce hybridize extensively (Manley 1972). Spruce hybrids tend to dominate poorly drained sites with a frequent history of disturbance (Manley and Fowler 1969; Osawa 1989). Rigorous procedures for determining the degree of hybridization (Manley 1971) are rarely applied. Trees with characteristics of both species are usually called "red" spruce, perhaps causing the abundance of genetically pure red spruce to be overestimated.

Disturbance history

Spruce budworm

During the 20th century, the most profound natural influence on growth and development of red spruce forests in Maine has been the eastern spruce budworm, a native insect. Documented outbreaks resulting in extensive tree mortality have occurred on two occasions: ca. 1913-19 and recently from 1972-86. A third outbreak that reached epidemic status in the boreal forest during the late 1940s caused only non-fatal defoliation in northern Maine (Irland et al. 1988). This insect arguably has been studied more than any other forest pest in North America, and a voluminous literature has accumulated from comprehensive investigations of these outbreaks (Swaine and Craighead 1924; Morris 1963; Sanders et al. 1985). Most research on budworm dynamics has been carried out in boreal forests where fir occurs in mixture with white and black spruce. The red spruce-fir forests in Maine and the southern Maritimes are more diverse in composition and structure, and have more complex and less well understood response patterns to uncontrolled budworm attack.

Controversy remains about the causes and periodicity of budworm outbreaks. Intensive studies of budworm population dynamics and observations of forest age structures suggest a natural cycle of 30-40 years, controlled by climate, availability of extensive areas of mature host foliage (balsam fir), and other natural limiting factors (Morris 1963; Royama 1984). Blais (1985), on the other hand, has argued that outbreaks were less common and less extensive in virgin forests than during the current century, based on studies of radial growth suppression of old white spruce in trees mostly in eastern Canada. Blais attributed the more frequent outbreaks to an increased abundance of balsam fir (the favored budworm host) resulting from extensive cutting of spruce, insecticidal protection of mature fir stands, and fire protection that has gradually reduced the area of non-host hardwood forests. The argument that forest exploitation has rendered spruce-fir forests more vulnerable to budworm attack is not new or unique; Swaine and Craighead (1924), Mott (1980), and many other writers have advanced similar arguments in support of silvicultural control strategies aimed at reducing fir abundance (Blum and MacLean 1985).

No definitive studies have examined the presettlement history of budworm outbreaks in the red spruce-dominated forests of Maine, so what follows is somewhat speculative. Mott's (1980) review of historical evidence reveals the possibility of two outbreaks during the 1800s: one in coastal Maine during the 1870s which did not reach tree-killing status inland, and a speculative one during the early 1800s. Evidence for an early 1800s outbreak comes from eyewitness accounts of extensive spruce mortality by early explorers and a distinct pattern of growth suppression during 1810-14 described by Austin Cary (1894a) as "reduced in some cases to almost microscopic." Although Cary attributed this growth suppression to a period of unusually cold weather, it does coincide with an outbreak postulated by Blais (1968) nearby in Quebec. The extent of this outbreak throughout northern Maine is unclear. Cary stated that "this zone of rings has been found in spruce trees in all parts of the State", whereas Lorimer (1977) found little evidence of extensive budworm-caused tree mortality in surveyors' notes dating from the 1820s, when such evidence would have been quite apparent. One can conclude that Maine's red spruce region escaped a tree-killing budworm outbreak for a minimum of ca. 100 years (i.e., from the early 1800s to 1913), a much longer frequency than the nine outbreaks reconstructed by Blais (1985) throughout the boreal fir region during the same period. Further evidence that budworm was not an important influence during the late 1800s comes from Hopkins (1901) monograph on "insect enemies of spruce in the Northeast", and Cary's (1900) accounts, which concentrate on bark beetles and do not even mention any budworm-like defoliators.

Uncontrolled budworm outbreaks exert a controlling influence on stand development in forests dominated by balsam fir. After several years of complete defoliation, mature stands invariably are completely killed, while immature stands suffer partial mortality analogous to a heavy crown thinning. The outbreak then collapses as a result of foliage depletion, and surviving trees develop without further defoliation until the next outbreak ca. 40 years later. Between outbreaks, advance seedlings that originated beneath mature stands prior to defoliation develop into vigorous immature stands, while surviving 40-year-old trees develop into large-crowned, highly vulnerable 80-year-old individuals. In this manner, the budworm effectively perpetuates a two-aged forest structure by periodically thinning the 40-year age class and killing the 80-year age class (Baskerville 1975a; MacLean 1984).

The budworm's effect on stand development in forests with a strong red spruce component is more complex. Studies of uncontrolled outbreaks during the 20th century demonstrate that some spruce can be killed, but that this species is much less vulnerable to mortality than fir (Craighead 1924; MacLean 1985; Osawa et al 1986). Post-outbreak stand structures thus depend on the relative abundance of red spruce and fir, as well as associated non-host species. One common pattern involved immature mixedwood or spruce-dominated stands that originated during the late 1800s after a heavy sawlog cuttings. Many such stands also had a residual overstory of fir left after the sawlog harvests, and were thus two-storied when the budworm outbreak developed ca. 1910. In these mixed-species stands, budworm tended to kill only the fir in both the residual and regenerating strata, thereby creating even-aged stands dominated by spruce or hardwoods (Seymour 1980). Sixty years later, these stands at age 90-100 were less vulnerable to the 1970s outbreak than were younger, 50-60-year-old pure-fir stands that originated after the 1913-19 attack. The logging-origin, spruce-dominated and mixedwood stands experienced only partial mortality and patchy regeneration, whereas the budworm-origin, fir-dominated stands were again completely killed and regenerated. Over time, in the absence of logging or catastrophic disturbances, the lower vulnerability of red spruce to budworm attack, coupled with its greater inherent longevity than fir, would eventually tend to promote multi-aged structures. Stands would become increasingly dominated by red spruce, as the more vulnerable fir was repeatedly purged from mixed stands.

Other natural disturbances

The preoccupation with spruce budworm as a dominant natural influence on stand development is relatively recent. When professional and scientific attention was first directed at the red spruce forest in late 1800s, the dominant concern was damage by the spruce bark beetle (originally *Dendroctonus piceaperda* Hopk., now *D. rufipennis* Kirby). Hopkins' (1901) investigations showed that, like other *Dendroctonus* species, this insect caused serious damage primarily to old, large-diameter trees. The scale of mortality apparently varied widely, from scattered mortality of individual stems to up to heavy losses over several adjoining townships. Peak mortality evidently occurred during the 1880s, coincident with large-scale spruce sawlog cutting operations on all major river systems of the State (Cary 1900). Severe bark-beetle mortality evidently regenerated "dense thickets of fir and other young growth" (Cary 1900); photographs in Hopkins (1901) show radial growth response of suppressed balsam fir that had formerly occupied understory status but was released by complete mortality of overstory spruce.

Wind damage is also an important disturbance agent, because of the shallow-rooted habit of spruce and susceptibility of fir to heart rots. Unlike insect epidemics, however, wind damage is usually a chronic phenomenon. Early surveyors noted large-scale windfalls (over 0.5 km in length) along only 2.6% of township lines, virtually all of which was spruce and fir on stony flats and swamps (Lorimer 1977). Assuming that surveyors would dependably record windfall for ca. 30 years, Lorimer calculated a recurrence interval for major windthrow (>25 ha) of 1150 years.

Fire was a concern of early foresters in Maine as it was elsewhere, but it is virtually impossible to separate a few, very large and probably man-caused fires from purely natural events. Extensive spruce logging, especially of pure stands on upper slopes and near railroads, appeared to increase both fire frequency and severity during the late 1800s in comparison to the presettlement era (Cary 1894b; Weiss and Millers 1988). Estimates of the natural recurrence interval of fire in northeastern Maine vary from a minimum of 800 to over 1900 years, depending on whether the 1803 fire (which covered 80,000 ha in the survey area alone) is included (Lorimer 1977).

Lethal disturbances affecting hardwood species also have greatly influenced stand development of mixed spruce-hardwood stands. Reams and Huso (1990) documented radial growth increases during 1935-55 in 54% of stands sampled ca. 1980 throughout northern Maine. They noted that this period coincides with extensive mortality and top-kill of *Betula* spp. resulting from the birch dieback epidemic, and the "killing front" of the introduced beech bark disease that eliminated *Fagus grandifolia* from the overstory of many stands in eastern and central Maine (Millers et al. 1989). Lethal disturbances affecting other conifer species that potentially have affected spruce-fir stand development include the larch sawfly (*Pristiphora erichsonii*) outbreak of the late 1800s that virtually eliminated tamarack from poorly drained sites throughout the spruce-fir region, and the pine leaf adelgid (*Pineus pinifoliae*) which can kill eastern white pine without damaging the alternative host red spruce (USDA Forest Service 1985).

Effect of natural disturbances on stand structures

The largely historical evidence reviewed above suggests that, unlike the boreal spruce-fir forest, large-scale stand-creating disturbances were much less common in the red spruce region prior to extensive logging of spruce. Rather, the evidence appears to support a regime of disturbances that were perhaps quite frequent relative to the life span of red spruce, but which rarely resulted in complete overstory mortality. The typical origin of virgin spruce appears to be a gradual response to a series of releasing disturbances, until reaching the overstory at relatively advanced ages. Evidence of this pattern comes from comparing the age-size relationships of virgin spruce from Cary (1894a) and Graves (1899) with those of both managed (Seymour and Lemin 1988) and unmanaged (Meyer 1929) even-aged stands (Fig. 1). Cary aged 1050 spruce logs from all major river systems of Maine during the early 1890s and found that 72% of all trees fell into the 100-year age class between 150-250, and less than 5% were under 125 years old. On the average, a spruce required nearly 200 years to reach a stump diameter of 36 cm (14 inches). Graves' age-size data from the Adirondacks follow a very similar pattern. In contrast, Meyer's normal yield tables show that spruces will reach 25 cm (10 inches) dbh at ca. age 100 on average site land; with early spacing, this time can be shorted to ca. 70 years -- 40-60% of the ages of comparably sized trees from virgin stands.

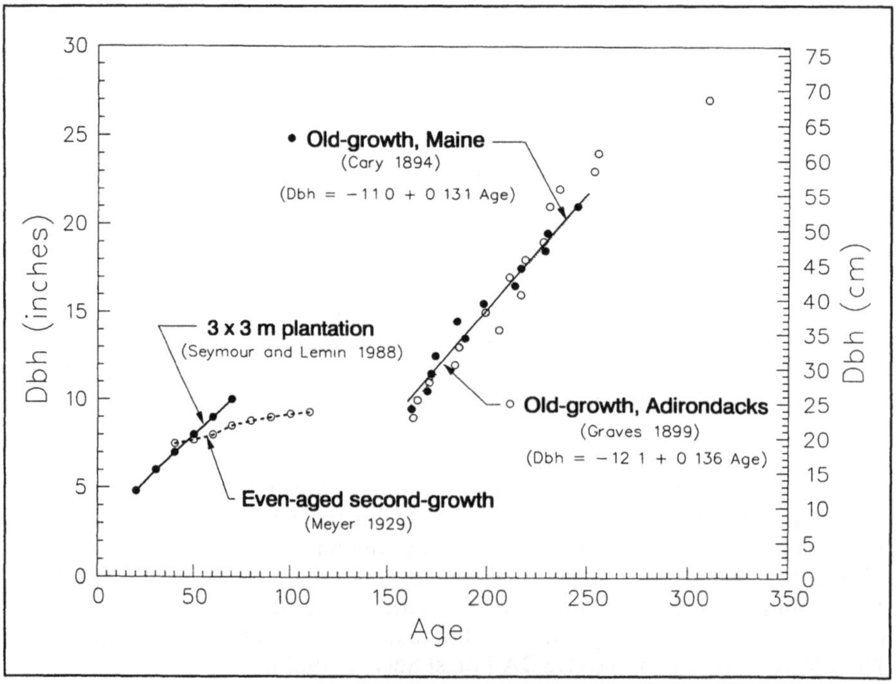

Fig. 1. Comparison of diameter-age relationships of virgin old-growth red spruce with managed and unmanaged even-aged stands.

Graves (1899) also measured recent periodic growth, and found that trees averaged 0.25-0.30 cm (0.10-0.12 inches) of diameter per year, with little pattern by diameter class other than a slight increase in the 28-36 cm (11-14 inch) classes. These growth rates are slightly less than the diameter-growth rate implied by the least-squares regression slope of 0.33 cm (0.13 inches) per year, but are significantly greater than the *average* dbh growth of 0.15-0.23 cm (0.06-0.09 inches) for all dbh classes over their entire lives. The fact that mean annual growth was still increasing, even for spruce over 50 cm (20 inches) dbh, is further evidence that such trees initially developed through a period of suppression where they grew at a below-average rate. More recent studies of age structure of intact, old growth stands also document the presence of several age classes, and an overall relationship between age and size such that the largest diameter, dominant red spruces were usually the oldest trees in the stand (Oosting and Billings 1951; Leak 1975).

The relative proportions of even- and uneven-aged stands in the presettlement forest are uncertain. Cary (1896) described height-growth patterns of free-growing spruces on two different soils, but noted that "but a small proportion of the trees that make up our spruce lumber have grown in any such way. Most have grown under a shade, often a dense and overpowering one." Cary (1896) further described a common forest type of "small and thick spruce timber" growing on "rocky knolls or ridges alternating with swampy ground" as "very old second growth which started up after some primeval fire or blowdown." He also encountered a 110-year-old fire-origin stand in which spruce was still in a subordinate stratum relative to the dominant white pine and declining paper birch and aspen, and commented "plainly, the prehistoric forest was by no means free from fire", but the clear implication was that such an origin was quite uncommon. Graves (1899) observed that pure spruce-fir stands, which occurred mainly on poorly drained flats or shallow, organic soils on upper slopes, tended to exhibit even-aged structures, as a result of their shallow-rooted habit and high risk of complete blowdown there. When Meyer (1929) sought even-aged stands for the first red spruce normal yield tables, he relied heavily on old-field stands; natural even-aged red spruce stands in Meyer's data base throughout northern New England occurred mostly on poorly drained sites.

It is possible that even-aged stands were common, but overlooked by early foresters simply because they were either immature, small-diameter, or dominated by balsam fir and hence of little value. Most even-aged stands probably originated after pure spruce-fir stands on poorly drained soils were subjected to severe windstorms; outbreaks of bark beetles or spruce budworm would usually not kill entire stands of spruce unless all trees were very old. Another probable origin is a heavy partial disturbance followed by windthrow of isolated surviving trees with unstable height:diameter ratios resulting from having grown in dense, undifferentiated stands (Oliver and Larson 1990). Post-disturbance blowdown was a common pattern of development after early partial cuttings, which led Cary (1899, 1902) to recommend clearcutting such stands. Fires were uncommon, and usually created stands of intolerant hardwoods which can require over a century to develop a composition of pure spruce-fir. The apparent lack of truly even-aged stands does not necessarily imply the existence of more than two age classes. Nevertheless, if extended suppression were the characteristic origin of dominant, mature spruce trees, then small-scale releasing disturbances must have been the dominant influences on stand development and structure.

Historical logging practices

While study and analysis of structure and development of virgin spruce stands offers many interesting and provocative ideas, over a century of logging in Maine's spruce-fir region has greatly reduced the possibility of further research. Nearly a century ago, Cary (1896) estimated that only 14% of the entire Kennebec River drainage had never been cut for spruce. Hosmer (1902) also commented that the virgin stands he studied "represents a class of forest of which very little is now left in Maine." Oosting and Billings (1951) found abundant examples of virgin old-growth in the southern Appalachians, but were able to locate only four northern examples in the late 1940s, all in the White Mountain National Forest. Nearly all the present commercial forest in Maine has developed in response to some kind of harvesting, and likely exhibits structures and compositions that may be quite different than those of virgin stands. The following chronological narrative, patterned after Seymour (1985), is offered on the premise that an understanding of how forest structures influenced logging practices, and vice-versa, is valuable background for discussing more intensive silvicultural systems for today's forests.

Although Maine is probably most famous for its logging of old-growth white pine, there is little doubt that red spruce, not pine, has been the staple of both early and present industries in Maine. Logging in the early 1800s initially concentrated solely on pine, and the easily accessible pine resource was depleted within a few decades. By ca. 1870, spruce supplanted pine as the dominant species in the annual river drives (Cary 1896). The first spruce harvests during this period tended to cut only spruce over 30-40 cm (12-16 inches) dbh in a highly selective fashion, probably from mixed stands on deep well-drained soils where the largest trees grew. As markets improved and large-diameter stands became scarce, many stands were cut repeatedly for sawlogs, each time to lower diameter limits, until all timber over 25 cm (10 inches) stump diameter was removed. In the 1890s, the rapidly developing pulp and paper industry built several mills on major rivers and began large-scale acquisitions of timberland formerly cut over for pine and spruce sawlogs. Unlike other regions such as the Lake States and parts of the Appalachians that were liquidated during the same period, Maine's spruce-fir forest remained well stocked with merchantable trees after the initial wave of exploitation for sawlogs. As a result, many stands were again harvested to even smaller diameter limits. Continued lowering of merchantability standards allowed not only stands cut previously to be re-entered, but also rendered operable "bunches" of small-diameter stands that had been skipped over during earlier entries into the same area (Cary 1896).

The increased severity of cutting in response to pulpwood demands was perceived with concern. Cary (1896) described an early pulpwood operation near Berlin, N.H. as follows:

> "It is the hardest cutting ever seen by the writer...the surface of the ground was an almost unbroken brush-heap...Plenty of ground that started with fifty hadn't more than two or three cords[1] of wood of any kind standing on it... A hundred years will not suffice to grow another crop of spruce logs ... and at two hundred it could not fail...to be much smaller than the original stand."

[1]One cord of spruce equals approximately 85 cubic feet or 2.4 cubic meters of solid wood, excluding bark.

In the first report of his extensive regeneration research, Westveld (1928) also described the transition in cutting practices:

> "Increased demands for spruce pulpwood, making possible utilization of small-sized trees, have resulted in a gradual increase in the severity of cuttings, until in recent years the practice is generally being followed of clearcutting lands of all pulpwood species. Prior to the adoption of this cutting method a rough selection system was being practiced in which only trees of large size were removed. Under this method of cutting little difficulty was experienced in keeping the forest in a productive condition."

These early foresters recognized that repeated cuttings had been made possible only by the incompleteness of earlier entries. During this era, stand productivity was measured in terms of how much residual growing stock was left for future cuttings and did not consider future yields from regeneration.

By 1910, fifty years of preferential cutting of old-growth spruce had left extensive areas stocked with old balsam fir of all sizes to respond to release. The epidemics of spruce bark beetle, while probably less extensive, also had essentially the same effect. By early in the 20th century, Maine probably had far more mature balsam fir than had ever existed in the virgin forest (Zon 1914), and the state's spruce-fir resource was subjected to a extensive spruce budworm outbreak that killed an estimated 27 million cords by the early 1920s. While the timing could have been coincidental, early entomologists attributed the unprecedented severity of this outbreak to the unnaturally high fir component of the forest during the early 1900s (Swaine and Craighead 1924). By all accounts, the 1913-19 budworm outbreak left the spruce-fir forest seriously depleted of merchantable trees, and pulpwood shortages apparently were regarded as inevitable. In an assessment of Maine's future pulpwood supply, Clapp and Boyce (1924) wrote:

> "The outlook..is probably an enforced curtailment of pulp and paper production...which will hit first and hardest the pulp mills without available timber supplies of their own. The cut of many other mills will probably be shifted in much greater degree than at present to their own inadequate holdings, with still more serious overcutting. It is very doubtful if immediate application of the most intensive forestry measures over the entire spruce-fir type of the State can produce results soon enough to prevent such a curtailment."

Historical evidence reviewed by Seymour (1985) shows volumes per acre standing and harvested during the 1930s as low as 18-30 cubic meters per ha (3-5 cords per acre). Harvesting operations covered larger areas than formerly, removing scattered remnants of the virgin forest that survived the outbreak in very understocked stands.

Beginning ca. 1950, stands regenerated after early pulpwood cuttings and the 1913-19 budworm outbreak began to reach merchantable size over large areas. Growth rates and stocking levels increased dramatically, and the first official assessment of the Maine forest (Ferguson and Longwood 1960) showed a large surplus of periodic growth over harvest. Cutting practices from the late 1940s through the 1960s were dominated by diameter limit cuttings that varied by species (Hart 1963). These cuttings tended to remove the larger spruces from two-storied stands, most of which had survived and responded to release as saplings or small poletimber from the earlier sawlog cuts and budworm attack. Periodic growth continued to exceed harvest during the 1960s (Ferguson and Kingsley 1972) as the early-1900s-origin age class began to mature into high-volume, single-canopy stands. Diameter-limit cuttings in this age class became increasingly unsatisfactory. As removal rates rose, windthrow of residual stands became more severe, just as they had after similar cuttings in old-growth stands several decades earlier. Then in the mid-1970s, a massive budworm outbreak infested the entire resource, necessitating large-scale annual insecticidal protection programs. Within a short time during the late-1970s, clearcutting became an important (although not necessarily dominant) harvesting practice to pre-salvage dying stands dominated by fir. Pre-salvage clearcutting continued through the early 1980s until the outbreak subsided, and some landowners continue to rely primarily on clearcutting.

Evolution of stand and forest structure, 1860-1990

While the primary purpose of harvesting in Maine's industrial spruce-fir forest has always been pulpwood production, not silvicultural treatment, there are nevertheless important lessons for silviculturists in the patterns of stand response. For example, merchantability limits for spruce pulpwood have not changed greatly since the turn of the century, yet "clearcuts" designed to harvest all merchantable pulpwood have produced very different results depending upon the particular stand structure(s) that were common during each era. The original, scattered sawlog cuttings removed only modest volumes from irregular, old-growth stands; their main silvicultural effect was probably to encourage establishment, and partial release, of large advance regeneration in the understory or in small gaps (Fig. 2). Subsequent re-entries, first for smaller logs and then for pulpwood, probably tended to release this regeneration more or less completely, as well as any residual, older balsam firs that were usually not merchantable. Such heavy cutting was often followed on poor sites by extensive windthrow, and virtually all stands were then subjected to the 1913-19 spruce budworm outbreak, further reducing the representation of the older age classes and giving complete occupancy to the already well established regeneration. Cutting in the decades following the outbreak probably continued this pattern. Over time regeneration established during or prior to the outbreak continued to develop, but merchantable volumes remaining in unharvested stands continued to decline as higher-volume stands were harvested. The budworm's effect was so pervasive that virtually all spruce-fir stands that had not already been released by early cuttings were at least partially regenerated by 1925. Indeed, most spruce-fir harvesting from before 1900 until the 1960s could be characterized as a staged liquidation of original members of the old growth forest that had survived the early partial sawlog cuts and the budworm attack and had not blown down, gradually releasing somewhat irregular, but essentially even-aged stands. This history has created a forest with a seriously unbalanced age structure, dominated by even- or two-aged stands that originated in the decades surrounding the peak budworm mortality ca. 1920 (Fig. 3).

When the more uniform, even-aged stands regenerated during the early part of the century became merchantable during the late 1960s, new and different regeneration patterns began to emerge when excessive losses to windthrow and budworm risk led some landowners to substitute clearcutting for the previous diameter-limit prescriptions. Stand structures were quite uniform and had matured to the point where small-diameter, unmerchantable stems had mainly died from suppression or budworm defoliation. As a consequence of their younger age, higher stand density, and total lack of previous disturbances, advance regeneration beneath these stands was either absent or poorly established. Like earlier commercial clearcuts, these harvests left essentially no residual growing stock, but unlike previous cuts, large well established advance growth was not present to occupy the cutover sites immediately. Hence, stand development tended to revert to an earlier successional stage dominated by pioneer vegetation, with spruce and fir relegated to substrata or shaded out completely except on the poorest sites. Unlike earlier "clearcuts" which merely released well established advance growth, harvesting operations began to regenerate truly new age classes on a significant scale.

The "New Forest"

The past century has witnessed a dramatic change in the structure of Maine's spruce-fir forest. These large-scale, episodic removals of mature trees, by logging in combination with spruce budworm, bark beetles and windstorms, have transformed Maine's spruce-fir forest from one dominated by mixed-aged, old-growth stands, to a forest dominated by younger, more uniform stands that may be more extensive in area and lack the within-stand height and age diversity of the old growth. This change in stand structures has, in turn, been accompanied by a change in silvicultural emphasis by large industrial landowners. Partial cutting systems designed to exploit the ability of residual trees and large advance regeneration have gradually given way to even-aged plantation-like silviculture that now characterizes practice in most intensively managed conifer forests worldwide. This rapidly expanding age class is known as the "New Forest", not only on account of its young age, but because it exhibits patterns of development in response to cutting and management that may well be unprecedented.

The maturation of uniform, second-growth stands and the related onset of clearcutting as a widespread harvesting practice have led to an oversimplified view of the current spruce-fir forest and its silvicultural challenges. Much silvicultural research on "intensive management" currently focuses narrowly on creating and culturing uniform, even-aged stands. Recent evidence suggests that actual harvesting and silvicultural practices encompass a much broader spectrum of activities and stand structures. For example, if clearcutting had become the dominant regeneration measure during the 1970s, then the 1-10-year old age class in 1980 (Fig. 3) should comprise well over 0.4 million ha. The actual area is less than 20% of this total (Seymour and Lemin 1989), suggesting that practices other than complete clearcutting predominated during this decade. Some of this apparent discrepancy can be explained by spruce-fir clearcuts that reverted to the early successional aspen-paper birch forest type; however, most is likely a result of continued use of various types of partial harvests which leave significant numbers of residual trees, similar to most historical practices. Such stands would fall in the "two-storied" and "uneven-aged" categories (Fig. 3), which contain all plots that do not clearly exhibit an even-aged structure including the very common two-aged stands. While few, if any of these uneven-aged stands contain any semblance of a balanced age structure, they represent an important category that cannot be ignored in any comprehensive treatment of the resource.

The apparent diversity in silvicultural practices and stand structures is further supported by 1986 mid-cycle remeasurement of Maine's spruce-fir resource (Maine Forest Service 1988). In an attempt to avoid the confusion inherent in classifying age structures of stands that contain trees with suppressed origins, stands were

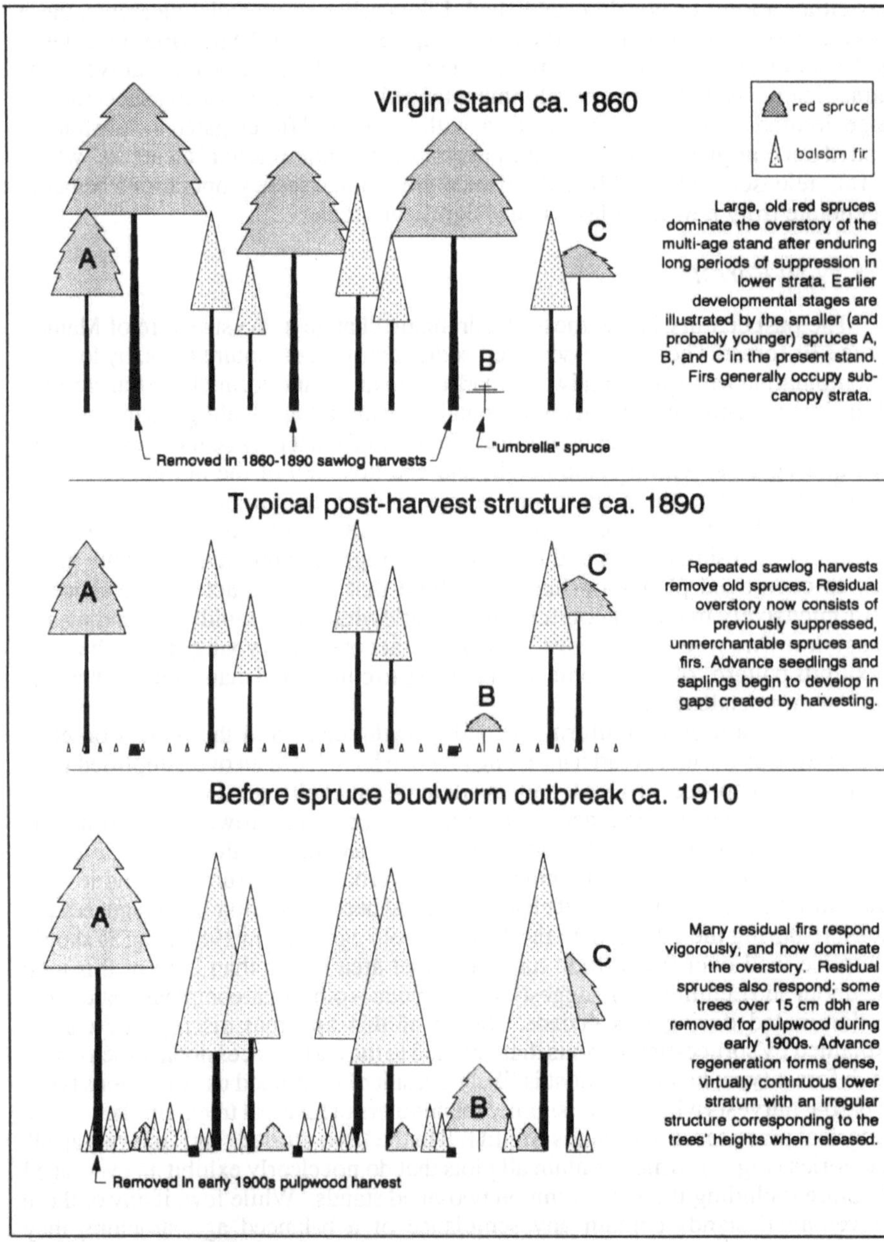

Fig. 2. Development of typical spruce-fir stand after logging and budworm attack ca. 1860-1970.

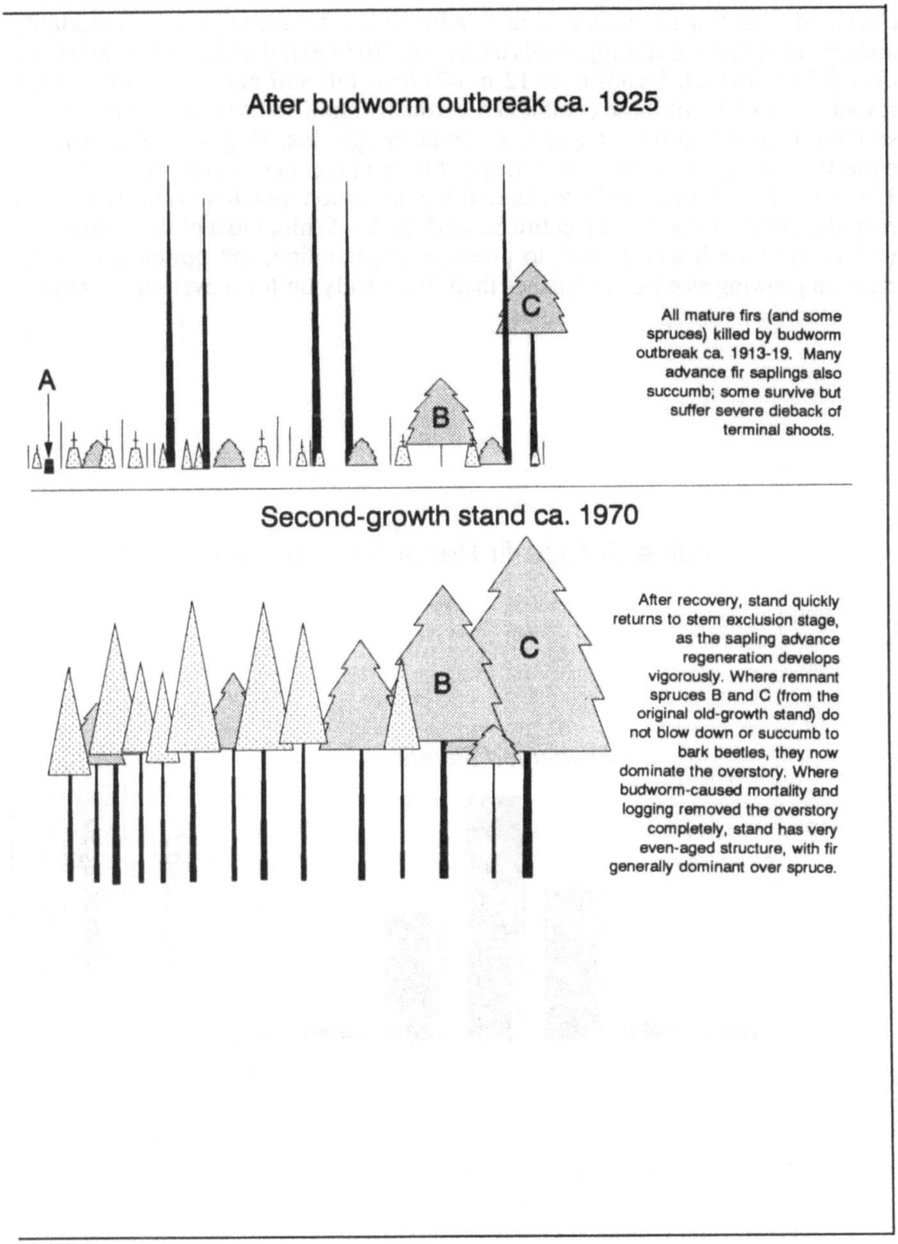
After budworm outbreak ca. 1925
A
B
C
All mature firs (and some spruces) killed by budworm outbreak ca. 1913-19. Many advance fir saplings also succumb; some survive but suffer severe dieback of terminal shoots.
Second-growth stand ca. 1970
B
C
After recovery, stand quickly returns to stem exclusion stage, as the sapling advance regeneration develops vigorously. Where remnant spruces B and C (from the original old-growth stand) do not blow down or succumb to bark beetles, they now dominate the overstory. Where budworm-caused mortality and logging removed the overstory completely, stand has very even-aged structure, with fir generally dominant over spruce.

classified according to height structure. After over a decade of budworm mortality and intensive salvage cutting, stands under 3 m (10 feet) tall accounted only for just over 0.2 million ha. Stands over 12 m (40 feet) tall still comprised 74% of the resource, and 59% of these exhibited a vertically stratified structure with a lower stratum of either a different species or younger age class (Fig. 4). These data are entirely consistent with 1988 summary of harvest practices (Maine Forest Service 1989) which show that complete clearcutting is less common than various forms of partial cutting throughout the entire State (Fig. 5). While most of the partial cuts undoubtedly are heavy enough to promote regeneration, the presence of some residual growing stock distinguishes them from truly uniform even-aged stands.

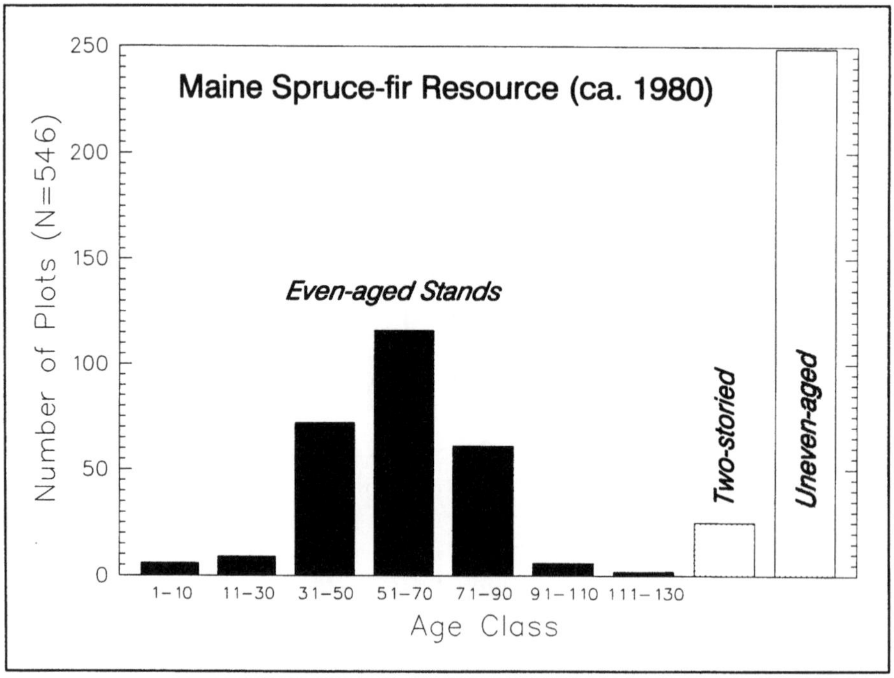

Fig. 3. Approximate age structure of Maine's spruce-fir resource as classified by the USDA Forest Service ca. 1980. [Source: unpublished data used by Powell and Dickson (1984)]

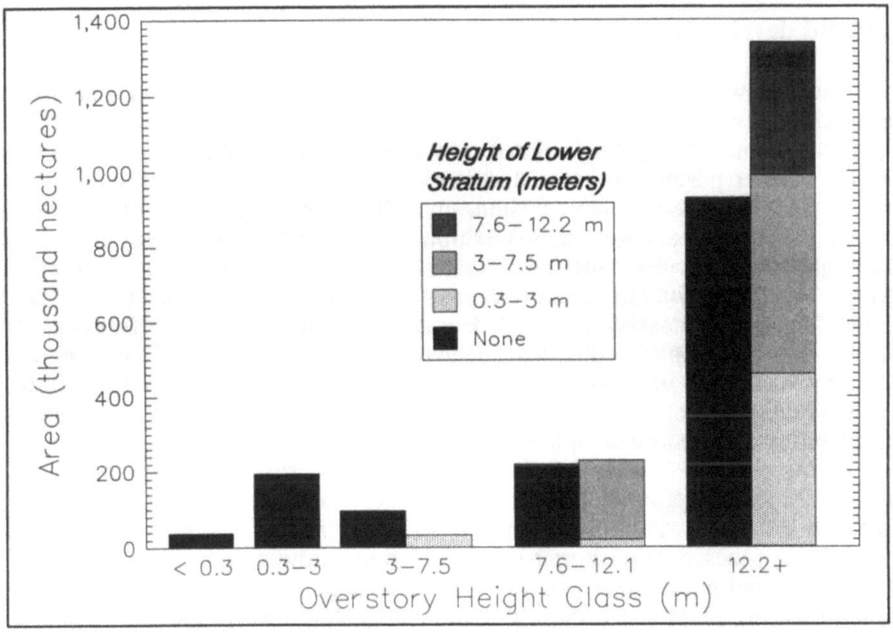

Fig. 4. Height structure of Maine's spruce-fir resource, 1986. [Source: Maine Forest Service (1988)]

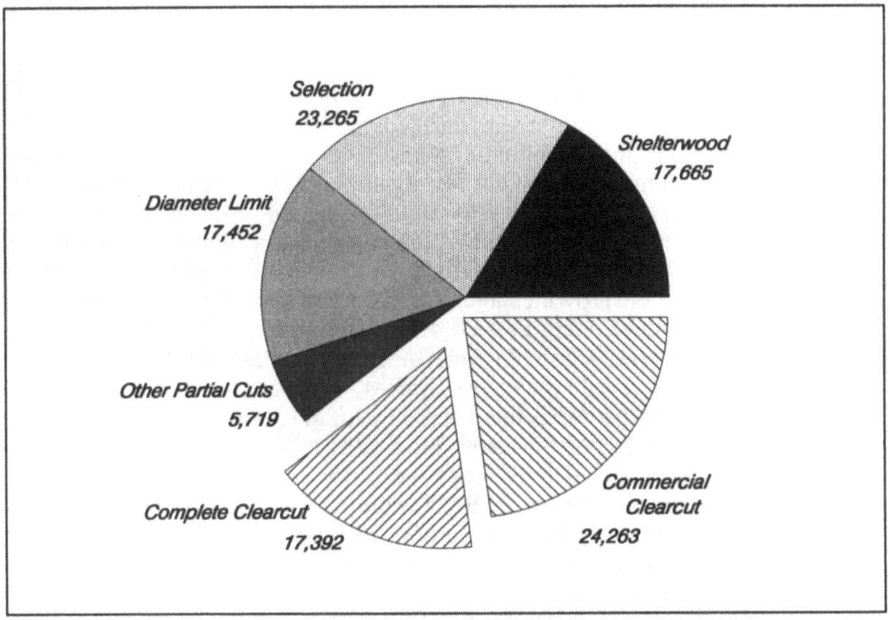

Fig. 5. Areas (hectares) treated by various harvesting practices in Maine, 1988. [Includes entire state; data for spruce-fir type only not available. Source: Maine Forest Service (1989)]

A stand development model

The following section attempts to assemble the historical and scientific information reviewed above into a working hypothesis of stand development for both natural and logging-origin red spruce-fir stands, using the model of Oliver (1981). Oliver postulated that disturbance-origin forests develop through four stages: stand initiation, stem exclusion, understory reinitiation, and old-growth. During the first three stages, stands exhibit a more or less even-aged structure, although their size structure may be quite stratified if species composition is mixed. Prior to exploitation, natural stand structures were probably dominated by the "old growth" developmental stage, in which new age classes may be continually released but which never completely dominate the overststory (Lorimer 1977). Many-aged, old-growth stands were probably most common on well drained soils supporting mixedwood stands, or pure spruce-fir stands in sheltered topographical positions which rarely experienced complete mortality from windstorms or other agents.

Pure spruce-fir stands on shallow soils are more susceptible to large-scale overstory mortality from "releasing disturbances" that release any younger age classes that may be present. When the overstory begins to break up gradually (for example, by senescence of shorter-lived intolerant hardwoods or fir, or budworm defoliation) during the understory reinitiation stage, abundant spruce and fir advance regeneration can occupy the understory completely prior to complete destruction of the overstory by windthrow or insect attack. In this case, the stand initiation stage is bypassed completely, and stem exclusion (defined as the time after which no new trees are recruited into the stand) begins immediately. Stem exclusion may prevail for 50 years or more, depending on the specific fir-spruce composition. As fir begins to drop out, either from episodic budworm outbreaks or rot-induced wind breakage, the stand again enters understory reinitiation, during which small advance seedlings, mainly of fir, begin to develop in the understory. Another 50-100 years of stem exclusion may elapse during which a two-aged structure prevails, with 50-150-year-old spruce dominating younger fir except in large gaps where a pure-fir composition may develop. As fir again begins to drop out of the stand, spruce seedlings and saplings that may have originated at the same time as the declining fir (Davis 1989), respond and become a more important stand component. After 150 years, stands may eventually develop a true, multi-aged structure indicative of Oliver's "old growth" stage, with spruce of several classes and younger fir opportunistically developing in gaps of varying scale. Once this structure is attained, it could conceivably be maintained indefinitely by small-scale disturbances or senescence that affect only the oldest spruces.

If this successional pattern is interrupted by logging or major disturbance, the resulting outcome depends on the status and fate of the advance regeneration and formerly suppressed residual trees from the older age classes, if any survive. Response is also influenced by site quality; the proportion of hardwood species; and where advance growth is sparse, the amount of disturbance to the forest floor. Extremes would be an understocked old-growth spruce stand with a completely regenerated understory of large spruce saplings that is killed by bark beetles, and an immature (age 40) even-aged stand still in stem exclusion killed by wildfire. The former would almost certainly respond promptly and maintain more or less the same pure-conifer composition, while the second would likely revert completely to intolerant hardwoods. An important lesson from recent experience is that superficially similar harvesting operations, which differ only slightly in their timing of overstory removal relative to the status of advance regeneration, can have a major effect on subsequent stand composition and resulting development.

Historically, stands reached understory reinitiation prior to (or at least coincidental with) their becoming merchantable for pulpwood. The type of commercial clearcutting practiced from the early 1900s to the 1960s rarely resulted in inadequate regeneration if logging damage was controlled; the main issue was an increased proportion of fir in the second-growth stands (Zon 1914; Westveld 1931). More recently, however, predicted wood scarcities throughout the spruce-fir region (Seymour and Lemin 1989) have led to cutting smaller-diameter, younger stands that may not have reached, or have only begun, understory reinitiation. Such complete clearcuts midway during the stem-exclusion stage have virtually no natural analogue. Spruce-fir stands rarely collapsed completely at this relatively young age, except perhaps when subjected to very uncommon hurricane-force windstorms which may completely blow down immature stands.

Unlike historical cuttings where the stand initiation stage occurred in the understory, complete "clearcuts" in such immature stands cause stand initiation to occur after harvest where much growing space is left vacant. Although small advance seedlings of spruce and fir may be present, they become rapidly overtopped by a vigorous growth of pioneer species (*Rubus* spp., *Betula papyrifera, Prunus pensylvanica, Populus* spp.) that do not develop nearly as vigorously if advance spruce and fir seedlings and saplings fully occupy the growing space prior to overstory removal. Site quality also strongly influences competing vegetation; on poorly drained soils, the main problem is overtopping by low-vigor intolerant hardwoods which may suppress growth but usually does not reduce stocking. On better soils, however, vigorous development of tolerant hardwoods and *Rubus* may suppress and eventually kill overtopped seedlings, causing a major forest type change. Small red spruces appear to be even more vulnerable than firs. Harvesting disturbance can further exacerbate regeneration problems by stimulating germination of buried seeds of weeds, by destroying advance seedlings, or by removing residues that provide "dead shade" for fragile advance seedlings (McCormack 1984; Seymour 1986).

Even in cases where casual observation suggests that conifer advance regeneration is well established, different patterns may emerge in the relative development of red spruce and balsam fir. Fir begins frequent seed production at a younger age than spruce, and seedlings are more robust because of their more vigorous root development (Place 1955). In dense even-aged stands early during the understory-reinitiation stage, advance growth will consist of nearly pure fir, even if the overstory is predominantly spruce. Even after both species are present as advance growth, fir usually continues to outnumber spruce and tends to respond more aggressively to release; in general, firs less than a half-meter tall will outgrow spruces of equal size during the decade immediately after release (Westveld 1931; Davis 1989). Later in development, rates of height growth become similar, but the initial superiority of fir conveys a lasting competitive advantage. Spruce saplings with a flat-topped "umbrella" morphology respond similarly to shorter spruces with full crowns, but may retain an advantage over fir because of their initially taller stature. Westveld (1931) found an optimum height growth response in spruces 1.0-1.5 meters tall that averaged 30-48 years old when released. Thus, spruce must attain a height advantage of 1-2 meters over fir in the understory prior to release in order to assure its continued presence in the overstory of the developing stand. Since light, partial disturbances favor small firs as much or more than small spruces, it is only after several decades that spruce eventually gains an advantage because of its greater ability to withstand extended suppression (Davis 1989).

Partial cutting can alter natural development patterns by selectively removing spruce from mixed stands, resulting in potentially long-lasting changes in stand composition. Although there is no definitive evidence, it is commonly believed that much of Maine's northern hardwood type, and even some mixedwood stands, have been converted to essentially pure hardwood stands via preferential high-grading of spruce (Weiss and Millers 1989). Concentrated partial cutting of spruce from pure spruce-fir stands over large areas has also likely changed spruce:fir proportions in favor of fir. This change may be even more dramatic and long-lasting after stand regeneration. Complete overstory removal cuttings in biologically immature spruce-dominated stands may cause a radical change in subsequent composition to fir, because of the greater initial abundance and establishment ability of advance seedlings of this species. The dominance of red spruce in virgin forests was because of its great longevity relative to fir. Over a century of harvesting concentrated until recently on spruce has effectively removed this critical natural advantage, and has led to increasing dominance of fir in young stands.

The above model of spruce-fir development does not attempt to describe subtle variations in development caused by regional differences in site quality, interactions with other species, responses of competing vegetation, effects of insects and disease, and many other ecological and socio-economic factors. For example, the mixtures of spruce and fir with hemlock and eastern white pine typical of northern Washington County respond quite differently to disturbance than the fir-dominated stands of the St. John Valley. The model is offered as a point of departure to structure further studies designed to clarify these important regional patterns.

Silvicultural systems

This section attempts to outline how silviculturists have applied knowledge of stand development gleaned from over a century of experience and research. Before considering specific systems, it is helpful to summarize some general principles of spruce-fir silviculture that apply to any system. The main even-aged regeneration methods currently in use are outlined in Fig. 6; variants of these general procedures and specific treatment options are detailed below.

Stand establishment

Avoiding regeneration failures requires that adequate stocking of advance seedlings be achieved prior to the final removal of the mature overstory. This simple but effective principle was advocated by Westveld (1931) and has been stressed repeatedly, most recently at a regionwide symposium (Needham and Murray 1991). Prior to 1950 when logging was confined mostly to winter with snow cover using relatively benign technology, in stands that were well-regenerated with large saplings as a result of prior partial entries, there was little need for concern about understocking. Indeed, Smith (1981) characterized this ease of regeneration as a "magic" property of the eastern spruce-fir forest in comparison to many other spruce forests worldwide where lower rainfall and vagaries of seed production make

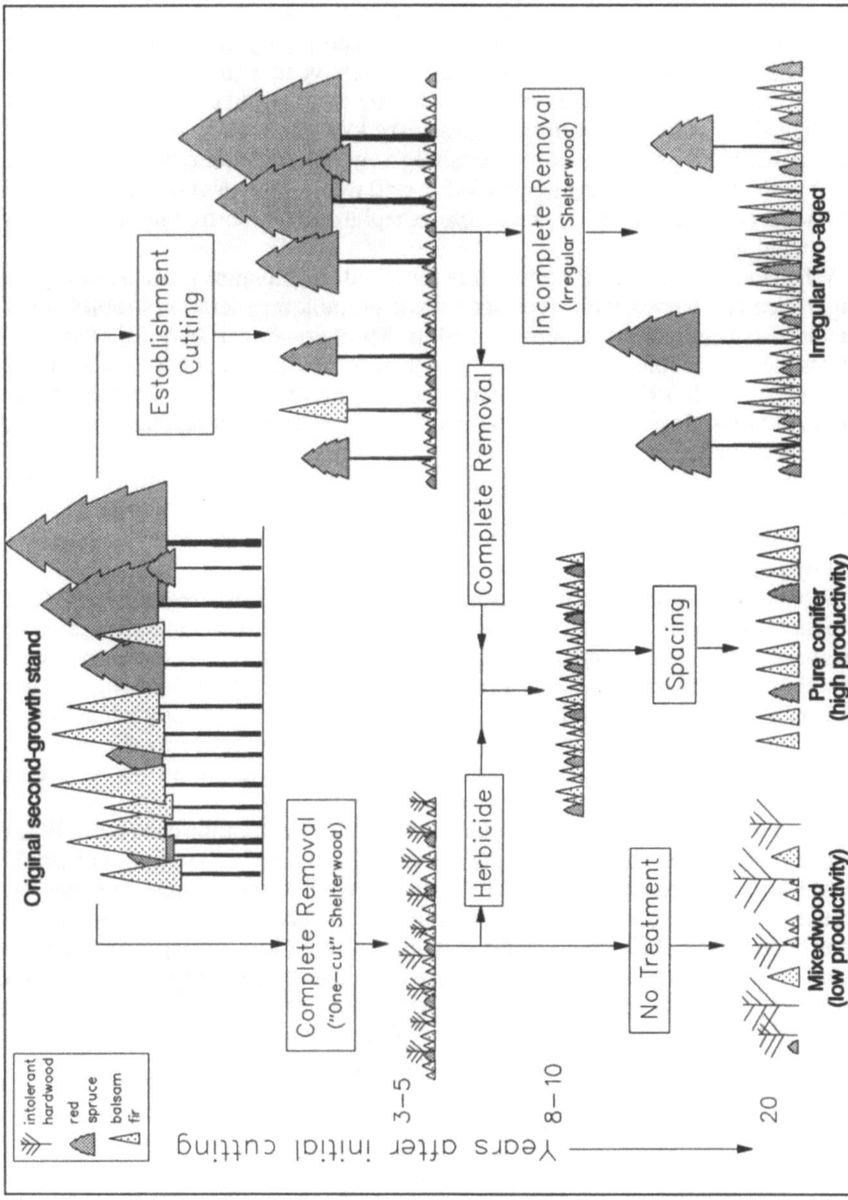

Fig. 6. Even-aged regeneration options for typical second-growth spruce-fir stands.

natural reproduction much less dependable. Since then, development of mechanized logging, year-round operations, and dense, even-aged stands with no history of disturbance and sparse advance regeneration, force close attention to the regeneration process.

Mature second-growth stands of today typically have adequate stocking of small (under 30 cm) advance growth that developed naturally in the understory. The first consideration is whether seedlings can withstand abrupt exposure. The traditional rule that trees must be 15-30 cm tall (Frank and Bjorkbom 1973) has been found to be conservative; smaller seedlings can survive in great numbers if they are rooted in mineral soil or if they are protected by dead shade of logging residues. Excessive site disturbance during logging risks loss of advance growth and exacerbates subsequent problems from competing vegetation. Winter cutting on frozen ground with snow cover using controlled-layout tree-length or shortwood systems with on-site delimbing is vastly preferable to haphazard whole-tree skidding during the growing season.

When the mature stand is completely removed and advance growth released in a single "clearcut" harvest without prior entries to promote regeneration establishment, silviculturists have adopted the terms "one-cut shelterwood" and "overstory removal" to stress the crucial role of advance seedlings in successful regeneration (Seymour et al. 1986; Smith 1986). True silvicultural clearcutting, where new seedlings develop in exposed microsites after the final harvest, is a very undependable method for regenerating spruce-fir stands. Like most non-serotinous conifers, little red spruce or fir seed remains viable beyond the year of dissemination (Frank and Safford 1970). If there is no seed crop in the year of cutting, or if a "catch" is not obtained within two years or less, clearcut sites invariably are completely dominated by pioneer vegetation.

Whether subsequent treatments are necessary to release regeneration from competition depends on the height and density of the advance seedlings when released, as well as the site quality which governs composition and development of competing vegetation. Seedlings under ca. 30 cm tall generally will become overtopped by pioneer vegetation except on the poorest sites, and early release using a selective herbicide is required to ensure continued dominance by the conifers. Larger advance growth (1-2 meters tall) will usually outgrow all but sprouting hardwoods without treatment.

Often advance growth is so abundant that severe overstocking can result. While stands usually do not stagnate and may eventually produce high volumes of small-diameter trees, rotations can be seriously delayed if early thinnings are not conducted. Concerns about species conversion, from spruce to fir and from conifers to hardwoods, are difficult to address with regeneration cuttings alone, since virtually all practices (such as conventional shelterwood cutting) that favor red spruce also favor fir or tolerant hardwoods. Thus, species composition is best controlled by recruiting adequate densities of the favored species and eliminating competitors with subsequent cultural treatments.

Growing stock manipulation

Most recommendations and experimental attempts at spruce-fir management have involved marginal changes in the common logging rules of the time, usually in the form of leaving more residual spruce growing stock. These partial cuttings tended to be a combination of thinning and release cutting, applied to the common, irregular stand structures of the time that defied generalization. Conventional density management regimes involving repeated commercial thinnings in uniform, even-aged stands, have been applied only experimentally. Virtually none of the large age class of second-growth even-aged stands was ever treated this way, even though stocking guides and suggested thinning schedules (Frank and Bjorkbom 1973) have existed for nearly 20 years. Poor windfirmness has been the major constraint to applying any sort of thinning or partial cutting system to middle-aged or older stands, and most attempts have found that blowdown losses nearly offset accretion of residual trees. Unlike most hardwoods, pine, and hemlock, spruce and fir are not windfirm as individual trees. The challenge is to make partial entries as light and uniform as possible, while taking care to leave fully stocked residual stands that resist wind as an integral unit. Lack of windfirmness is strictly an economic problem; chronic losses of individual trees or small patches are ideally adapted to perpetuating these species.

Precommercial thinning of densely stocked natural regeneration, to create uniformly spaced crop trees similar to a plantation, has received increasing interest during the last two decades (Murray and Cameron 1987). Studies of pure fir stands on good sites in northern New Brunswick established by G. L. Baskerville in the late 1950s show impressive growth responses, and suggest an optimum spacing of 2 meters to maximize volume production of pulpwood-sized (10 cm dbh) trees (Ker 1987). Ten-year results from the only study in mixed red spruce-fir stands (Frank 1987) show that fir responds more than red spruce, but that all species benefit from reduced competition. Commercial thinnings removing 10-20 cm trees are also contemplated as the "New Forest" reaches merchantable size, using single-grip harvesters developed to thin similar small-diameter spruce stands in Scandinavia. Such thinnings are still experimental, and no consensus has been reached about issues such as timing, thinning method, removal rates, residual stocking levels, or whether such treatment is even desirable.

Manipulating stand composition via partial cuttings that remove vulnerable fir to increase stocking of resistant red spruce and non-host species has long been recommended to reduce vulnerability to spruce budworm damage (Blum and MacLean 1985). Repeated trials show that such a strategy may be viable only when begun early in stand development, and where it is possible to create pure spruce stands (Dimond et al. 1984). Even pure spruce stands, while demonstrably less vulnerable than fir, will likely suffer economically significant growth reduction and tree mortality without insecticidal protection (MacLean 1985; Osawa et al. 1986) so this lower vulnerability may not be an exploitable advantage in today's forest economy with projected shortages. A further problem is that multi-storied stands, recommended by some authors (e.g., Westveld 1946) to reduce budworm vulnerability, may suffer nearly complete loss of the lower regenerating strata, since they are subjected to heavy feeding from dispersing large larvae and cannot be protected with conventional insecticide treatments that do not penetrate the overstory canopy. Hence, virtually any stand structure that can be used to grow significant volumes of spruce and fir also risks large losses from an uncontrolled outbreak (Baskerville 1975b). Currently, the best hope is to grow vigorous stands that will respond to protection when future outbreaks develop.

Selection cutting

Foresters who carried out pioneering studies of spruce-fir growth and management originally focused their attention on devising various diameter-limit prescriptions (generally, 36 cm or 14 inches dbh) that would increase future growth in comparison to the heavier cutting rules of the time that removed most trees over 23-25 cm (9-10 inches) dbh. Since the large-diameter spruces were very old (Fig. 1), cutting them could easily be justified on the basis of financial maturity, as well as to release the ostensibly younger, smaller-diameter poletimber and saplings. Using stand-table projection techniques, they estimated future growth and speculated about volumes removed in future cutting cycles of 20-40 years. A crude sort of selection management emerged as a paradigm, designed to take advantage of low logging costs of large trees and the usual abundance of small-diameter growing stock with a potential to respond to release. Westveld (1953) and Hart (1963) continued to recommend a slightly more refined type of selection cutting, implemented through differential diameter limits and careful individual tree marking based on risk and vigor classes. There is little evidence, however, that marking to a particular diameter-class structure was applied during this period; such attempts are more recent, and limited to the Penobscot Experimental Forest (Frank and Blum 1978).

As described earlier, actual harvesting practices from the beginning of spruce cutting ca. 1860 until the budworm outbreak 50 years later involved relatively frequent re-entries into the same stands. This may have created a false perception that stands would indeed yield periodic volumes indefinitely. Close examination of these studies reveals that repeated cutting was made possible largely by continued reduction in merchantability standards, not by accretion of residual trees. Both Graves (1899) and Hosmer (1902) found that a small proportion (18-20%) of merchantable residual trees actually increased in growth in response to earlier sawlog cuttings. Dramatic growth responses were more typical of saplings (Westveld 1931). Such frequently repeated cutting actually had the effect of narrowing within-stand age diversity in comparison to the virgin structures, not enhancing it. Initially, cuttings removed the oldest age classes without creating truly new ones; later entries then were forced to cut much of the remaining growing stock, regenerating far *more* area than a sustained yield calculation would dictate. At best, cuttings tended to create and maintain a two-aged stand structure, by periodically removing much of the merchantable growing stock while leaving some undersized, old residuals to respond along with abundant advance regeneration.

Even-aged systems

The transition to even-aged silvicultural practice in spruce-fir parallels that in other forest types in the Northeast (Seymour et al. 1986) and United States (Smith 1972), as large areas of even-aged stands regenerated early in the century began to mature in the 1960s. Recent summaries of spruce-fir silviculture (Frank and Bjorkbom 1973; Blum et al. 1983) tend to emphasize conventional treatment of uniform, even-aged stands, but as noted above, such practices apply to less than half the resource. The most common practice appears to be heavy partial cuttings which regenerate much of the stand, but which leave significant numbers of residual trees that often respond to release and that materially influence development of the reproduction. These stands usually have an irregular, two-aged structure, falling between the extremes of plantations and balanced selection stands. In general, stand development probably follows a pattern similar to that described by Marquis (1991) for Allegheny hardwood stands with residual sugar maple poles over black cherry (*Prunus serotina*) advance seedlings. In this case, red spruce, white pine or hemlock represent the "two-rotation" trees analogous to sugar maple; balsam fir would be the main short-rotation species (analogous to black cherry), along with an occasional, minor component of paper birch and aspen (Smith 1986; Davis 1989). Superficially, such cuttings resemble the seed-tree method. However, since most of the new age class originates as advance growth, they are best thought of as incomplete removal cuttings in an irregular shelterwood system.

Irregular shelterwood. Most two-aged management described above has been quite extensive in application; often, two-aged stands result merely because merchantability differences dictated different diameter limits by species, or because high logging costs of small trees have made them difficult to remove economically. Intensive application of two-aged irregular shelterwood management is more recent, and no consensus has emerged. Attention is focused on early selection of the two-rotation trees that will be retained after the final, incomplete removal cutting of the main age class, and that will be allowed to grow at least part way through the next rotation of the regenerating age class. For example, many even-aged spruce-fir stands contain a few white pines which develop high-quality, knot-free boles as a result of competing with densely stocked spruce-fir. At the end of a 50-70-year spruce-fir pulpwood rotation, such pines reach 20-30 cm (8-12 inches) dbh as codominants, but unlike similar spruces, have the ability to remain windfirm and grow as emergents until 60-76 cm(24-30 inches) dbh or larger, thereby earning very high rates of return. Leaving such pines as residuals involves little sacrifice in harvest value (since often they are only pulpwood or small-sawlog size) and does not interfere with the developing reproduction for at least several decades. Ideally, the final incomplete removal cutting would be preceded by at least one uniform shelterwood establishment cutting entry during a pine seed year, to establish pine advance reproduction along with the normally abundant spruce and fir seedlings.

Another common application of irregular shelterwood cutting involves retention of spruces or hemlocks from the lower crown classes. To respond well and not blow down, such trees must be much shorter than the maximum upper height for the site, so that they can expand their crowns upward after release and develop windfirmness as individual trees (Oliver and Larson 1990). Future commercial thinning schedules must consider developing such holdover trees. Unlike stand structures of the past, it will not suffice to expect a population of small-diameter residuals to "be there" at the final removal cutting. For example, early, heavy dominant or crown thinnings would tend to release partially spruces that had lapsed into lower strata, while at the same time favoring development of large-crowned dominant and codominant spruces or firs. Such treatments may be especially appropriate in stands on good sites where the upper stratum is pure fir, but where the lower stratum contains some intermediate or overtopped spruces of the same age. This practice could help to reverse the prevailing trend towards higher fir composition by maintaining or increasing the spruce component of fir-dominated stands. In contrast, heavy low thinnings in such stands would be inappropriate, since they would probably eradicate spruce from the stand.

Many foresters avoid creating two-storied stands under the belief that irregularity implies a less productive stand structure that also complicates management. Irregular stands may actually be preferable under many circumstances, however. First, several-stage removal of the overstory, culminating with retention of well distributed residual trees, promotes an uneven height structure in the sapling regeneration. This structure accelerates natural expression of dominance and may obviate precommercial thinning. Several decades after the final removal, residuals should develop into large-diameter stems that can be removed along with smaller stems in the first commercial thinning, perhaps making the thinning more profitable or allowing it to be conducted sooner. The naturally pruned, branch-free residuals provide a low-cost way to grow high value, knot-free sawlogs. Carefully chosen residuals also can greatly improve the appearance of harvested sites, and offer a simple alternative to the selection system where aesthetics are important. Finally, and perhaps most importantly, two-storied stands provide a way to use simple, even-aged practices to maintain and favor species with slow juvenile growth patterns such as red spruce and hemlock, which are at a great competitive disadvantage relative to more aggressive species such as fir or most hardwoods when grown in mixed-species, even-aged stands.

Intensive management of the New Forest. Since clearcutting has become a dominant harvesting technology on many industrial timberlands and markets have developed for very small stems of all species, very uniform, even-aged stands have become the focus for high-yield silvicultural systems. Typically, stands are regenerated by one-cut shelterwood removal cuttings, followed at ages 3-5 by an aerial herbicide release treatment to control pioneer vegetation and sprouting hardwoods. Stands also respond to release at older ages, but the current recommendation is to apply herbicide before any overtopping or suppression is evident (Newton et al. 1987). After some initial complacency during the 1970s, herbicide release is now regarded as virtually essential to prevent understocking of conifers or even complete type conversion to hardwoods after clearcutting. If density is high and stocking uniform, motormanual precommercial thinning is sometimes done by contract workers using Scandinavian spacing saws. When this technology was first applied in the early 1980s, efforts were made to favor spruces over firs, even where spruces were quite subordinate. More recently, crop-tree selection rules emphasize the most dominant stems regardless of species. This practice can create nearly pure fir stands in the typical case where spruce lapses somewhat behind fir early in development. Carefully targeted herbicide strip treatments, carried out early in the growing season by helicopters equipped with special booms and precise guidance systems, has also been tested experimentally (McCormack and Lautenschlager 1988), but has not become operational. Since herbicide release yields substantial benefits at modest costs (under $125/ha), it accounts for a much greater area treated annually in comparison to the more expensive motormanual spacing that can cost over $500/ha (Seymour and Gadzik 1985). Herbicide release has become by far the dominant silvicultural investment, and is expanding annually (Fig. 7).

A small amount of the area regenerated annually is planted (Fig. 7), mainly to convert productive soils dominated by previously high-graded, low-quality hardwood stands to monocultures of black or white spruce, red pine, or exotic larches. Site conversion planting has been greatly facilitated by development of wood energy markets and biomass harvesting technology which permits very effective site preparation to be done at no cost. Neither balsam fir nor red spruce are planted operationally for timber production. Some planting is done to remedy understocking resulting from failed attempts at natural regeneration, but the current tendency is to accept marginal stocking, and virtually no fill planting (i.e., in voids) is done.

Future yields and treatments of these intensively managed stands remain uncertain. As the old second-growth stands diminish and the new forest reaches merchantability, scarce wood supplies will likely encourage widespread commercial thinning in 25-40-year-old stands. Future options for natural regeneration will range from a repetition of the one-cut shelterwood in extensive practice, to intensive application of the uniform or irregular shelterwood methods involving several partial entries prior to the final removal. Unquestionably, the spruce budworm will return, and unless new population control intervention techniques are developed, insecticide spraying will be required on all managed, infested stands to prevent large-scale mortality.

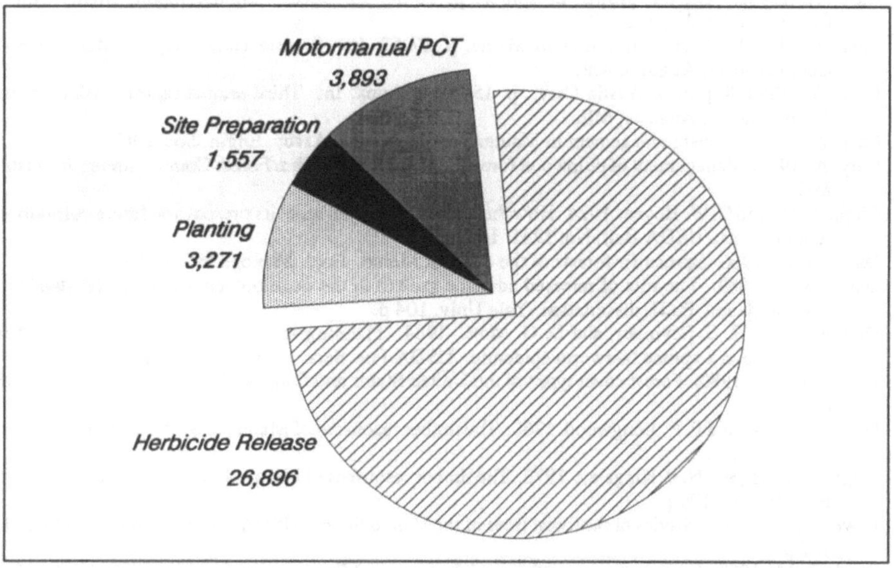

Fig. 7. Areas (hectares) treated by various intensive silvicultural practices in Maine, 1988. [Includes entire state, but virtually all are applied to spruce-fir stands. Source: Maine Forest Service (1988)]

Acknowledgments

This paper benefitted greatly from thorough reviews of Chadwick Oliver, Matthew Kelty, Robert Frank, Barton Blum, Morten Moehs, and Mary Ann Fajvan. I am especially indebted to D. Gordon Mott, Maxwell McCormack, Jr. and Gordon Baskerville for freely offering their ideas and advice as friends and colleagues since I arrived in Maine in 1977. Most of all, this paper stands as a humble tribute to David M. Smith, who not only provided the intellectual framework for this effort, but helped me to formulate many of the details based on his long-term observation and hands-on management of Maine's forests.

Literature cited

Baskerville, G. L. 1975a. Spruce budworm: super silviculturist. Forestry Chronicle 51:138-140.

Baskerville, G. L. 1975b. Spruce budworm: The answer is forest management: Or is it? For. Chron. 51:157-160.

Blais, J. R. 1985. The ecology of the eastern spruce budworm: a review and discussion. p. 49-59. In: Sanders, C. J., et al. (1985).

Blum, B.M, J. W. Benzie, and E. Merski. 1983. Eastern spruce-fir. p. 128-130. In: Burns, R.M., compiler. Silvicultural systems for the major forest types of the United States. USDA Agr. Handb. 445. 191 p.

Blum, B.M., and D.A. MacLean. 1985. Potential silviculture, harvesting and salvage practices in eastern North America. p. 264-280 In: Sanders, C. J., et al. (1985).

Cary, A. 1894a. On the growth of spruce. p. 20-36. In: Second annual report, Maine Forest Commissioner, Augusta, ME.

Cary, A. 1894b. Early forest fires in Maine. p. 37-59. In: Second annual report, Maine Forest Commissioner, Augusta, ME.

Cary, A. 1896. Report of Austin Cary. p. 15-203 + Appx. In: Third annual report, Maine Forest Commissioner, Augusta, ME.

Cary, A. 1899. Forest management in Maine. Reprint from: J. Assoc. Engin. Soc. 23(2).

Cary, A. 1902. Management of pulpwood forests. In: 4th Rept. Maine Forest Commissioner, Augusta, ME.

Clapp, E. H., and C. W. Boyce. 1924. How the United States can meet its present and future pulp-wood requirements. USDA Bull. No. 1241. 100 p.

Davis, R. B. 1966. Spruce-fir forests of the coast of Maine. Ecol. Monog. 36(2): 79-94.

Davis, W.C. 1989. The role of released advance growth in the development of spruce-fir stands in eastern Maine. Ph.D. dissertation, Yale Univ. 104 p.

Dimond, J. D., R. S. Seymour, and D. G. Mott. 1984. Planning insecticide application and timber harvesting in a spruce budworm epidemic. USDA For. Serv. Agr. Handb. 618. 29 p.

Eyre, F. H., ed. 1980. Forest cover types of the United States and Canada. Soc. Amer. Foresters. 148 p.

Ferguson, R. H., and F. R. Longwood. 1960. The timber resources of Maine. USDA For. Serv. NE For. Exp. Sta. 75 p.

Ferguson, R. H., and N. P. Kingsley. 1972. The timber resources of Maine. USDA For. Serv. Resource Bull. NE-26. 129 p.

Fowells, H. A. 1965. Silvics of the forest trees of the United States. USDA For. Serv. Agr. Handb. 271. 762 p.

Frank, R. M. 1987. Growth response of potential spruce and fir crop trees by precommercial thinning treatments and by fertilization treatments. Unpublished progress report for the period mid-growing season of 1976 to the mid-growing season of 1986. USDA For. Serv., Orono, ME. 9 p.

Frank, R. M., and L. O. Safford. 1970. Lack of viable seeds in the forest floor after clearcutting. J. Forestry 68(12):776-778.

Frank, R. F., and J. C. Bjorkbom. 1973. A silvicultural guide for spruce-fir in the Northeast. USDA For. Serv. Gen. Tech. Rep. NE-6. 29 p.

Frank, R. F., and B.M. Blum. 1978. The selection system of silviculture in spruce-fir stands — procedures, early results, and comparisons with unmanaged stands. USDA For. Serv. Res. Pap. NE-425. 15 p.

Graves, H.S. 1899. Practical forestry in the Adirondacks. USDA Bull. 26. 86 p.

Hart, A.C. 1963. Spruce-fir silviculture in northern New England. p. 107-110 in Proc. 1963 SAF Annual Convention, Boston, MA.

Hopkins, A.D. 1901. Insect enemies of the spruce in the Northeast. USDA Bull. 18 (new series). 48 p.

Hosmer, R. S. 1902. A study of the Maine spruce. In: 4th Rept. Maine Forest Commissioner, Augusta, ME.

Irland, L. C., J. B. Dimond, J. L. Stone, J. Falk, and E. Baum. 1988. The spruce budworm outbreak in Maine in the 1970s — assessment and directions for the future. Maine Agr. Exp. Sta. Bull. 819. 119 p.

Jacobson, G.L., Jr., T. Webb, and E. C. Grimm. 1987. Patterns and rates of vegetation change during the deglaciation of eastern North America. p. 277-288. In: Ruddiman, W. F. and H. E. Wright, Jr., eds. The Geology of North America, v. K-3. North America and adjacent oceans during the last glaciation. Geol. Soc. Amer., Boulder, CO.

Ker, M. F. 1987. Effects of spacing on balsam fir: 25-year results from the Green River spacing trials. p. 58-75. In: Murray, T. S. and M. D. Cameron, eds. (1987).

Hart, A. C. 1963. Spruce-fir silviculture in northern New England. p. 107-110 In: Proc. Soc. Amer. Foresters, Boston, MA.

Leak, W. B. 1975. Age distribution in virgin red spruce and northern hardwoods. Ecology 56:1451-1454.

Leak. W. B. 1982. Habitat mapping and interpretation in New England. USDA For. Serv. Res. Pap. NE-496. 28 p.

Lorimer, C. G. 1977. The presettlement forest and natural disturbance cycle of northeastern Maine. Ecology 58:139-148.

McCormack, M. L., Jr. 1984. Interaction of harvesting and stand establishment in conifers in northeastern North America. p. 233-239. In: Corcoran, T. J. and D. R. Gill, eds. Proceedings – Recent advances in spruce-fir utilization technology. Soc. Amer. For. Publ. 83-13. 222 p.

McCormack, M. L., Jr. 1985. Vegetation problems and solutions – Northeast. p. 315-326. In: Proc. S. Weed Sci. Soc. Houston, TX.

McCormack, M. L., Jr. and R. A. Lautenschlager. 1989. An aerial technique to adjust conifer stocking– modifications and results. Suppl. to Proc. NE Weed Sci. Soc. 43:27-29.

MacLean, D. A. 1984. Effects of spruce budworm outbreaks on the productivity and stability of balsam fir forests. For. Chron. 60:273-279.

MacLean, D. A. 1985. Effects of spruce budworm outbreaks on forest growth and yield. p. 148-175 In: Sanders, C. J. et al. (1985).

Maine Forest Service. 1988. Report of the 1986 midcycle resurvey of the spruce-fir forest in Maine. Maine Dept. Conservation, Augusta, ME. 51 p.

Maine Forest Service. 1989. Silvicultural practices report for 1988. Augusta, ME. 6 p.

Manley, S.A.M. 1971. Identification of red, black and hybrid spruces. Can. For. Serv. Publ. No. 1301.

Manley, S.A.M. 1972. The occurrence of hybrid swarms of red and black spruces in central New Brunswick. Can. J. For. Res. 2:381-391.

Manley, S.A.M., and D.P. Fowler. 1969. Spruce budworm defoliation in relation to introgression in red and black spruce. For. Sci. 15:365-366.

Marquis, D. A. 1991. Stand development patterns in Allegheny hardwood forests, and their influence on silviculture and management practices. (this volume)

Meyer, W. H. 1929. Yields of second-growth spruce and fir in the Northeast. USDA Tech. Bull. No. 142. 52 p.

Millers, I., D. S. Shriner, and D. Rizzo. 1989. History of hardwood decline in the eastern United States. USDA For. Serv. Gen. Tech. Rep. NE-126. 75 p.

Morris, R. F., ed. 1963. The dynamics of epidemic spruce budworm populations. Entom. Soc. Can. Memoirs 31. 332 p.

Mott, D. G. 1980. Spruce budworm protection management in Maine. Maine Forest Review 13:26-33.

Murray, T.S., and M. D. Cameron, eds. 1987. Proceedings of the precommercial thinning workshop. Can. For. Serv., Maritimes. Fredericton, N. B. 105 p.

Murphy, L. S. 1917. The red spruce – its growth and management. USDA Bull. 544. 100 p.

Needham, T.D., and T.S. Murray. 1991. Proceedings of the conference on natural regeneration management. Fredericton, N. B., March 17-19, 1990. Forestry Canada – Maritimes, Fredericton, N. B.

Newton, M., M.L. McCormack, Jr., R.L. Sajdak, and J.D. Walstad. 1987. Forest vegetation problems in the Northeast and Lakes States/Provinces. p. 77-104. In: Walstad, J.D. and P.J. Kuch, eds. Forest Vegetation Management for Conifer Production. Wiley and Sons, NY. 523 p.

Oosting, H.J., and W.D. Billings. 1951. A comparison of virgin spruce-fir forests in the northern and southern Appalachian system. Ecology 32:84-103.

Oliver, C. D. 1981. Forest development in North America following major disturbances. For. Ecol. Mgt. 3:153-168.

Oliver, C. D. and B. C. Larson. 1990. Forest Stand Dynamics. McGraw-Hill, Inc. 467 p.

Osawa, A. 1989. Causality in mortality patterns of spruce trees during a spruce budworm outbreak. Can. J. For. Res. 19:632-638.

Osawa, A., C. J. Spies, and J. B. Dimond. 1986. Patterns of tree mortality during an uncontrolled spruce budworm outbreak in Baxter State Park, 1983. Maine Agr. Exp. Sta. Tech. Bull. 121. 69 p.

Place, I.C.M. 1955. The influence of seedbed conditions on the regeneration of spruce and balsam fir. Can. Dept. North. Affairs and Nat. Res., For. Branch Bull 117. 77p.

Powell, D. S., and D. R. Dickson. 1984. Forest statistics for Maine, 1971 and 1982. USDA For. Serv. Resource Bull. NE-81. 194 p.

Reams, G. A., and M. M. P. Huso. 1990. Stand history: an alternative explanation of red spruce radial growth reduction. Can. J. For. Res. 20:250-253.

Royama, T. 1984. Population dynamics of the eastern spruce budworm *Choristoneura fumiferana*. Ecol. Monogr. 54:429-462.

Sanders, C.J., R.W. Stark, E.J. Mullins, and J. Murphy. 1985. Recent advances in spruce budworms research. Proceedings of the CANUSA spruce budworms research symposium. Bangor, Maine, Sept. 16-20, 1984. Can. For. Serv., Ottawa. 527 p.

Seymour, R. S. 1980. Vulnerability to spruce budworm damage and 100-year development of mixed red spruce-fir stands in north central Maine. Ph.D. dissertation, Yale Univ. 160 p.

Seymour, R. S. 1985. Forecasting growth and yield of budworm-infested forests. Part I: Eastern North America. p. 200-213 In: Sanders, C. J. et al. (1985).

Seymour, R. S. 1986. Stand dynamics and productivity of northeastern forests – biomass harvesting considerations. p. 63-68. In: Smith, C. T., C. W. Martin, and L. M. Tritton, eds. Proc. 1986 Sympos. on the productivity of northern forests following biomass harvesting. USDA For. Serv. Gen. Tech. Rep. NE-115. 104 p.

Seymour, R. S., and C. J. Gadzik. 1985. A nomogram for predicting precommercial thinning costs in overstocked spruce-fir stands. N. J. Appl. For. 2:37-40.

Seymour, R.S., and R. C. Lemin, Jr. 1988. SISTIM —A new model for simulating silvicultural treatments in Maine. Abstr. p. 1147. In: Ek, A. R., S. R. Shifley, and T. E. Burk. Forest growth modelling and prediction. USDA For. Serv. Gen. Tech. Rep. NC-120. Vol. 2., p. 580-1149.

Seymour, R.S., and R. C. Lemin, Jr. 1989. Timber supply projections for Maine, 1980-2080. Maine Agr. Exp. Sta. Misc. Rep. 337. 39 p.

Seymour, R.S., P.R. Hannah, J.R. Grace, and D.A. Marquis. 1986. Silviculture – the next 30 years, the past 30 years. Part IV. The Northeast. J. Forestry 84(7):31-38.

Smith, D.M. 1972. The continuing evolution of silvicultural practice. J. Forestry 70:89-92.

Smith, D.M. 1981. The forest and Maine's future. p. 79-87. In: Proc. Blaine House Conf. Forestry. Augusta, ME, Jan. 21-22, 1981.

Smith, D.M. 1986. The Practice of Silviculture (Ed. 8). Wiley and Sons, NY. 578 p.

Swaine, J.M., and F.C. Craighead. 1924. Studies on the spruce budworm (*Cacoecia fumiferana* Clem.). Can. Dept. Agric. Tech. Bull. 37 (new series). 91 p. + Appx.

USDA Forest Service. 1985. Insects of eastern forests. Misc. Publ. 1426. 608 p.

Weiss, M.J., and I. Millers. 1988. Historical impacts on red spruce and balsam fir in the northeastern United States. p. 271-277. In: Proc. US/FRG research symposium: effects of atmospheric pollutants on the spruce-fir forests of the eastern United States and the Federal Republic of Germany. USDA For. Serv. Gen. Tech. Rep. NE-120. 543 p.

Westveld, M. 1928. Observations on cutover pulpwood lands in the Northeast. J. Forestry 26:649-664.

Westveld, M. 1931. Reproduction on the pulpwood lands in the Northeast. USDA Tech. Bull. 223. 52 p.

Westveld, M. 1946. Forest management as a means of controlling the spruce budworm. J. Forestry 44:949-953.

Westveld, M. 1951. Vegetation mapping as a guide to better silviculture. Ecology 32:508-517.

Westveld, M. 1953. Ecology and silviculture of the spruce-fir forests of eastern North America. J. Forestry 51:422-430.

Zon, R. 1914. Balsam fir. USDA Bull. No. 55. 67 p.

Temperate zone roots of silviculture in the tropics

13

FRANK H. WADSWORTH

Introduction

It should come as no surprise that modern silviculture in the tropics has had its roots in the temperate zone. From Sir Dietrich Brandis' beginnings in Pegu, Burma, in 1856 to the present, most of the fundamentals and many of the silvicultural practices have been adaptations from earlier temperate zone experience, mostly in Europe. The subject is relevant to the rising temperate-zone interest in the growing, globally significant forest problems of the tropics.

Early temperate zone impressions

The complexities of the forests of the moist tropics produced a profound impression on early foresters of the temperate zone. Broun (1912), who in 1912 wrote one of the earliest books on silviculture in the tropics, was so influenced by the ecology of such forests described in Schimper's *Plant Geography* (1903) that he dedicated a third of his book to factors governing and influencing the existence of forests.

The complexity of these forests was again brought to light in 1921 in Troup's (1921) *The Silviculture of Indian Trees*. In three volumes it described some 600 tree species that foresters faced in India. *The Silvics Manual of the Forest Trees of the United States* (Fowells 1965) appeared 44 years later and covered less than one-quarter as many species.

Application of silvicultural systems

The degree of application in the tropics of the silvicultural systems that had their origins in the temperate zone provides a basis for relating forestry practice in the two zones. Troup's (1928) *Silvicultural Systems*, written primarily for the temperate zone, and Trevor's "Silvicultural Systems," contained within Champion and Trevor's (1938) *Manual of Indian Silviculture*, describing 10 years later experience where it had developed to a higher degree than in most other tropical areas, make possible such a comparison.

Troup introduced his subject as follows:

> "Certain European countries, faced centuries ago with the problem of future timber supplies, as a result of long experience have evolved methods of treatment, termed "silvicultural systems," which are an object-lesson to the whole world.
>
> "Lest it may be held that systems which have been evolved in Europe are not applicable to other parts of the world where totally different conditions prevail, the fact may be mentioned that more than fifty years past the silvicultural systems of Europe, with suitable modifications, have been applied successfully in many parts of India under a variety of

M. J. Kelty (ed.), The Ecology and Silviculture of Mixed-Species Forests, 245–255.

> conditions and in many types of forest; and it may be truly said that the great progress which forestry has made in that country during the past half-century has been due to a large extent to the fact that the officers of the higher branch of the Forest Service have received their practical training in the forests of continental Europe."

Trevor adopted verbatim for India Troup's definition of a silvicultural system and stated that every forester should have a complete knowledge of the systems practiced in Europe and their application.

Both Troup and Trevor described clearcutting systems, including variants by strips. Trevor described tropical variations of clearfelling with natural reproduction of sissoo (*Dalbergia sissoo*), coppice of sal (*Shorea robusta*), and "taungya" interplanting with food crops. Trevor related experience with clearfelling and artificial regeneration in India to that much earlier in Europe.

Trevor described the use in India of the five shelterwood systems recognized by Troup: uniform, group, irregular, strip, and wedge. The uniform method has been used in India with chir pine (*Pinus longifolia*), deodar (*Cedrus deodora*), sal, teak (*Tectona*), and mixed forest, including dipterocarps.

Trevor reported that group shelterwood was tried in India with sal but grass became excessive in the openings, and fires destroyed the vegetation. He concluded that the system was not suitable to Indian conditions.

A system allied to Troup's irregular shelterwood, which Trevor nevertheless considered distinct, was applied in India to deodar, chir pine, and sal. Trevor concluded that the value of the immature trees retained generally does not compensate for the complications resulting from a wide range of tree ages in the next crop.

Trevor stated that selection was the system mostly applied in India, to sal, deodar, mixed deciduous, and evergreen forests. Trevor foresaw continuation of the system on the slopes of the Himalayas, the hill sal forests, the teak forests, most of Burma, and in water catchment areas generally. The system, however, has not achieved all that is expected of it, such as the development of a normal series of age-classes.

Trevor described simple coppice, shelterwood coppice, and selection coppice, as well as coppice with standards. He stated that vast areas of India are incapable of producing more than small timber yet they contain teak and can be managed only under some coppice system. Coppice has been widely used there also with *Eucalyptus*, *Eugenia*, *Prosopis*, and *Acacia*.

The recognized systems capable of sustaining forest productivity and new crops, with local adaptations, thus have all been applied in the South Asian tropics. For those who question today whether it is possible to "manage" forests in the tropics, whether anyone knows how, or whether anyone has ever done it, Trevor's reporting of nearly 9 million hectares under silvicultural systems in India as early as 1935 (Table 1) may be a revelation.

Table 1. The use of silvicultural systems in 10 Indian states, 1935.

System	Forest Area	
	10^3 Ha	Percent
Clearfelling	1,359	15
Shelterwood	1,493	17
Selection	3,596	40
Coppice	2,471	28
Total	8,919	100

Related silvicultural practices, with local variations, but generally on a lesser scale, have been applied in Indonesia (Indonesia 1975), Malaysia (Burgess 1975), Uganda (Dawkins 1958), Ghana (Mooney 1963), Nigeria (Lowe 1984), Francophone West Africa (Catinot 1965), and tropical Australia (Queensland 1983), guided by foresters trained chiefly in the United Kingdom, France, Germany, and The Netherlands. Elsewhere in the tropics, silvicultural practice in natural stands is still in its infancy, and no totally new systems appear in prospect.

Application of silvicultural treatments

Few silvicultural practices are unique to the tropics. One, more widely practiced there than in the temperate zone, is described as follows, by Troup (1921) (p.lvi):

> "A provisional method of treatment, the main objective of which is to utilize the available stock of mature and over-mature trees of marketable species, while endeavoring to safeguard the future stock as far as possible, was adopted in the early days of forest administration: it has continued to be the principal method of treatment. It consists of working over the forests under a definite felling cycle and removing trees which have reached exploitable size, seed-bearers being left where natural reproduction is deficient.
>
> "The fellings in question have been termed "selection fellings," though they differ materially from those of the true selection system. These fellings cannot be regarded as anything but a provisional means of working the forests pending the introduction of more scientific systems of management. Their most serious defects are: first, they do not tend towards the establishment of a normal forest; Second, they do not ensure adequate reproduction; and third, that in many cases they do not sufficiently take into account the silvicultural requirements of the species, particularly in the case of light-demanders."

More recently, under the terms "refinement" and "liberation thinning" such treatments have been more sharply focused on providing growing space for selected crop trees (Dawkins 1961, Hutchinson 1980).

More purely tropical is "taungya," the use of food and forage crops interplanted with trees to defray the costs of initial weed control. This practice, apparently applied centuries earlier in Europe (King 1968) reappeared independently in the tropics as a result of land hunger and high alternative weeding costs. Apparently begun in Burma in 1856 (King 1968), by 1891 it had been used to establish some 40,000 hectares of teak in Java (Becking 1951).

Underplanting trees beneath existing stands probably has been practiced more extensively in the tropics than in temperate zone forests. The practice has been a response to a particularly tropical forest condition, with most of the tree species perceived to be much less useful than the rest. One variant of the practice, "enrichment," begun by the French in West Africa (Ford-Robertson 1971), consisted of planting only unregenerated areas within forests. Providing the needed early tending of trees so planted proved impractical because they were so scattered as to be difficult to locate. Enrichment therefore was abandoned in India and Africa long ago (Griffith 1941, Wood 1970) and was replaced by underplanting in systematic corridors, a practice that remains promising where needed to upgrade forest composition.

Practically all that is silviculture in the tropics either has been or could be practiced in the temperate zone. It thus appears to be but one end of a continuum, a commonality obscured by the prevailing penchant of many to accentuate minor differences among technical practices and the significance of latitude to forestry. Writers who incessantly apply the word "tropical" to practices that in truth are universal exacerbate this questionable dichotomy.

More recent temperate-zone concepts for the tropics

Sixteen years after Champion and Trevor's *Manual of Indian Silviculture* was published, the sixth edition of *The Practice of Silviculture*, by Hawley and Smith, appeared in the United States (Hawley and Smith 1954). Although directed specifically toward North America, it presented many new concepts applicable in the tropics. A selection of these with my personal opinion as to their tropical relevance, follows:

> "The forester should work for the good of the forest as an entity, ... to ensure that it will remain a permanently productive source of goods and benefits." (p. 5). Here appears to be what in the temperate zone we are now hearing called "The new forestry," holistic ecosystem rather than single-resource emphasis in management. In the tropics this concept may prove to be the only possible savior of both the intangible values and the commodity value productivity of the forests.
>
> "Silviculture (may be) most profitable on sites of intermediate quality, where intolerant trees can be maintained with less effort." (p. 8) In the tropics this relationship is exemplified by the much higher costs of tending regeneration on the better sites.
>
> "The best opportunities for intensive silvicultural practice lie in ... accessible second-growth stands on good sites, which are now so often in outwardly unpromising condition." (p. 496) In the tropics vast and almost nondeclining areas of secondary forest are the main heritage for forest managers.
>
> "As a general rule, skillfully managed second-growth forest differs markedly from the climax association, being more productive, more vigorous, and safer from injury." (p. 7) In the tropics these concepts probably are as true as in the temperate zone. Much more is already known as to how to manage them than is applied.

"The balance achieved by long-continued natural processes, operating more or less at random, is (not) necessarily more favorable to the trees than to the organisms that feed upon them." (p. 44) This concept is apparently of universal applicability, suggesting great prospects for increasing useful productivity of tropical forests by silvicultural suppression of the many such "organisms" they contain. Notwithstanding this, these prospects are commonly either unrecognized or perceived to be ecologically hazardous or economically unrewarding.

"The most valuable commercial species in any region tend to be relatively intolerant trees representative of the early or intermediate stages of natural succession." (p. 7) In the tropics this relationship is exemplified by such outstanding species as mahogany (*Swietenia*), cedar (*Cedrela*), teak, and obeche (*Triplochiton*).

"There has been a tendency for... foresters to limit their silvicultural outlook to partial cuttings aimed primarily at indefinite extension of the lives of existing stands. While this policy has advantages as a temporary expedient, it must ultimately collapse because it evades rather than solves the basic problem of renewing existing stands by attainment of reproduction. The easy doctrine that nature will somehow provide a new and desirable stand regardless of how the old stand is removed is tenable only if one is ready to welcome whatever drifts in on the tide of natural succession." (p. 496) In the tropics unproven, hopeful assumptions made regarding natural regeneration commonly have proven unfounded.

"Most of the answers to problems of securing natural regeneration are to be found in existing forests." (p. 40) In the tropics this source of information could become a keystone to the silviculture of secondary forests, yet has received attention only sparingly, even by scientists.

"Most violent objections to (artificial reproduction) can usually be traced to lack of success with pure plantations of species naturally occurring in mixed stands, use of exotic species or poorly chosen strains of native species, and improper selection of sites." (p. 49) In the tropics the complexities of the environment and forest composition accentuate these risks. Errors are still being made, some on a large scale, and at the expense of temperate-zone donors.

"Introduced species and strains should not be used on a large scale until their superiority to the indigenous has been demonstrated in trials over a period long enough for the trees to reach merchantable size." (p. 222) In the tropics disregard for this precaution has led to numerous unnecessary plantation problems.

"The future attractiveness of silviculture lies in the fact that, properly applied, it can produce relatively large returns on small investments." (p. 25) In the tropics this attractiveness, even greater because of the social rewards from needed rural employment is, nevertheless, constrained by option for shorter-term economic gain.

"The closer (the owner) carries the product to the ultimate consumer, the greater is his ability to capture the values added by increases in intensity of practice in the woods." (p. 18) In the tropics the virtues of this vertical linkage are of national significance. Processing of forest products before export, unlike logging alone, creates economic and social commitments that are compelling as motives for reinvestments in future timber crops.

"There is an unfortunate tendency for foresters to adopt extreme opinions on such questions as whether to resist or follow natural succession, to employ even-aged or unevenaged management, to foster pure or mixed stands, or to regenerate stands by artificial or natural means. The unreasonable attitudes of one school of thought too often goad adherents of opposing doctrines to seek equally untenable positions." (p. 50) In the tropics these irrational extremes exist, commonly among professionally trained expatriates from the temperate zone.

Smith's introduction to forestry in the tropics

The 1986 edition of *The Practice of Silviculture*, authored by Smith alone, retained Hawley's primary focus on North America but was significantly broader in scope than the 1954 edition. Seventeen major references on forestry in the tropics are cited. Of more significance are new text passages that give inklings of tropical forests and timber production practice, such as these:

"Stratified mixtures of species were first recognized in the wet evergreen and moist deciduous tropical forests, which are practically incomprehensible without this means of analyzing their structure." (p. 492)

"Production rates in the evergreen tropical rainforest (are, due to) the high temperatures that induce profligate respiration,... not clearly higher than those of the most favorable parts of the temperate zone." (p. 44)

"It is often stated that the production of primary forest in the humid tropics is very low. Only the production by few scattered trees of cabinet wood may be counted. The situation is also complicated by... the difficulty of knowing ages of trees." (p. 505)

"(With) strongly leached soils of tropical rainforests,... most of the nutrients are in the vegetation." (p. 227) As a generality this now appears extreme, but the high proportion of the nutrients in the vegetation remains a special concern for the sustainability of forest productivity on such soils.

"Even in the humid tropics, the main forestry problems... are, as has commonly been the case with forestry, how to rehabilitate forests and lands that have been damaged by hard or unwise use." (p. 505)

"Under (dry tropical) conditions the silvicultural practices are not different in principle from those applicable in most of the temperate zone." (p. 505)

"Poorly aerated (tropical) swamps (are) conducive to stands of pure or simple composition. Silvicultural management of these has often been comparatively uncomplicated." (p. 504)

"The... closest approach to success in (achieving natural regeneration of moist tropical forest) is in Southeast Asia, where under the "tropical shelterwood system" one or more lower strata are removed to make way for regeneration." (p. 504)

"Many species of moist tropical forests are so thoroughly adapted for immediate germination after seedfall that their seeds are virtually impossible to store or even to transport." (p. 285)

"In the humid tropics seedlings (in the nursery) remain succulent while growing very large, (so) there may be no alternative to containerized planting stock." (p. 305)

"In humid tropical climates it is often necessary to do cleaning (of plantations)... even several times during the first year, as well as afterwards." (p. 154)

"The perennial grasses that normally appear after site preparation probably serve to inhibit hardwoods, (an) effect deliberately used to inhibit invasion by the native angiosperm forest (into stands of planted Caribbean pine) in the Jari Valley of Amazonian Brazil." (p. 235)

"The taungya method (integrates) agriculture and reforestation. Just before the land is taken out of cultivation it is planted with both trees and (herbaceous crop) plants. The weeding of the crop also keeps the tree seedlings free." (p. 506)

"The concept of the stratified mixture is... applied in agroforestry, in which the production of crops of trees is combined with that of agricultural plants or forage for domestic animals. These practices are especially important for subsistence agriculture in the humid tropics (but also in semiarid forests)." (p. 494/5) "(They may include) overstories of timber trees over middle strata... of fruit trees, with food plants in the lowest strata." (p. 506)

Smith's counsel for silviculture in the tropics

Smith's recent edition of *The Practice of Silviculture* also describes state-of-the-art silvicultural practices, the utility of which is not limited to the temperate zone, including salvage and sanitation cutting, liberation, site preparation, use of herbicides, selection of species, seed handling, genetics, direct seeding, planting stock production, planting, fertilization, drainage, irrigation, thinning, and pruning. More fundamental to forestry in the tropics, however, is the broader counsel Smith presents, such as the following (again, with my opinion as to their tropical relevance):

"The duties of the forester are to analyze the natural and social factors bearing on each stand and then devise and conduct the treatments most appropriate to the objectives of management." (p. 1) In the tropics, giving social factors parity with those considered "natural" is belatedly seen as crucial to the success of forestry.

"Studies of total production tell much about the fundamental biological factors controlling forest production. They also show the tantalizing possibilities of any technology that might enable closer utilization of the massive production of organic matter by forests." (p. 45) In the tropics the differences between total and marketable forest productivity have generally been much greater than in temperate zone forests, and thus, so are these possibilities.

"(Forest animals), whether they be defoliating insects or carnivores that feed on herbivorous mammals... can exert a major influence on the nature of the vegetation." (p. 15) In the tropics the role of animals in forest welfare, as yet little understood, may well be more crucial than in the temperate zone.

"Habitat" more (than) "site" fully connotes the idea that the place is one in which trees *and other living organisms* subsist and interact." (p. 263) This term is equally desirable in the tropics.

"Understory shrubs and other lesser vegetation are of high importance (for site diagnosis)." (p. 268) In the tropics, where forests tend to be heterogeneous in composition, this is a promising field for research.

"(Sound policies about the care and use of) areas set aside for wilderness, scenery, or scientific study inevitably involve something other than leaving them entirely alone." (p. 5) Present perceptions in the tropics generally to the contrary, this conclusion can be expected to come to light as the full costs of "neglective" management become apparent.

"Repugnant as it may be to certain influential segments of public opinion, useful forests are created and maintained chiefly by the destruction of judiciously chosen parts of them." (p. 15) In the tropics this concept is clearly applicable but is subject to challenge, and so needs more effective demonstration.

"Silviculture ... is not any less "practical" than forest management." (p. 22) In the tropics this concept, although apparently applicable, is not widely accepted because it has been demonstrated only locally.

"The boundary between topographic convexity and concavity sometimes defines differences in species composition and productivity." (p. 269) In the tropics site differences as apparently minor as these are showing up as significant to plantation performance.

"It is entirely within the realm of possibility to conduct forestry permanently without the degradation that is almost inevitable in agriculture." (p. 8) In the tropics this fact is fundamental to land use allocation for sustainability, yet there is little support for forests in the face of pressures for agricultural expansion.

"It is easy to argue that one should do nothing to the forest for fear of doing something wrong. This charge can never be entirely refuted, but society requires practitioners of applied science to act in the absence of full knowledge. (p. 9) Decisive action cannot await absolute proof of validity nor can it be evaded indefinitely by fence-straddling. The forester must... proceed as far as possible on the basis of proven fact and then complete plans for action in the light of the most objectively analytical opinions that can be formed." (p. 363) A lack of recognition of these principles threatens to delay rational forest management in parts of the tropics.

"The necessity that nature should be understood and emulated does not mean that silviculture should slavishly follow either the reality of natural processes or abstract theories about them." (p. 8) In the tropics lack of understanding of this concept has led to indecision delaying the testing and application of silviculture.

"It appears that wood utilization tends to change faster than foresters can change the forest composition (so) it is perhaps best to let the site determine which inherently useful species to grow." (p. 270) In the tropics there is need for more recognition that future markets will accept timber heretofore disdained, thus greatly increasing secondary forest productivity potentials.

"Most efforts to explain tree growth in quantitative terms seem to work best (on the basis) of direct relationships between amounts of foliage and wood volume growth." (p. 70) In the tropics also crown size is showing up as a good growth rate predictor, opening exciting new potentials for forest productivity.

"In ... multiple regression analysis ... as many factors as possible that might govern growth ... are tested as independent variables." (p. 125) In the tropics these techniques are even more revealing because of forest complexities.

"The basal area of a tree is fairly well correlated with the cross-sectional area of the crown." (p. 119) In the tropics this has been found true by species groups (Dawkins 1963), and makes possible the estimation of individual tree efficiency: the rate of useful wood increment relative to the forest area occupied, as indicated by stem basal area.

"Some shade-tolerant trees can lapse into long stages of postponed development such that it may take them twice or even thrice as long to grow to mature size as some more quickly growing associate. (Such trees), left as being too small to cut, (can become valuable components)." (p. 501) In the tropics many important timber species have this capability, with important implications for silviculture.

"There is no need to have seedlings present at all stages in the life of a stand." (p. 337) In the tropics this is equally true, yet not understood by many temperate-zone critics of timber production there.

It is seldom a good idea to expect natural regeneration as the unearned or unplanned by-product of treatments done for other purposes at vaguely defined times during the rotation." (p. 337) This applies equally to the tropics, where desired regeneration generally requires positive measures to assure survival and development.

"Because of the high energy requirements of substitute materials (mostly from nonrenewable resources) solid-wood products represent the best way to use wood for energy conservation." (p. 21) In the tropics this concept should have been recognized in the recent emphasis on species purely for fuelwood, whereas other species, capable of producing solid-wood products as well, could have served both purposes.

"(In the selection of tree species) the first step ... is analysis of limitations imposed by environmental factors that collectively constitute the site. The second step ... is the choice of species that will most nearly meet the humanly ordained objectives of stand management. A third step is consideration of the degree of artificial control that will be exerted over the genetic constitution of the species selected." (p. 262/3) In the tropics the search for the best species to plant, following these steps, is more critical than in the temperate zone because of the larger number of species from which to select.

"When one sees fine ... hardwoods in stratified mixtures ... it is prudent to resist the temptation to assume that a pure plantation of the species should be better yet." (p. 497) In the tropics, anomalies in the performance of timber tree species planted pure corroborate this need to know in advance the autecology of the species in its native environment.

"(Stratified mixtures of species with) storied structure can have the... advantages that intolerant overstory species get plenty of room in which to grow, and the understory species can usually make effective use of the leftover growing space that might otherwise fill up with undesirable vegetation." (p. 489) In the tropics mixed plantations generally reflect only a belief that pure plantations are ecologically unsound. Most mixed plantations do not come up to expectations because mixtures both compatible and productive, such as are called for here, are generally inadequately understood.

For international technical assistance agencies, donor governments, nongovernmental organizations, policy makers, planners, and decision makers concerned with the future of tropical forests Dave Smith, perhaps not inadvertently, presents fundamentals of production forestry applicable worldwide. For foresters and other students, even those concerned primarly with the tropics, he achieves his stated purpose, "to stimulate thought."

Literature cited

Becking, J.H. 1951. Forestry technique in the teak forests of Java. In: Proceedings, United Nations Scientific Conference on the Conservation and Utilization of Resources; 1949; Lake Success, N.Y. United Nations; 5:106-114.

Broun, A. F. 1912. Sylviculture in the tropics. Macmillan, London. 309 p.

Burgess, P.F. 1975. Silviculture in the hill forests of the Malay Peninsula. Research Pamphlet 66, Kepong, Malaysia: Forest Research Institute 100p.

Catinot, R. 1965. Sylviculture en foret dense Africaine. Bois et forets des Tropiques 100:5-18; 101; 3-16; 102:316; 103:3-16; 104:17-29.

Champion, H. G. and Sir Gerald Trevor, 1938. Manual of Indian Silviculture. Humphrey Milford, Oxford University Press, Oxford. 374 p.

Dawkins, H.C. 1958. The management of tropical high forest with special reference to Uganda. Institute Paper 34, University of Oxford: Imperial Forestry Institute. 155p.

Dawkins, H.C. 1961. New methods of improving stand composition in tropical forests. Caribbean Forester 22(1/2):12-20.

Dawkins, H.C. 1963. Crown diameters, their relation to bole diameter in tropical forest trees. Commonwealth Forestry Journal 42:318-333.

Ford-Robertson, F.C. (ed.) 1971. Terminology of forest science, technology, practice, and products. The multilingual forestry terminology series, Society of American Foresters, Washington D.C. 349 p.

Fowells, H. A. (ed.) 1965. Silvics of Forest Trees of the United States. Agricultural Handbook no. 271. U. S. Department of Agriculture, Washington, D.C. 762 p.

Griffith, A.L. 1941. Notes on gap regeneration with the selection method of exploitation. In: Proceedings, 5th Silvicultural Conference; Dehra Dun, India: Forest Research Institute. p. 161.

Hawley, Ralph C., and David M. Smith, 1954. The Practice of Silviculture (Sixth edition). John Wiley, New York. 525 p.

Hutchinson, I.D. 1980. Processes of optimizing and diversifying uses of resources: approach adopted to define interim guidelines for silviculture and management of mixed dipterocarp forest in Sarawak. In: Papers International Forestry Seminar; 1980 November; Kuala Lumpur, Malaysia: University of Malaysia. 24 p.

Indonesia Directorate General of Forestry. 1975. Forests and forestry in Indonesia. Prepared for Technical Conference on Tropical Moist Forests. 21p.

King, K.F.S. 1968. Agri-silviculture (the taungya system). Bulletin 1. Lagos, Nigeria: University of Ibadan, Forest Department. 109p.

Lowe, R.G. 1984. Forestry and forest conservation in Nigeria Commonwealth Forestry Review 63 (2): 129-136.

Mooney, J.W. 1963. Silviculture in Ghana. Commonwealth Forestry Review 42 (2): 159-163.
Queensland Department of Forestry. 1983. Rainforest research in north Queensland. Brisbane, Australia 52p.
Schimper, A. F. W. 1903. Plant Geography upon a Physiological Basis. P. Groom, and I. B. Balfour (eds.) Clarendon Press, Oxford.
Smith, David M. 1986. The Practice of Silviculture. (Eighth edition) John Wiley, New York. 527 p.
Troup, R.S. 1921. The Silviculture of Indian Trees. 3 vols. Clarendon Press, Oxford. 216 p.
Troup. R. S. 1928. Silvicultural Systems. Clarendon Press, Oxford.
Wood, P.J. 1970. Silvicultural notes on a tour in West Africa. University of Oxford, Department of Forestry, Commonwealth Forestry Institute. Oxford. 88 p.

Forest analysis: Linking the stand and forest levels

14

GORDON L. BASKERVILLE

Introduction

Analysis of forest level dynamics is similar to the analysis applied in arriving at an understanding of the dynamics of development in a stand. While stand analysis is used to evaluate stand dynamics and to define alternative silvicultural treatments, forest analysis is used to examine the biological and practical bases of alternative forest management regimes. Forest analysis is not something to replace contemporary forest management design tools, but rather a process of organizing the thoughts of the manager who uses those powerful tools.

In arriving at an appropriate choice of silvicultural treatment in a stand, the silviculturist analyzes the futures that could reasonably be achieved, given current stand condition and the tools at hand. The silviculturist forecasts stand performance by means of biological reasoning, commonly carried out while actually on the ground in the stand under discussion. The intent is to arrive at a logical definition of the stand development problem and of the treatments that might be used to correct that problem, in the context of their probable impacts on the dynamics of stand development. The thesis offered here is that application of a biologically-based analysis of development at the forest level would facilitate the design of credible management at the forest level.

The idea of forest level analysis is offered as an application of reasoning in the process of management design, and not as innovative science. The intent is to illustrate the concept in a simple manner, yet with sufficient detail to show the potential for this approach to improve the ability of the manager for reflection before action.

Stand level analysis

To put forest analysis in perspective it is appropriate to begin with a discussion of analysis at the stand level. In the broadest sense, there are two approaches to stand silviculture. At one extreme (the 'cookbook' approach), prescriptions are made by a book, often without even going to the woods to see the stands in question. At the other extreme, stand prescriptions are designed 'on the ground' from analysis of the unique dynamics of each stand under consideration.

In the cookbook approach to silviculture, the desired response is chosen, and the associated treatment is prescribed. It requires only a leap of faith to believe the result in the forest will be as illustrated in the book. In reality, this works only when the dynamics of the actual stand in question are similar to those stated or implied in the book. Because the actual response of a stand to a treatment is dependent on its unique history and its unique current structure, cookbook silviculture has a high frequency of failed recipes. Perhaps no silviculturist would admit to prescribing by the book. On the other hand, the public and the media have a strong attachment to these simple prescriptions.

M. J. Kelty (ed.), The Ecology and Silviculture of Mixed-Species Forests, 257–277.

The second approach to silviculture, that of stand analysis, requires careful definition of the dynamics of change in a particular stand, and analysis of these dynamics to arrive at a prescription. My introduction to this process was at Yale University, where every course, every seminar, and every field trip developed into an analytical exercise. No matter what the topic, the challenge was to reason through a set of cause-effect connections. In every stand visited, the same series of questions came up: what is the dynamic structure of the stand?, how have these dynamics driven past history of the stand?, and how will these dynamics condition what the stand will look like in the future? In every cutover visited, a similar set of questions emerged: what was the nature of the stand that was here before the harvest?, what was the means by which that previous stand was removed?, and, what will the dynamics of stand development on this area be in the future? The *process of reasoning* was always the same, and the outcome was always different in that it related to the unique situation of the stand in question.

The chief value of stand analysis is that discussion centers on stand dynamics, and where these dynamics will lead stand development in the future. The discussion seldom founders on what treatment to apply. In fact, when the analysis of dynamics is thorough, the silvicultural need and silvicultural treatment become obvious and self-standing. The analytical approach to silviculture does not presuppose an objective, and, while it is handy if one exists, a key power of the approach is to outline a set of futures that could reasonably be reached from the present stand condition. Stand analysis invites examination of a range of possible objectives each associated with a way of getting there. That allows easy modification of any initial objective into one that is both desirable *and* attainable.

Along with its considerable advantages, the analytical approach to silviculture has what some would consider substantial disadvantages: it can be complex, it requires considerable thought in the context of a particular situation, it requires that the prescriber acquire woods experience with respect to stand development, it requires that the prescriber actually go to the woods to make prescriptions, it does not give a single answer, and different people can come to different answers, and perhaps worst of all, it recognizes that *the right answer* cannot be known.

Forecasts of future stand performance arrived at by reason do not have the elegance of a fitted yield curve (complete with standard error) but in forecasting the future of stand development, a reasoned analysis of where dynamics in the stand at hand will lead in the future can be more reliable than a line fitted to a set of data from other stands in the past. The difference is between accuracy with respect to characterizing nature, and precision with respect to fitting data. From a biological point of view it boils down to whether the silviculturist wants to be approximately right, or exactly wrong.

Forest level analysis

Boothroyd (1978) argued that the most important ingredient in the design of system control is what he called reflection before action. The aim of forest level analysis is to provide reflection before action in a manner that maximizes learning by the forest manager. Of equal importance in the 1990's is the aim of providing the public with an opportunity for reflection before action, and for learning. This is achieved: (i) by examination of the dynamics of forecasts used for management planning before a choice is made; and (ii) by displaying the probable outcomes of management so that the manager, and the public, can visualize the forest future.

A forest manager should know the dynamics of the forest he is trying to control, just as a silviculturist should know the dynamics of the stands he tries to control. It is harder to 'know' a forest, and clearly doing that will require experience in the forest as well as considerable skill in data collection and in the conversion of data into information.

To achieve biologically realistic forest management it is essential to characterize a forest and its dynamics in a format that facilitates biological analysis. That means a model of some sort. Analysis of the model forest should explore weaknesses in the biological logic of proposed management strategies, identify risks of management failure due to errors in information, and risks due to errors with respect to the practicality of implementation. In short, forest analysis is used to expose weaknesses in the model representation of biological and operational reality in the forest. While all forest models represent a substantial compression of biological reality, a model has one convenient feature—in a model it is always possible to discover exactly what the cause is for every effect generated. The question then becomes: are these cause-effect relationships credible in nature. Perhaps the most important thing to verify in a forecast is that the dynamics that cause an outcome in the forecast are also operating in the real forest in a manner that would deliver the same result. Forest analysis establishes an understanding of the dynamic response at the forest level and helps identify the feasible range for management strategies in both biological and practical terms.

Limited resources and the prevalence of interactions at the forest level mean that not everything can be done everywhere all the time. If one stand is being cut another is not being cut, and if one stand is treated another is not treated. In these circumstances, the spatial and temporal logistics at the forest level make averaging dangerous to the success of management—yet only by some form of averaging can the forest manager hope to understand the dynamics of change in his forest.

Holling (1978) emphasized the importance of 'compression for understanding' in dealing with natural resource and environmental systems. To many, the phrase 'compression for understanding' appears to be an internal contradiction. What Holling meant is that in systems that are large spatially, or that change slowly in time, or that are rich in dynamic connections, some compression is an essential step towards allowing human understanding of system dynamics. The compression he advocated requires care with respect to mimicking the dynamic processes of the subject system. Only by such careful compression of the temporal and spatial dynamics in forests will it be possible to design forest level management that is biologically realistic and practically implementable.

The goal is not to build a model that *looks* like the subject forest, but rather to build a model that *acts* like the subject forest, *reacts* like the subject forest, *and* which does not misrepresent the spatial and temporal variability of such issues as forest access and cover type pattern as these must be faced by the manager of the subject forest. The emphasis in such models must be on the processes that drive change. The manager needs a model with which it is possible to logically examine future situations for which no observational experience exists, or can exist. The construction of such models is almost entirely analytical, with data needs being identified by the processes involved. Fitting a model to data from the past may not be particularly helpful, because management is concerned with the dynamic development of the forest as it changes in the future under a different set of interventions than have prevailed in the past. Starting with a data set describing past states of the resource, and trying to extract the dynamics retrospectively is impossible for all but the simplest forest systems.

Walters (1986) extended the idea of 'compression for understanding' and provided a rich array of examples. Although neither has a forestry background, Holling and Walters would make fine designers of forest resource management. Their work represents the definitive description of analysis of natural systems dynamics. They acknowledge that the dynamics in a real system with the size and complexity of a forest will forever be beyond our capability to fully measure or to fully comprehend. At the same time, rather than give up or use overly simple tools, they argue persuasively for systematic attempts to capture the key dynamics of large systems in simulation models that are not themselves so large and mysterious as to be incapable of human comprehension.

Simulation models are in common use in forest management design. Simulation of forest dynamics consists of preparing a characterization of the initial condition of the forest and applying a set of rules of change to this condition (illustrated in general form by the age-class structure and the arrows in Fig. 1). Application of the rules mimics a step forward in time, and each application results in a new condition of the forest at the end of that step in time. Repetition of this process by applying the rules of change to each successive new condition allows forecasts for any length of time into the future.

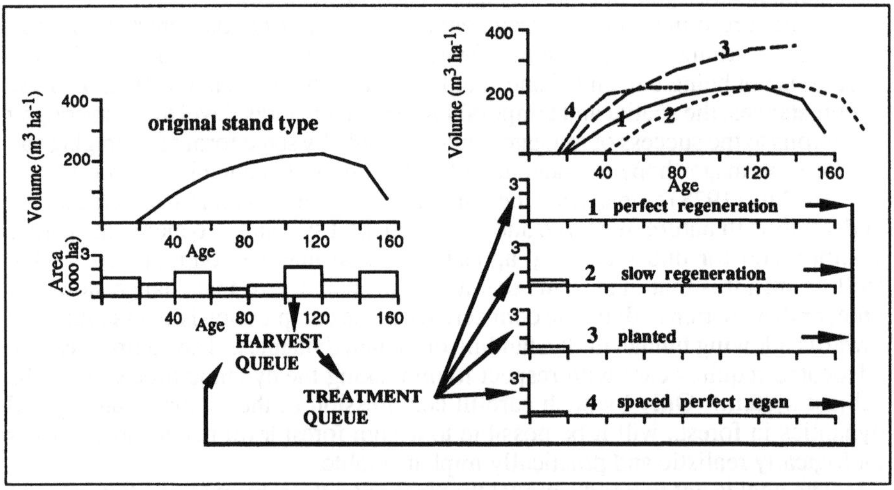

Fig. 1. Overview of the format of a forest level simulation. Age-class structures and yield curves represent the initial conditions, and solid lines represent the rules which govern the repetitive processes. See text for detail.

The rules of change specify such things as: how each stand type responds to harvest; how each annual harvest will be chosen in terms of area and type of stand to be harvested; what area and type of treatments will be applied and to which stands; what the responses of treated and untreated stands will be, in the form of a yield curve for future stand development; and what the cost of carrying out the various operations will be. There is also a rule about losses to insects, disease, and fire. The forecasts can be no better than the degree to which the initial conditions and the rules of change mimic reality in the forest.

All wood supply forecasts either state, or imply, the assumptions outlined in the previous two paragraphs. That is, no matter how simple the forecast, the initial conditions and rules for change may be either explicit, or they may be implicit, ***but they are always there.***

Forest analysis examines these assumptions employing the widest possible array of forest level indicators to define the forest management problem in terms of the causes of forest dynamics. The forest management problem is defined in terms of interactions among four key processes that drive forest development (i.e., stand growth, harvesting, stand treatment, and protection). Understanding how these operate in the temporal and spatial domain of a particular forest allows management design to concentrate on influencing the correct causal mechanisms as their relative importance in the dynamics of the forest development shifts over time. The process is to forecast future development for one management strategy at a time using an array of indicators of forest performance, compare these to existing knowledge of the forest, and analyze them with respect to their causes.

Analysis is used to verify that objectives are feasible; that is, feasible in the sense that the yield curves are believable, the responses are attainable, the harvest queue is economic, the treatment queue is implementable for the existing budget and for existing forest characteristics, and that there is consistency across all four of these features. The issue here is not precision, rather it is purely one of accuracy in reflecting what will happen in the real world as the future unfolds. *The validity of a forecast depends entirely on how accurately the initial conditions, and the rules of change reflect reality in the forest.*

Analysis should precede prescription. That is true for stand silviculture where the appropriate approach is stand analysis, and it is true for forest management where the appropriate approach is forest analysis.

A simple exercise

The example of forest analysis that follows is simple so that logical connections can be followed. There are no technical limitations to viewing more complex (realistic) situations in the same detail.

The basics of the forecast

The initial conditions. Grouping stands to set up a forest level simulation assumes a common developmental pattern for all stands in each group (see age-class structure in Fig. 1). This uniformity of development of all stands within a type is biologically crucial to the realism of a forecast of forest performance: first, because in the simulation every stand in a group will follow the same developmental pattern; and second, because in the simulation every stand in a group will respond similarly to treatment, whether this consists of silvicultural intervention, allowing the stand to break up in over-maturity, or allowing natural regeneration after harvest.

The grouping of stands should be based on similarity of stand dynamics, and not just on similarity of stand appearance. That is easier said than done, and it is common to group stands by their present cover type and age. These features may, or may not, reflect how the stands have grown in the past, and how they are developing at present. *When stands grouped in a type for forest level forecasting do not have the same dynamic characteristics, the biological realism of the forecast is reduced.* For purposes of forest management design, any loss of biological realism at this point cannot be offset by mathematical elegance in subsequent calculation of the simulation. The only protection against grievous error here is experience in the forest being forecast. A forecast of forest development will be no better than the biological realism in the definition of the initial state.

Stand yield rules. Forecasts of change in stand characteristics in the example forest over time are shown in Figure 2. Figure 2a shows yield curves for the volume of primary species. To the extent these yield curves are not an accurate representation of how these stands will grow in the actual forest in the future, the forest level forecasts will be in error. There is no way to bypass this critical assumption, and it exists in all forest level forecasts.

Figure 2b shows the proportion of the volume in each stand type that will be in the form of sawlogs. The remainder will be in pulpwood form. Here again, the accuracy of forecasts of sawlog availability from the whole forest can be no better than these stand level proportions in the context of the associated yield curves.

Figure 2c shows cost per cubic meter of harvesting and putting the wood at roadside. Costs are a crucial element in forest management design, largely because actual harvest pattern as it emerges in the forest tends to be strongly conditioned by the relative cost of various possible sources of wood regardless of "forest management rules".

Figure 2d shows the volume of secondary species growing in the stand types. This volume is not part of the commercial volume shown in Figure 2a; rather, it is unusable volume that must be dealt with in forest management. In the present example, the response yield curves are built on the assumption that secondary volume is removed at the time of harvesting the primary volume.

The harvest rule. The harvest rule used in the example is as follows: (i) harvest stands that are forecast to lose volume in the next five years; (ii) if no mortality is forecast, harvest stands with the lowest forecast 5-year growth; and (iii) if potential growth among stands is equal, those with the lowest harvest cost are taken, and if costs are equal, those with the most sawlog material are taken.

The harvest is even-flow, and the level of harvest must maintain the equivalent of 10 years harvest in the growing stock at all times.

Stand dynamics rules. All stands grow exactly on their assigned yield curves. When the original type is harvested, 50% of the cutover area regenerates to the same type on the same yield curve. The remaining 50% of the cutover regenerates to the same shape of yield curve but with a 20-year delay in regeneration (shown as 'slow regen' in Fig. 2a). When stands of the original type are allowed to break up due to overmaturity they regenerate to follow the original yield curve with a 5-year overlap to mimic advanced regeneration under an over-mature stand. Spaced stands that breakup also follow the original yield curve. Planted stands that are allowed to break up suffer a 20-year delay of regeneration, but thereafter follow the original yield curve shape.

For each stand type, the period during which the stand is eligible for harvest is defined by a minimum first volume and a minimum last volume that are operable. Stands that are younger eventually grow into the operable range. Stands that are older do not become operable until the old stand has broken up and been replaced with a new stand which has grown to exceed the lower operability threshold. Stands

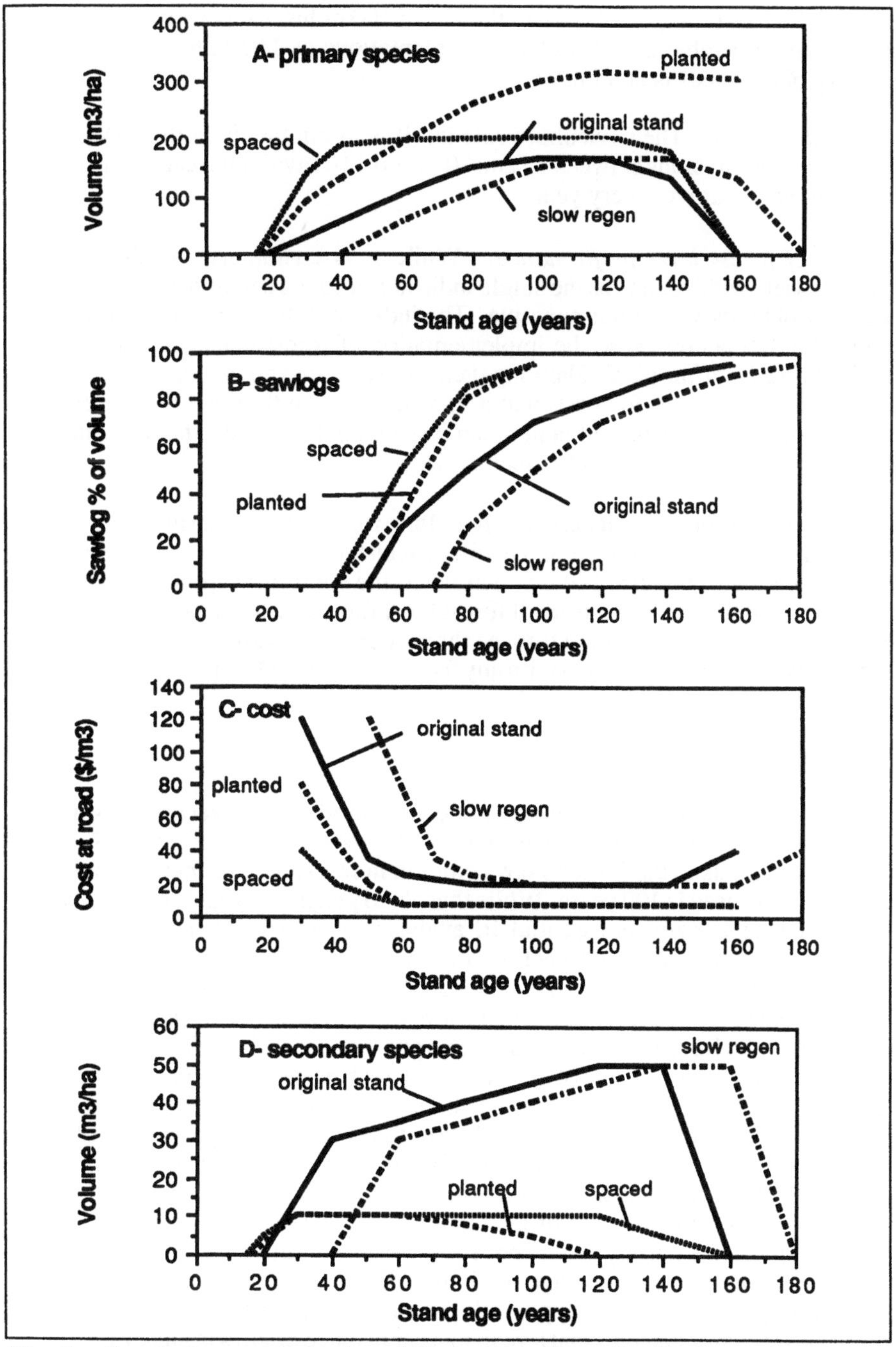

Fig. 2. Stand characteristics in relation to stand age for the forest used in the example forecast: (a) operable volume, (b) proportion of operable volume that is in sawlog form, (c) cost of harvesting and putting the wood to roadside, and (d) volume of secondary species. See text for detail.

are eligible for harvest (operable) when they have 100 m^3ha^{-1} except spaced stands which are first operable at 120 m^3ha^{-1} and plantations at 140 m^3ha^{-1}. The last operability of all stands is at 45 m^3ha^{-1}.

The silviculture rule. The silviculture rule is: (i) pre-commercially thin 50 ha of natural regeneration stands between 15 and 30 years old (if available) each year; and (ii) plant 50 ha of cutover every year.

Indicators of forest level performance. While it is customary to talk about sustainable yield with volume as the single indicator of forest performance, a dozen indicators will be shown for this example. The indicators shown relate to the kind of issues that frequently cause the implementation of forest management in the forest to deviate from 'the plan'. Showing them here required no innovative science – the tool used to make the forecasts was a standard one, but the other indicators of forest performance are displayed rather than just stating the harvest that is sustainable while remaining silent on other features.

Caveat. The yield curves, cost curves, silviculture rule, and harvest rule described above and used here are examples, and it will be seen they *completely condition* the forecast results that are outlined below. Similarly, the yield curves, costs, and rules are the fundamental determinants in all forest level forecasts, and in all cases these assumptions exist either in explicit form as shown here or they are implicit. For this reason, the assumptions associated with any forecast deserve clear presentation and intense review.

Examination of the forecast

While forest analysis involves the review of an array of forecasts, a single example is used here for ease of illustration. The forecast was generated in 5-year time steps analogous to the 5-year steps common in forest management planning. The time horizon used is 100 years. While that clearly exceeds what is practical for planning, a long horizon is essential in analysis to ensure that early steps in management do not accidently foreclose important options.

Using the initial conditions and rules for change described above a harvest level of 25 000 m^3 per year is sustainable from this forest through 100 years into the future. Figure 3 shows the forecast change in operable growing stock for this harvest level. The steady decline in the volume of growing stock with partial recovery in the future is common in forests where management is being introduced. Reference to the initial condition for this forest (Fig. 1) will show that a substantial proportion of the stands are over mature. That offers an attractive proposition for industrial development in terms of the quality raw material on the stump; however, this structure is not a highly productive one in terms of m^3 sustained per year since significant areas support stands that are already in decline. It can be seen intuitively that an age-class structure with a maximum stand age of about 120 years would give a lower total growing stock, but would also support a higher annual harvest.

The above example shows that a year-to-year balance of growth and drain is an inadequate basis for planning or explaining forest management, especially during the transition from a wild forest structure to a managed state. There is an important distinction between a forest structure that has a high operable volume now but is

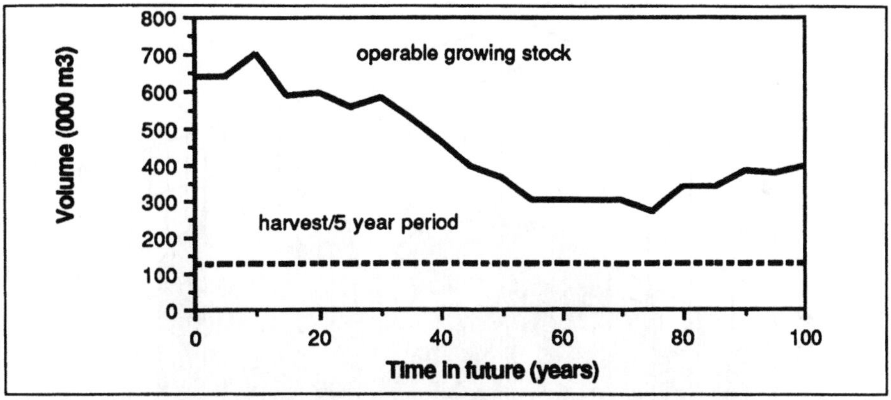

Fig. 3. Harvest level and volume of operable growing stock, plotted over time in the future for the example forecast.

but is limited at some point in terms of long run harvest, as compared to a forest structure with a lower current growing stock that supports a higher long run annual harvest. The point is that forest structure, not the total level of growing stock, makes the difference in productivity. While a decline in growing stock may indicate over-harvesting (as the media and public are quick to assume) it is not necessarily a bad sign; in the present case it will be shown the decline in growing stock is associated with rejuvenation of this forest into a more productive age-class structure for purposes of growing timber. On the other hand, it also shows the elimination of old growth stands and reduced productivity of non-timber values associated with such stands.

If the manager wishes to increase the minimum growing stock buffer (reached in year 75, and equal to 10 annual harvests here) to be more conservative, or decrease the buffer to get a greater harvest although at greater risk, a new forecast will give different values for all the indicators discussed below. Changes in other indicators of forest performance are rarely linearly related to the growing stock, as will become more evident in what follows. The point, however, needs emphasis – *each forest level forecast is a whole, and when either (i) a change is desired in one or more of the indicators, or, (ii) a decision is taken to operate in the real forest differently than specified in the forecast in order to 'correct' one of the indicators in real time, the entire forecast is invalidated!* The only way to retain logical consistency among the indicators, and between the forecast and what happens in actual forest, is to rerun the forecast building in the necessary alterations to the rules to cause the change in the indicator of concern. This will result in a new forecast for *all* indicators.

Figure 4 displays the change in growing stock over time for harvest levels of 0, 10 000, 20 000, 30 000, 40 000, and 50 000 m^3 per year – each with the same initial conditions and rules as in the example. The example forecast lies midway between "cut 20" and "cut 30" in the figure. The figure illustrates the discontinuous response of growing stock to increases in harvest level. There is relatively little impact on the pattern of growing stock volume over time at harvest levels below 20 000 m^3 per year. However, at harvest levels above 20 000 m^3 per year, there is a sharp drop in the growing stock over time, and only limited recovery. Harvest levels above 25 000 m^3 are *not sustainable*, and although the growing stock recovers slightly in each case, its *structure* does not support the higher harvest level.

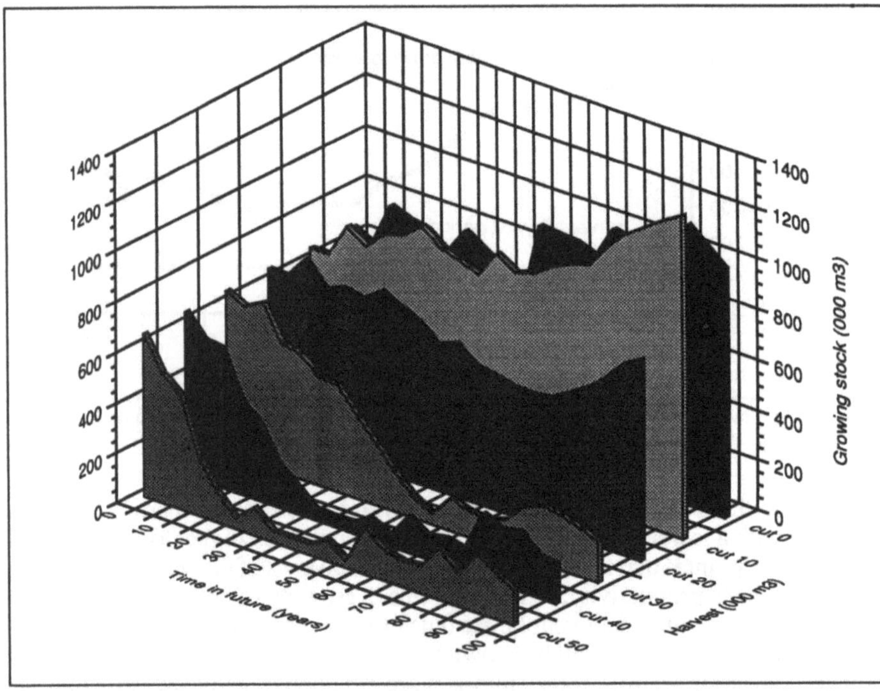

Fig. 4. Volume of operable growing stock over time in the future for six different harvest levels, using the same initial conditions and rules of change as in the example forecast. See text for detail.

The harvest level of 25 000 m^3 per year used in this example is seen to be a risky strategy. Since that level results in a growing stock pattern that borders on disaster about 60 years into the future, a minor mistake in either the accuracy of the forecast or the implementation of the harvest and treatment schedules in the forest could result in a collapse of the growing stock. For example, if the yield curves used in the forecasts happen to be slightly optimistic relative to the real forest, or, if the actual harvest in the forest did not target initially the stands most in danger of losing volume to overmaturity, the result in the real forest would *not* be as forecast, and in fact the 25 000 m^3 harvest would *not* be sustainable. The point is that a relatively small error in the forecast procedure (and therefore in the plan) could result in a major discrepancy between the forecast and the outcome in the forest. If the risk associated with a 25 000 m^3 harvest is not acceptable, then a lower harvest level can be chosen leading to a new forecast and a new set of indicator patterns.

Figure 5 shows the forecast cost per cubic meter at roadside for the initial conditions and rules for change in the 25 000 m^3 per year example. The cost varies with the type of stand being harvested from year to year, and with the stage of development of those stands (see Fig. 2c). The high cost in the early years results from the harvest rule targeting over-mature stands that are near breakup and which have relatively low volume per hectare, although relatively large individual trees. The low cost in the period 70-80 years in the future results from harvesting mature plantations which predominate in the harvest queue at that time. The silviculture costs are small in comparison to those for harvesting when expressed per cubic meter harvested.

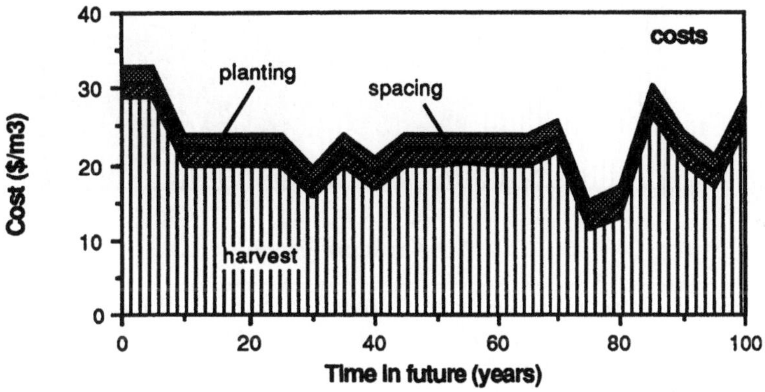

Fig. 5. Cost of harvesting (to roadside) for the example forecast, and annual cost of planting and spacing expressed per m^3 of harvest.

In the actual forest operations it is possible to avoid the high costs forecast in the early years by harvesting younger stands with higher volume (and which were not breaking up) but that action would invalidate the forecast given here. This is perhaps the most common manner in which a management plan is invalidated in the forest (i.e., when the planned harvest queue is abandoned in order to maintain a desired level of real wood cost). It is possible to design a plan (forecast) with a more even pattern of harvest cost over time by altering the harvest rule, but this will result in a lower harvest if the other sustainability guidelines are retained. There are two points which bear emphasis again: *(i) the forecast being reviewed is integral, and while any part of it can be changed by rerunning the forecast, this change will also alter all other elements of the forecast;* and *(ii) the actions in the forecast must be realistic mimics of the actions in the forest if there is to be consistency between the plan and reality.*

Figure 6 shows the forecast annual harvest in volume of sawlog material and pulpwood, as well as the volume of secondary (i.e., non-commercial) species that is generated by harvesting the primary species. The forecast change in sawlog material over time, did not show in the aggregate picture in Figure 3. The rather sudden transition from a high to a low proportion of log volume is due partly to the manner in which the harvest rule seeks to minimize mortality losses, and thereby incidentally targets stands with a high sawlog out-turn. The major cause of the decline in sawlog volume is that the harvest level is so high that regenerating stands are not allowed sufficient time to attain a high log volume (Fig. 2b). It is possible to have a harvest strategy that evens out the flow of sawlog quality material. However, this will also result in a change in harvest cost, and in all other indicators. Further, such a strategy is likely to result in a lower level of sustainable harvest in terms of total volume.

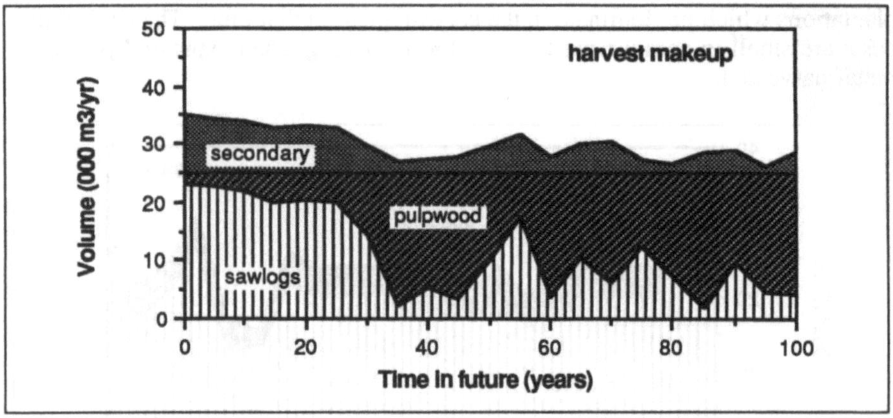

Fig. 6. Harvest volume for the example forecast in terms of the proportions in sawlog and pulp material, over time. Secondary volume is unusable material generated by the harvest.

Here as elsewhere, to the degree the actual harvest operation in the forest and the harvest rule in the forecast do not match, the forecast is rendered invalid. For example, the response curves in Figure 2, and the stand dynamics rules assume the secondary species are removed with the harvest. If the secondary volume is not removed, the real dynamics in stands in the forest *will not be as forecast*, and a new forecast adjusted for competition from the secondary species would give different results than shown for this and for all other indicators.

Figure 7 shows the source of the annual harvest by stand type. During the first two decades the harvest comes from the original stand type. By year 30 in the future, stands that were spaced at the start of the forecast become the dominant source of harvest. Since there is low risk of losses to mortality beyond year 30 when the last

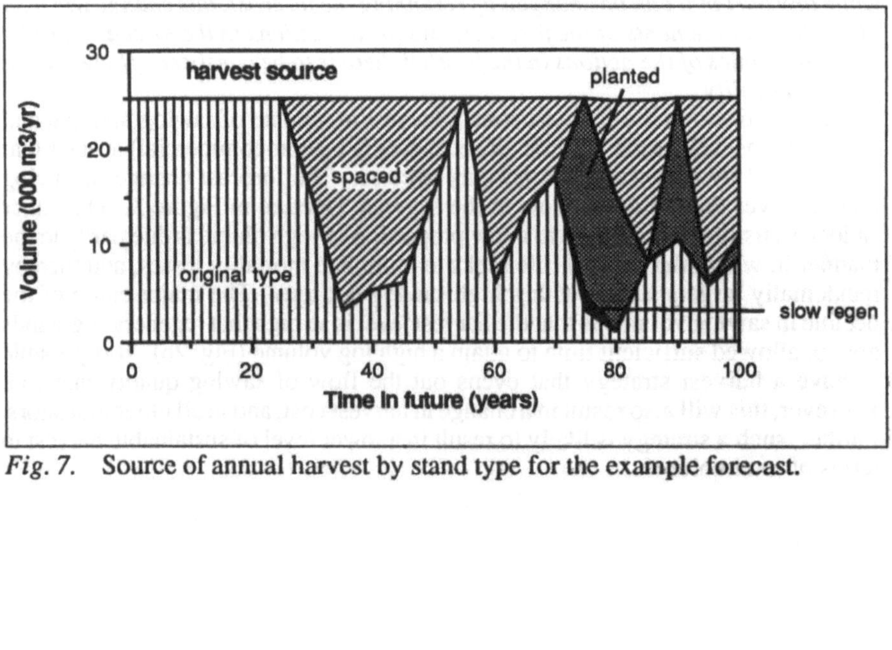

Fig. 7. Source of annual harvest by stand type for the example forecast.

of the over-mature stands have been harvested, the rule switches the harvest to spaced stands where growth has peaked. Plantations do not form a significant contribution to the harvest until about 70 years in the future. While plantations are operable much sooner than that, in this case, the harvest rule emphasis on current stand growth results in targeting the spaced stands where growth levels off earlier than in plantations (see Fig. 2a). The spaced stands also become operable sooner than plantations by virtue of a lower limit on first operable volume, and the large piece size in these stands makes harvesting from this source possible at relatively low cost (see the result by relating harvest source to Fig. 5). Although the slow regeneration stands have the same characteristics as the original stand type, the regeneration delay means they do not contribute to the harvest until the seventh decade, and then only slightly. These latter stands do become a source of harvest beyond 100 years in the future.

Figure 7 indicates several risks in this management strategy for this forest. If the spacing program is not implemented in the forest as planned, there will be a large shortfall of wood beyond year 30 in the future. The original stand type is being harvested at a rate that *anticipates* the early availability of wood from spaced stands. Thus, even if the spacing is carried out, but stand growth happens to be poorer than forecast by the response curve (Fig. 2a), the result equally will be a shortfall beyond year 30. Such risks are common in management plans, but by not showing the harvest source, they conveniently can be ignored. As with the other indicators demonstrated here, the purpose for showing harvest source explicitly is to invite comparison and test against what is happening in the real forest.

If implementation of the harvest in the actual forest chose to override the forecast dependence of the harvest queue on spaced stands by spreading the actual harvest over other stand types, the result would be invalidation of the forecast shown here. If the situation shown in Figure 7 is not acceptable, the entire forecast should be reworked to conform to an acceptable reality in the forest.

Figure 8 shows the source of the sawlog portion of the forecast harvest by stand type. Clearly the original stand type is the principle source of logs. While spaced stands have the potential for a high proportion of sawlogs, these stands are cut early

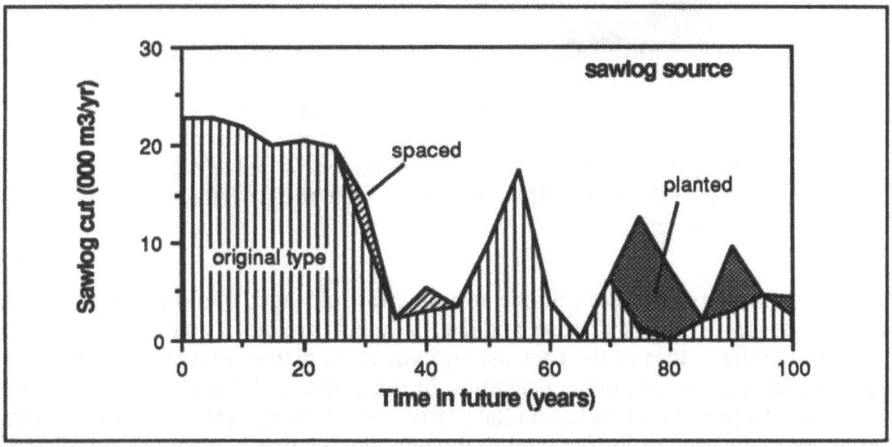

Fig. 8. Source of sawlog portion of harvest by stand type.

in their life before they have built a significant proportion of log volume because the harvest rule targets them due to their flat-topped yield curve. While the 25 000 m^3 harvest per year is sustainable it is clear that, with the harvest rule used here, this level will not allow stands to grow to an age where there is a significant proportion of sawlog material. The exception is in plantations where the delay in harvesting due to the nature of the harvest rule allows a buildup of sawlog material (not shown in the figures) although the area involved is not large enough to sustain the early log harvest levels.

Figure 9 illustrates the effectiveness of the element of the harvest rule designed to minimize losses to stand breakup in overmaturity. The harvest in the first decade is made up almost entirely of stands that would have become inoperable due to low volume had they not been harvested. The forecast losses to mortality are small in comparison. While this saving was made at some cost (see Fig. 5), it has a major impact on the magnitude of the long term sustainable harvest in this case, and failure to capture these potential losses to decadence would reduce the sustainable harvest. First, because the volume would have to be replaced by harvesting stands that were younger, thereby reducing the contribution of those stands to subsequent harvests, and secondly, when the old stands become inoperable due to low volume they continue to occupy area while they break up and gradually regenerate. Even with the edge in timing provided by the stand dynamics rules, the unavailability of these stands for harvest scheduling in the early years of the forecast period would severely constrain the level of even-flow sustainable harvest.

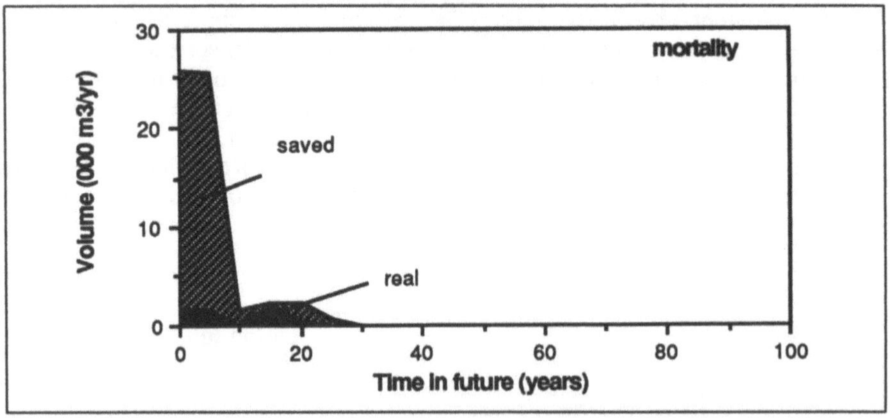

Fig. 9. Forecast of potential losses to overmaturity that were preempted by harvest, and of real losses due to stand breakup in over-mature stands that were not harvested.

The mortality discussed above results from old age; this does not include episodic losses to insects, disease, and fire. While it is relatively straightforward to include a volume loss in the forecast equivalent to historical averages for these factors, this can be seriously misleading in the context of forest dynamics. The impact that a particular loss has on long term sustainability relates to the age-class structure at the time of impact, and to the age-classes that happen to be depleted, more than it does to the volume of growing stock that happens to be lost. Further, the age-classes associated with the greatest risk to sustainability change over time.

For example, near the beginning of the forecast period in this example, long term sustainability is sensitive to any reduction to the growth of the spaced stands. Nearer the end of the forecast period, the greatest sensitivity is to loss of stands, or of stand growth, in the 60-100 year old range.

The most appropriate manner to handle periodic losses to insects, disease, and fire is to use the buffer in the growing stock (held at the lowest point in the forecast period) as security, and to rerun forecasts after each episode of losses specifically altering the initial conditions to reflect the actual damage experienced in the forest. Only in this manner can the impact of a loss to these agents on the integrity of the management strategy be logically determined.

Figure 10 shows the volume of growing stock over time with reference to the amount in sawlogs and in pulpwood, as well as in the secondary species. The basis of the problems discussed earlier with respect to the availability of log volume is clear, in that beyond 50 years in the future the growing stock contains relatively little log material.

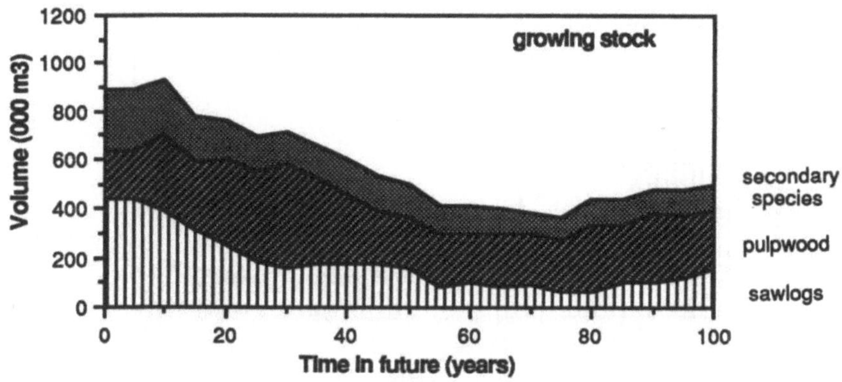

Fig. 10. Volume of operable growing stock in terms of sawlog and pulp material, and volume of secondary species.

Figure 11 shows the volume of primary species growing stock (logs and pulpwood) by stand type. The original stand type does not contribute to the operable growing stock beyond year 70 in the future. There are stands of this type in the forest at that time, but they are either young, or they are harvested as soon as they acquire an operable volume (see Fig. 7). The growing stock from spaced stands does not build up because these stands are harvested at a relatively young age due to the nature of the harvest rule.

Figure 12 shows the volume of sawlog growing stock by stand type. Here again, it is clear the original type is the principle source of log volume. The other types all have sawlog potential, but at a harvest level of 25 000 m³ per year those stands do not get old enough to show a substantial amount of log volume.

Figure 13 shows the area harvested and the area treated, over the forecast period. The area planted and the area of natural regeneration that is spaced are each 50 ha per year. The area harvested is high in the early years when the low-volume over-mature stands are being cut to preempt mortality. The area harvested drops near the end of the forecast period when harvesting focuses on high volume

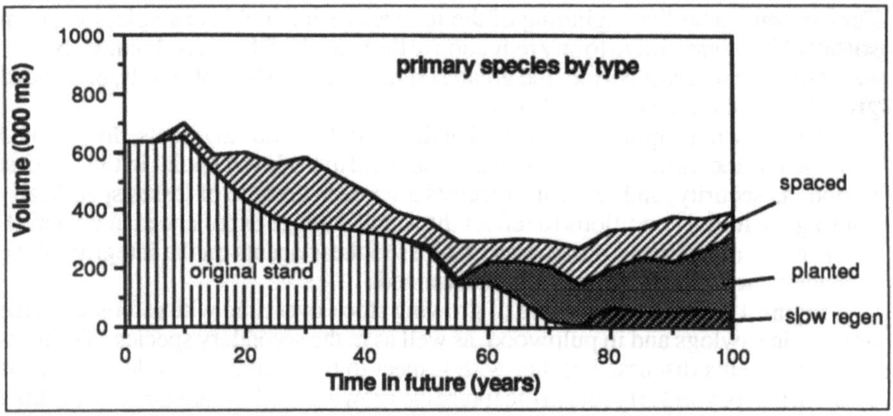

Fig. 11. Operable growing stock (primary species) by stand type.

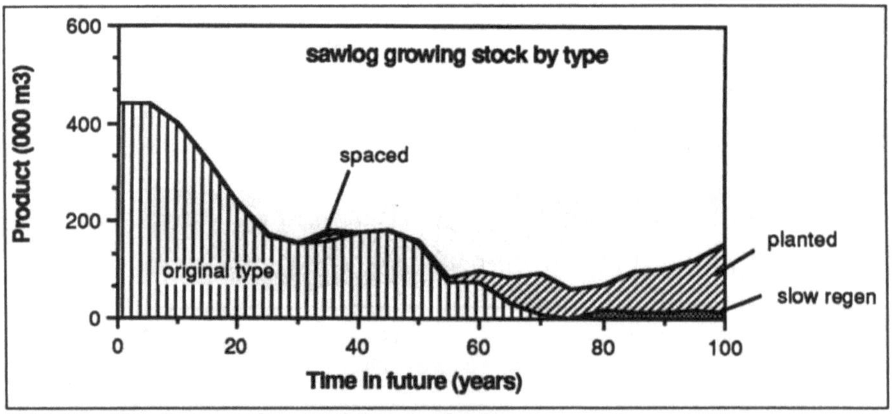

Fig. 12. Sawlog growing stock by stand type.

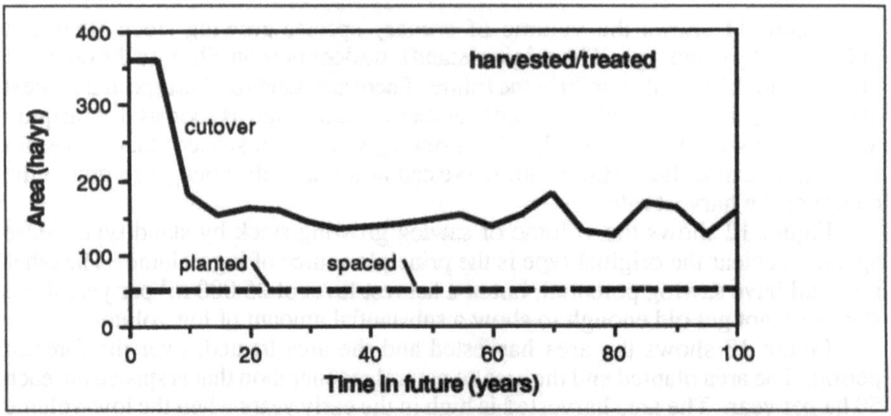

Fig. 13. Area harvested and treated annually.

plantations. In the context of the foregoing analyses it is clear that even in this simple case 'the management problem' has several dimensions, and the problem changes over the forecast period; thus, the simple treatment rule is an inefficient approach to management in this forest.

The change in area of the forest occupied by the various stand types over time is shown in Figure 14. It is obvious that a significant area has regenerated on the original yield curve but, as noted above, in the period 70-100 years in the future these stands have not yet reached ages that support operable volume, or are cut as soon as they reach operability. There is operable volume from these stands when the forecast is extended beyond 100 years. The figure shows the transition from a wild forest to a managed forest, and it is worth noting that stands of natural origin make up more than one-half the forest area 100 years in the future, and that there is little change in the proportions at this point. The figure does show a growing dominance of slow regeneration stands. These accumulated: (i) when the original stand type was cut (50% of each cutover); (ii) when plantations are cut (100% of each cutover); and (iii) when the slow regeneration type is cut (100% of each cutover). The slow regeneration stands become a target for harvest shortly after year 100.

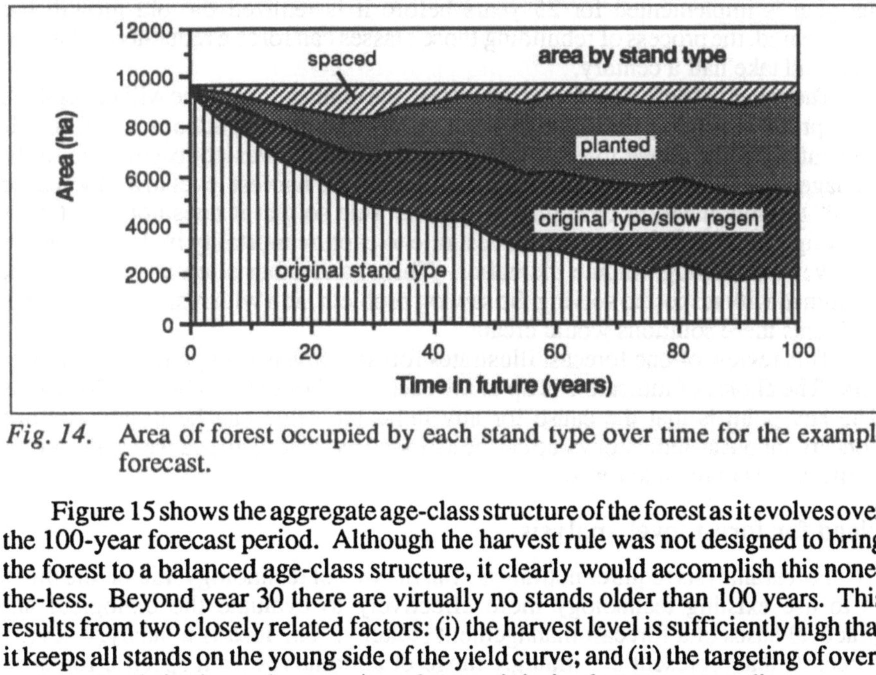

Fig. 14. Area of forest occupied by each stand type over time for the example forecast.

Figure 15 shows the aggregate age-class structure of the forest as it evolves over the 100-year forecast period. Although the harvest rule was not designed to bring the forest to a balanced age-class structure, it clearly would accomplish this none-the-less. Beyond year 30 there are virtually no stands older than 100 years. This results from two closely related factors: (i) the harvest level is sufficiently high that it keeps all stands on the young side of the yield curve; and (ii) the targeting of over-mature stands in the early years in order to minimize losses to mortality.

The balanced age-class structure shown in Figure 14 is across all stand types. From the perspective of raw material availability the crucial issue is the age-class structure by stand type (available but not shown). Also, this age-class pattern may not be acceptable for reasons of wildlife habitat and some older stands may have to be retained. *While any mix of age classes can be achieved over time, each would result in a different harvest level and a new set of indicator values for the associated forecast.* In this latter case, it is crucial to see the problem in the model before it is created in the forest. It is possible to retain older age classes from the outset, but if

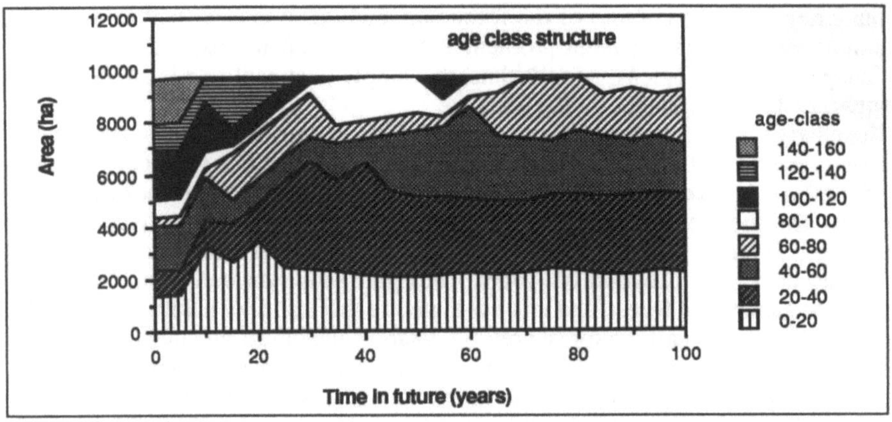

Fig. 15. Evolution of age-class structure of the example forest over the forecast period.

the plan is implemented for 25 years before it is realized the old growth has disappeared, the process of rebuilding those classes can force a reduction in harvest level and take half a century.

The principal benefit of forest analysis to the manager is that he will understand what problem is being corrected by each set of management actions. As a simple illustration, if the above forecast of sawlog availability was not acceptable to the manager, he has several options, such as reducing the harvest level to allow spaced stands to grow older, or changing the harvest rule so that it does not target slow growing stands, or changing the harvest rule to capture mortality in the early years but even out sawlog volume thereafter. These, and other alternatives, could be examined with respect to solving the sawlog problem, and with respect to any other problems these solutions would create.

This review of one forecast illustrates forest analysis using a range of indicators. The choice of indicators is up to the user, but should be as broad as possible. The key point is that the cause for any indicator value can be discovered and concerns about an indicator value can be assessed by relating the causes in the model to the causes in the real forest.

Need for forest level analysis

Forest analysis requires nothing new in the way of technology; rather, the thrust is to use existing technology more effectively to evaluate the biological and practical bases of a forecast management strategy. The idea is to analyze several forecasts that represent the strategic range of alternatives for a specific forest, in terms of forest dynamics as these operate in the model in order to discover places where forecast results are not logical, either by reason of biology or practicality. There is a place for forest analysis in forest management design analogous to that described earlier for silviculture design, and, it is argued, such analysis is a proper base from which to begin the design of management at the forest level.

Two approaches to forest management design dominate the contemporary scene. Both are subject to challenge on the grounds that they run unnecessary risks with respect to failure in application. The dominant approaches to forest management are stand-by-stand silviculture, and forest level simulation/optimization.

Simulation is often used by itself, but optimization approaches embody a simulator and the presence of the forest level simulator makes the two approaches the same in terms of realism in forest management design.

When forest management is designed through stand-by-stand silviculture prescription, forest level management becomes a simple sum of the series of stand level decisions, often based on economic analysis. In any event, the assumption is that actions designed independently for various stands are additive in terms of impact at the forest level. Another way of stating this assumption is that treating one stand has no influence on the opportunities open in other stands in the forest. That assumption would not be fatal where funds are not limiting, and where all stands can receive the best treatment prescribed, at the time prescribed. In all other cases, the fact of non-additivity of stand level actions in addressing a problem of forest level dynamics would be more or less devastating to the realism of a stand-by-stand approach to forest management design. However elegantly the silviculture might be determined, the stand-by-stand approach to forest management is risky in application. The fact that it ignores interactions at the forest level renders it trivial as a tool for designing management at the forest level, and unlikely to provide a realistic chance of successfully controlling forest level dynamics towards forest level goals. This approach cannot be logically subjected to forest level analysis because forest level dynamics are ignored in the procedure.

With forest level simulation/optimization, the management problem is defined and solved at the forest level, and the problem of non-additivity across stands is avoided. In this case, the risk is that the forest level characterization may not be consistent with biological dynamics at the stand level, and in the ultimate sense, a forest management solution must be applied through actions at the stand level. Therefore, consistency of cause and effect between the forest level assumptions in these models and the real stand responses, is crucial to successful management design. While the risk of an inaccurate representation is considerable, the structure of simulation/optimization approaches at the forest level invites analysis of the assumptions, keeping that risk exposed.

The chief problem is a tendency to use forest level simulation/optimization tools in an almost antiseptically prescriptive manner. Some management plans produced with these tools read like a collection of prefabricated parts supplied in a kit for easy home assembly – take a known objective, screw in a set of silvicultural responses, attach a set of costs, and plug it in to get the optimum harvest strategy and silvicultural program.

In the above context, it is worth remembering that the great precision apparent in forest management plans generated by forest level simulation/optimization techniques is largely an illusion. The plans are precise in terms of the exact output of what is to be done, when, where, and at what intensity. However, the accuracy of these plans is usually untested, and frequently presented in a form that makes them untestable by the manager. Biological accuracy of the representation of the forest in the model in terms of its dynamics, and its response to interventions to control those dynamics, can be no better than the way the model captures stand level dynamics, and the realism of the rules used to generate forest level dynamics.

Related to the prescriptive use of contemporary computer models is an absence of built in learning about the forest system for the manager. The forest level optimization processes are usually treated as if they were uni-directional. The 'right' run is made and the plan is written. There is little motivation for reflection before action, and little review of the biological reality of the outcome—with optimization, after all, it can be argued that the best answer has been defined. In many cases where simulation/optimization approaches have resulted in non-implementable forest management, the problem is misuse of the tools by the kind of 'forest manager' who never goes to the the forest. While it would be a good idea to make it a law that anyone who would design forest management must spend some minimal time in the subject forest, the display of an analysis of the internal consistency of the tools used also should be mandatory.

In the ultimate sense, the person who implements management actions in the forest is 'the forest manager', and often this is not the same person who operates the computer model. The 'computer specialist' may be aware of all the indicators, constraints, etc., in the model but unaware of the biological importance of model coherence. In this case if the indicators are not displayed, the manager has the equivalent of a black box answer. If the manager and the person who operate the model are not the same, the need to facilitate forest analysis by the manager as described above is paramount.

The stand-by-stand approach to management design has a strong element of analysis, but only at the stand level. The assumption that stand effects are additive across the forest is rarely valid, especially when management is being introduced. The forest level optimization approach is heavy on mathematical analysis, but customarily begins with assumptions roughly equivalent to 'the biology works this way', and little or no analysis is carried out with respect to realism of cause-effect connections. In the end, what passes for forest level analysis in contemporary forest management design has been largely reduced to economic analyses to determine which of a set of fixed silvicultural tools, with fixed responses, should be applied to achieve a standard economic goal. *Review of the internal consistency of the underlying biology in the resultant plans is conspicuous by its absence.* It is rare to find a forest management plan with an analysis of the biological basis of change in the system being managed that is carried out at the same scale that management is being designed and implemented. Yet the degree to which the model mimics this change in the real forest will determine whether the plan can reasonably be implemented to reach the goal in the forest, and will determine the extent to which other forest values are influenced both directly and indirectly. Despite the power of existing computer models, there is evidence their unthinking use is regularly failing to deliver reasoned management on the ground, out there in the forest (Barber and Rodman 1990, Reed and Baskerville 1990).

A common cause of failure of forest management plans is simply that the plans are not being followed in the forest. That is, the plan exists in an office on pieces of paper or in a computer, but not on the ground in the forest. Since a plan should offer some edge to the manager in facing the unfolding future, it is paradoxical that forest management plans are to be found more frequently in the office than in the forest. The most frequent reason for failure to follow a forest management plan is that the plan is not implementable in some fundamentally practical manner. Since the emphasis in simulation/optimization tends to center on the mathematical problem statement, the accuracy of the reflection of the forest can easily go unchallenged. However, it does not take much of an error in aggregation of stands to result in a solution which *cannot* be implemented in the spatial domain of the real forest. Nor does it take much of an error in the statement of a harvest rule to arrive at a solution that *will not* be implemented in the temporal domain of the forest.

Plans that violate practicality by ever so little are quickly modified out of context, or discarded, by those who ultimately carry out the work in the forest. In any event, the need is for management plans that have biological integrity *and* that are implementable in the target forest with the tools available.

Underlying the growing public concern over forest resource management is an unthinking use of powerful tools like forest level simulation/optimization—and virtually no attempt to explain the basis of the related forecasts of forest development to the profession, let alone to the public. The result is an alarming proportion of forest management plans that are: (i) not implementable in the biological domain of the forest, and at the temporal and spatial dimensions of the forest; or (ii) are rejected by the public as obscure. A non-implementable plan is considerably worse than no plan at all; indeed, it has a negative impact in the public domain. Creating plans in a computer that cannot be made to work in the real world, as advertised, is foolishly dangerous to the advancement of forestry in a society as active as that of the 1990's. It is staggering to hear modern management technologists say they have 'implemented' their plan, only to discover that to them implementation means it is running on a computer. The plan, and its impacts on forest development, must be stated in a manner that can be followed by an interested non-professional. This can be as simple as providing a time line for a series of indicators as illustrated above.

There is no intent here to take anything away from forest level simulation/optimization approaches; rather, the need is to augment their obvious strengths by extending their use from prescription to analysis of the subject system. In short, linear programming approaches would be the appropriate approach to find the proper silviculture rule in the above example, *but only after forest analysis has shown the statement of the dynamics to be biologically reasonable, and resulting plan to be operationally implementable.*

What is missing in the design of contemporary forest level management is a forest analysis analogous to stand analysis – literally a systematic review of: (i) what is the current dynamic structure with recognition of spatial pattern in the forest; (ii) what dynamics drove the forest to its spatial pattern including such historic influences as fire, insects, and past harvesting; and (iii) where are forest dynamics taking the forest in the future, again with recognition of the spatial pattern. The important point is that, spatially and temporally, the impact of management actions on the forest will be non-average, and the response to these actions as it emerges in the forest will be non-average in these dimensions. In these circumstances, forest analysis is a prudent approach to forest management design.

Literature cited

Barber, K.H. and S.A. Rodman. 1990. FORPLAN: The marvelous toy. Journal of Forestry 88: 26-30.

Boothroyd, H. 1978. Articulate intervention. Taylor and Francis, London / John Wiley & Sons, New York. 154 p.

Holling, C.S. 1978. Adaptive environmental assessment and management. John Wiley & Sons, New York. 377 p.

Reed, F.L.C. and G.L. Baskerville. 1990. A contemporary perspective on silvicultural investments. Journal of Business Admininistration. Vol. 19, in press.

Walters, C. 1986. Adaptive management of renewable resources. MacMillan Publishers, New York. 374 p.

Concluding Remarks

V

Ideas about mixed stands

15

DAVID M. SMITH

Most of silviculture is based on having a working understanding of how forest stands become established and how they develop during the rest of the rotation. It is a scientific art that depends on ability to carry out constructive disturbances in forest stands and predict with some confidence what will take place. One of the best ways of achieving this ability comes from identifying the kinds of forest stands that we want and then working backwards in time to determine how they reached the desired condition.

The model of stand development to which we have too steadfastly clung is that of the even-aged single-canopied stand typified by the pure pine (*Pinus*) or spruce (*Picea*) plantation. Most of the attempts that have been made to accommodate complexity of stand structure have tended to preserve the basic idea of the pure, even-aged, single-canopied aggregation of trees. The classic J-shaped curve of the hypothetical all-aged stand is really nothing more than the sum of a series of little pure, single-canopied stands representing a series of evenly spaced age classes each occupying an equal area. The ideas about mixed stand structure to which I was exposed when I was a student almost 50 years ago were that such stands started as little pure groups that might ideally thin down to single trees over the course of a rotation. Probably these concepts of complexity developed as compromises between complexity and the simplicity of bookkeeping epitomized by the yield tables for pure, even-aged stands that have so often governed silvicultural practice.

When the ecological behavior of stands fits the model of the pure, even-aged, single-canopied stand our efforts go well and predictably. If a site factor such as soil moisture is sufficiently restrictive, nature usually provides us with simple stands that fit the model or the yield tables. If fires or similar regenerative disturbances are severe enough, we may also get such stands on less restrictive sites. However, we start getting into trouble and bewilderment when some combination of site conditions and history of regenerative disturbance gives us stands that do not fit the model. If we try to force the simple model on some situations, nature rebels and presents us with the kinds of mixed-species vegetation that have long since been termed "jungles."

What we need to do is to fit the silviculture to the stand development process rather than always trying to get the stand to fit the simple model.

The purpose of all this inquiry into the nature of mixed stands is to inflict some order on the chaos of forests that are highly productive in a biological sense but so complicated that they are sometimes regarded as hopeless from the silvicultural standpoint. A common approach to them has often been to clear them off, wipe the slate clean of competing vegetation, and establish pure stands of the sort with which we have long been familiar. While this kind of silviculture is seldon really harmful to the environment or the long-term productivity of the forest, its devastated appearance scares the public and costs much money. It is a kind of silviculture that has a very important place and we need to keep pointing out that its main drawback

M. J. Kelty (ed.), The Ecology and Silviculture of Mixed-Species Forests, 281–287.

is that it may be highly visible and look bad in the early stages. However, we should have better reasons to do it than that it merely creates forest stands that we can understand.

Some of the papers of this symposium have underscored the important and often indispensible role that advanced growth plays in mixed-species forests. The attitude about this source of regeneration has traditionally been colored by the view that understory seedlings, saplings, and poles are hopelessly stunted and will always be cripples if they ever recover. This can be true but is conditional on the shape of the stems, sizes of the crowns, and health of the small trees involved. It is also true that reliance on advanced growth can mean favoring shade-tolerant species over the intolerant, although usually it means that only some weedy pioneers are eliminated.

Perhaps the most important thing that we have learned recently is that some of the most desirable species of mixed forests *must* start as advanced growth in partial shade (Marquis 1973, Devoe 1989). They may have to be released by heavy cutting later on but if they do not get the right conditions initially they will not be there to release.

More generally, in regeneration practice the emphasis on the regeneration that comes *after* harvest has obscured the importance of that which is already present *beforehand*. Advance growth comes in several varieties. The several forms of sprout regeneration, especially in broadleaved species, are much more effective and useful than bad experiences with short-rotation coppicing have led us to believe. Much forest vegetation arises from seeds stored in the forest floor. There is also the closely related phenomenon of regeneration with delayed development, exemplified by but not limited to those of longleaf pine (*Pinus palustris*) and oak (*Quercus*) seedling-sprouts. It appears that the seedlings and saplings of many species are not adapted to commence their race for the sky until the root system or some other part of the plant has developed to an extent adequate to support that race. Mixed forests and forests in general contain many species which we cannot manage effectively unless we recognize that not all are adapted to grow most rapidly in height just after they germinate.

One virtue of such mixtures is that there is opportunity to give some upper-stratum trees plenty of space in which to grow. It is also possible that the lower stratum trees may increase total stand production (Kelty 1989). Regardless of whether they do or not, many of them retain capacity to grow well after the overstory is removed, althought reliance upon them can be merely high-grading if they are poor trees. The lower strata of stratified mixtures and well-established advanced growth not only provide a source of renewal of stands but they can also play an exceedingly important role in excluding undesirable intolerant plants (de Graaf 1986). This is especially crucial in those kinds of rich, mixed stands in which one must guard against opening up the lower strata so much that jungles of undesirable vines, shrubs, and pioneer tree species invade the site.

One of the key virtues of using advanced growth and lower stratum tree species is that it enables us to keep command of the growing space at all times. This not only keeps out certain undesirables but it can also give a headstart on the next rotation. If the released trees are fundamentally sound and straight, every unit of initial height is a gain in production for the new rotation (Davis 1989). If the additional height keeps the desirable tree ahead of something undesirable, so much the better.

If niches are left vacant for establishment of undesirable plants we lose twice; once in wasting productive growing space and then in costly releasing operations.

Too often we have opened doorways for unwanted vegetation by trying to remove it completely from extensive areas when having 1000 little spots that were suitable on every hectare would be enough. With forest-tree regeneration, more is often not better.

Much of this symposium has been about stratified mixtures of species. It may shed some light on the matter if I recounted my own introduction to the idea and on subsequent efforts to develop it.

It started about 1947 when I was metamorphosing from the instar of the graduate student into that of the silviculture instructor. The prevailing interpretation of complexity in mixed stands was that of the uneven-aged stand constructed with the J-shaped curve of diameter distribution. The silviculture that went with it was ostensibly that of the selection system which was standard operating procedure in, for example, the U.S. Forest Service. My mentors at Yale, Hawley, Chapman, and Lutz, were in the minority of foresters who were skeptical about the concept.

Shortly before then I worked for Leo Isaac (1956), the high priest of Douglas-fir (*Pseudotsuga menziesii*) silviculture in the Pacific Northwest, another person with doubts about selection management. My specific duties were to measure the outcome of attempts at uneven-aged selection management of old-growth stands that had come to be of mixed composition. It was such a fiasco that I speedily saw why Isaac had long condemned this kind of misplaced silvicultural enthusiasm. The old stands were of mixed species and they had diameter distributions of the sort associated with the all-aged condition. The biggest trees were being cut and this accelerated the deterioration of the smaller ones.

At the time we assumed that all the species other than the initial Douglas-fir of fire origin had not become established until after the Douglas-firs began to dwindle with old age. The decline of the smaller trees of the other species after partial cutting was interpreted as meaning that they became old and decrepit at ages much younger than would be the case with the Douglas-firs. Since then I have come to suspect that parts of the stands were even-aged stratified mixtures, but that is getting ahead of the story. Similar problems were common in so-called selection management in eastern hardwood forests. The basic idea was not working in spite of an almost idyllic plausibility. However, there remained the difference between knowing that the idea did not often work and knowing why it did not.

About the only important place in the country where the selection method was actually working was in the truly uneven-aged ponderosa pine (*Pinus ponderosa*) stands of the western interior (Munger 1941, 1950). The fact that these happened to be essentially pure stands of a single species is an important clue about the significance of the J-shaped diameter distribution. In such stands it can truly be a diagnostic criterion of the balanced uneven-aged structure. In mixed stands it is simply not.

At the time, the wood science people at Yale were engaged in a thorough federally financed study of the woods of tropical America. Their purpose was to find more diversified uses for the bewildering variety of species in moist forests. I thought it desirable to inquire into the potential silviculture of the forests involved so I did my own unfinanced library study of the matter. I concentrated on the moist forests around the Caribbean and learned much from the publications of the old British and French colonial forestry services as well as the botanical literature. Most importantly I encountered the idea of stratified structure, as in the papers that P.W. Richards later summarized in his 1952 book on tropical forests.

During the same winter, gypsy-moth mortality impelled me to set up a partial cutting in a mixed stand of hardwoods, hemlock (*Tsuga canadensis*), and white pine (*Pinus strobus*) at the family forest in Massachusetts. As I was marking the stand, it suddenly became apparent that I was in one of the stratified mixtures of the tropical literature. It was very easy to see because the main continuous canopy of leafless hardwoods separated a solid lower stratum of evergreen hemlocks from evergreen pine emergents soaring above the hardwoods. The next thought that came to mind was that the stand had to be even-aged. I knew that my great grandfather, who owned a sawmill and a large drafty house, had cut it very hard about 65 years earlier.

The stand came from the advanced regeneration that had been released by the clearcutting of a stand of old-field white pine. The new stand was even-aged, and a stratified mixture, but it had the J-shaped diameter distribution. Hawley and Goodspeed (1932) had already pointed out that this distribution could exist in even-aged hardwood stands but the significance of this had not previously registered on me.

The various species had started off together and, through differential patterns of height growth, sorted themselves into the strata to which they were physiologically adapted. Furthermore, the species that we had come to regard as the biggest and best had forged their way to the top unaided. Perhaps their ability to do so and become large was the reason why we deemed them the best species. This is the same process which Egler (1954) soon after termed *initial floristics*. What he called the *relay floristics* of the succession theory of Clements (1916) fits better with regenerative disturbances that nearly wipe the slate clean of pre-existing vegetation.

At about the same time I took a pioneer Danish forest geneticist, Syrach Larson, to see an American pioneer, Arthur Graves, and his trial plantation of hybrid chestnuts (*Castanea dentata*) near New Haven. The visitor briefly transferred his attention from chestnuts to an adjacent stand of Connecticut sprout hardwoods that had about a dozen species. He asked, "How do you ever go about thinning a stand like that?" I lamely responded that we cut trees here and there in what we construed as crown thinnings of the sort described in the Danish literature. Inwardly it dawned on me that to him the only proper hardwood stand was of pure oak or beech (*Fagus*) and that we had something which the well-established concepts of silvicultural practice did not fit. Therefore, it was better and cheaper to augment the concepts than to simplify the stands.

Since then I have seen the phenomenon wherever heavy cutting or disturbances have released small advanced growth and created mixed, even-aged stands. The northeast part of the United States is full of them because of very heavy cutting around the turn of the century.

However, true differences in age can superimpose more complicated structures. As shown by the studies of Marquis (1981) in northwestern Pennsylvania, the release of residual shade-tolerant species of the lower strata can create stands of two age-classes and circumstances in which such species may take two or more rotations of some quicker-growing associate to complete their development.

The existence of this kind of variation serves to warn that we should observe and analyse the patterns of stand development first and devise silvicultural systems to fit or modify them afterwards.

❧

The concept of the even-aged stratified mixture provides part of the answer to the question of why the so-called selection cutting schemes of 1930-1955 worked so poorly. It also helps explain why high-grading often works badly. In many mixed stands the problem is that the stands are seldom all-aged or even close to that condition; sometimes they are even-aged. Once the even-aged condition exists, it is very difficult to change it. The common failure to provide space for new regeneration is only part of the problem.

Much of the difficulty comes from trying to force stand development into a model that seldom fits. The model is the diameter distribution of DeLiocourt's so-called Law defined by J-shaped curves convertible to straight-line form with logarithmic transformation. Fascination with mathematics is such that this relationship is treated as if it were a law of physics. It is presumed that if one removes some prescribed number of trees at the large-diameter end, all the other diameter classes will adjust and the "system" will continue to feed out the same number of trees to cut at regular intervals of time in the future.

The equations involved are thus used as if they really described processes that go on in stands that are mixtures either of species or of age classes. They do not fit because the only process that they actually describe is the attrition of numbers from competition as *pure, even-aged* aggregations of trees get older. When they are used in connection with stands or forests, they are best thought of as describing conditions that might conceivably exist at single points in time. These conditions are not automatically self-perpetuating in nature or necessarily under deliberate silvicultural management. In other words, for complex stands the equations describe a kind of temporary condition but not a process. The process that they describe is appropriate to series of age-classes of pure stands. Their best practical use is in testing the degree to which the trees of a given species in a large forest unit have the appropriate diameter distribution required in a perfect sustained-yield unit.

The concept of the stratified mixture of species comes closer to accounting for what happens in mixed stands. It does not explain them all. True differences in age often complicate and obscure the structural patterns.

In silvicultural application, both the concept of the stratified mixture and that of the all-aged selection stand often derive much advantage from the release of small shade-tolerant trees growing underneath larger ones. Under one concept they are treated as younger trees; in the other, as trees of a lower stratum. It appears to me that their silvicultural behavior is understood best by treating them as lower-stratum trees. However, if we want to keep track of all the different sizes of trees of that single species for purposes of sustained yield on a whole forest, then we can usefully use the J-shaped curves of the DeLiocourt relationship. However, for that purpose we may find it prudent only to count trees that are free to grow.

Finally I present some sketches (Fig. 1) that suggest various pathways for the treatment or modification of stratified mixtures, starting with an essentially even-aged one. The sketch at the middle of the whole diagram shows the even-aged 5-species stand that is the starting point. The path to the upper left corner of the diagram shows how the same kind of stand could be created again by opening up all strata enough to establish a new mixed, essentially even-aged crop of advanced regeneration before a final heavy removal cutting. As an alternative it is also shown how, by accelerating regeneration in one part of the stand and holding it back in others, one might commence the very long process of building an uneven-aged stand in which each age-group is a stratified mixture.

The path to the upper right shows how some trees of a lower stratum can be reserved to become part of a new 2-aged stand. It may take two rotations of the faster-growing species for trees of the lower-stratum species to complete full

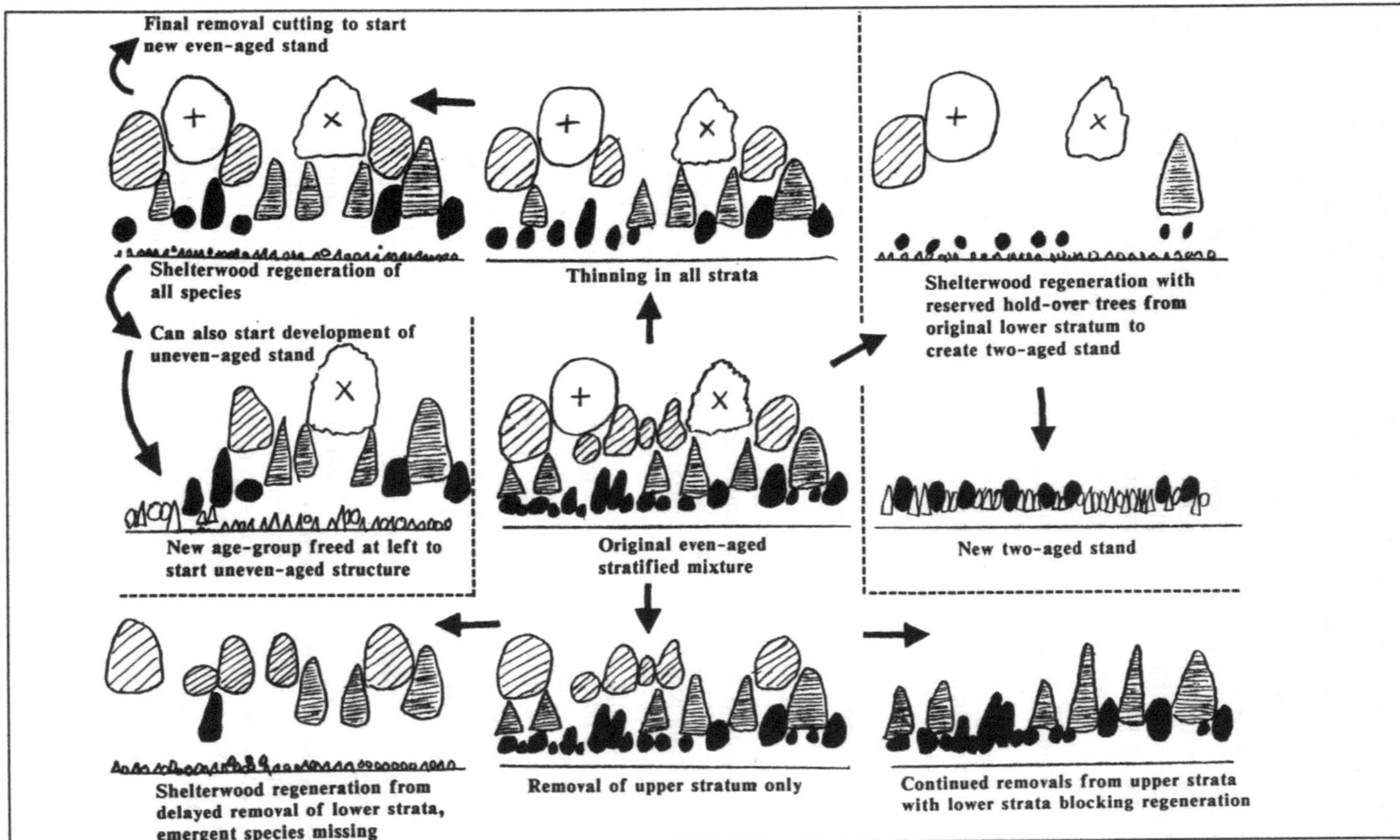

Fig. 1. Pathways for the silvicultural treatment of stratified mixtures; original even-aged stand containing five species is shown at center.

development. The path to the lower right is "high-grading" by stages. The impedence of regeneration by the uncut lower strata may be desirable to the extent that it keeps pioneer weeds out and undesirable if it prevents establishment of advanced growth of good species. The path to the lower left shows an attempt to halt the high-grading effect.

References

Clements, F.E. 1916. Plant succession: an analysis of the development of vegetation. Carnegie Institute of Washington. 512 p.

Davis, W.C. 1989. Role of released advance growth in the development of spruce-fir stands in eastern Maine. Unpublished Ph.D. dissertation, Yale University. 150 p.

de Graaf, N.R. 1986. A silvicultural system for natural regeneration of tropical rain forest in Suriname. Agricultural University, Wageningen, Netherlands. 250 p.

Devoe, N.N. 1989. Differential seeding and regeneration in openings and beneath closed canopy in sub-tropical wet forest. Unpublished D.For. dissertation, Yale University. 316 p.

Egler, F.E. 1954. Vegetation science concepts. I. Initial floristic composition–a factor in old-field vegetation development. Vegetatio 4: 412-417.

Isaac, L.A. 1956. Place of partial cutting in old-growth stands of the Douglas-fir region. Pacific Northwest Forest & Range Experiment Station Research Paper 16. 48 p.

Hawley, R.C., and A.W. Goodspeed. 1932. Selection cuttings for the small forest owner. Yale University School of Forestry Bulletin 35. 45 p.

Kelty, M.J. 1989. Productivity of New England hemlock/hardwood stands as affected by species composition and canopy structure. Forest Ecology and Management 28: 237-257.

Marquis, D.A. 1973. The effect of environmental factors on advance regeneration of Alleghany hardwoods. Unpublished Ph.D. dissertation, Yale University. 147 p.

Marquis, D.A. 1981. Survival, growth, and quality of residual trees following clearcutting in Alleghany hardwood forests. U.S. Forest Service Research Paper NE-277. 9p.

Munger, T.T. 1941. They discuss the maturity selection system. Journal of Forestry 39: 297-303.

Munger, T.T. 1950. A look at selective cutting in Douglas-fir. Journal of Forestry 48: 97-99.

Richards, P.W. 1952. The tropical rain forest; an ecological study. Cambridge (England) University Press. 450 p.

FORESTRY SCIENCES

1. P. Baas (ed.): *New Perspectives in Wood Anatomy*. Published on the Occasion of the 50th Anniversary of the International Association of Wood Anatomists. 1982 ISBN 90-247-2526-7
2. C.F.L. Prins (ed.): *Production, Marketing and Use of Finger-Jointed Sawnwood*. Proceedings of an International Seminar Organized by the Timber Committee of the UNECE (Halmar, Norway, 1980). 1982 ISBN 90-247-2569-0
3. R.A.A. Oldeman (ed.): *Tropical Hardwood Utilization*. Practice and Prospects. 1982 ISBN 90-247-2581-X
4. P. den Ouden (in collaboration with B.K. Boom): *Manual of Cultivated Conifers*. Hardy in the Cold- and Warm-Temperate Zone. 3rd ed., 1982 ISBN Hb 90-247-2148-2; Pb 90-247-2644-1
5. J.M. Bonga and D.J. Durzan (eds.): *Tissue Culture in Forestry*. 1982 ISBN 90-247-2660-3
6. T. Satoo: *Forest Biomass*. Rev. ed. by H.A.I. Madgwick. 1982 ISBN 90-247-2710-3
7. Tran Van Nao (ed.): *Forest Fire Prevention and Control*. Proceedings of an International Seminar Organized by the Timber Committee of the UNECE (Warsaw, Poland, 1981). 1982 ISBN 90-247-3050-3
8. J.J. Douglas: *A Re-Appraisal of Forestry Development in Developing Countries*. 1983 ISBN 90-247-2830-4
9. J.C. Gordon and C.T. Wheeler (eds.): *Biological Nitrogen Fixation in Forest Ecosystems*. Foundations and Applications. 1983 ISBN 90-247-2849-5
10. M. Németh: *Virus, Mycoplasma and Rickettsia Diseases of Fruit Trees*. Rev. (English) ed., 1986 ISBN 90-247-2868-1
11. M.L. Duryea and T.D. Landis (eds.): *Forest Nursery Manual*. Production of Bareroot Seedlings. 1984; 2nd printing 1987 ISBN Hb 90-247-2913-0; Pb 90-247-2914-9
12. F.C. Hummel: *Forest Policy*. A Contribution to Resource Development. 1984 ISBN 90-247-2883-5
13. P.D. Manion (ed.): *Scleroderris Canker of Conifers*. Proceedings of an International Symposium on Scleroderris Canker of Conifers (Syracuse, USA, 1983). 1984 ISBN 90-247-2912-2
14. M.L. Duryea and G.N. Brown (eds.): *Seedling Physiology and Reforestation Success*. Proceedings of the Physiology Working Group, Technical Session, Society of American Foresters National Convention (Portland, Oregon, USA, 1983). 1984 ISBN 90-247-2949-1
15. K.A.G. Staaf and N.A. Wiksten (eds.): *Tree Harvesting Techniques*. 1984 ISBN 90-247-2994-7
16. J.D. Boyd: *Biophysical Control of Microfibril Orientation in Plant Cell Walls*. Aquatic and Terrestrial Plants Including Trees. 1985 ISBN 90-247-3101-1
17. W.P.K. Findlay (ed.): *Preservation of Timber in the Tropics*. 1985 ISBN 90-247-3112-7
18. I. Samset: *Winch and Cable Systems*. 1985 ISBN 90-247-3205-0

FORESTRY SCIENCES

19. R.A. Leary: *Interaction Theory in Forest Ecology and Management.* 1985 ISBN 90-247-3220-4
20. S.P. Gessel (ed.): *Forest Site and Productivity.* 1986 ISBN 90-247-3284-0
21. T.C. Hennessey, P.M. Dougherty, S.V. Kossuth and J.D. Johnson (eds.): *Stress Physiology and Forest Productivity.* Proceedings of the Physiology Working Group, Technical Session, Society of American Foresters National Convention (Fort Collins, Colorado, USA, 1985). 1986 ISBN 90-247-3359-6
22. K.R. Shepherd: *Plantation Silviculture.* 1986 ISBN 90-247-3379-0
23. S. Sohlberg and V.E. Sokolov (eds.): *Practical Application of Remote Sensing in Forestry.* Proceedings of a Seminar on the Practical Application of Remote Sensing in Forestry (Jönköping, Sweden, 1985). 1986 ISBN 90-247-3392-8
24. J.M. Bonga and D.J. Durzan (eds.): *Cell and Tissue Culure in Forestry.* Volume 1: General Principles and Biotechnology. 1987 ISBN 90-247-3430-4
25. J.M. Bonga and D.J. Durzan (eds.): *Cell and Tissue Culure in Forestry.* Volume 2: Specific Principles and Methods: Growth and Development. 1987 ISBN 90-247-3431-2
26. J.M. Bonga and D.J. Durzan (eds.): *Cell and Tissue Culure in Forestry.* Volume 3: Case Histories: Gymnosperms, Angiosperms and Palms. 1987 ISBN 90-247-3432-0
    Set ISBN (Volumes 24-26) 90-247-3433-9
27. E.G. Richards (ed.): *Forestry and the Forest Industries: Past and Future.* Major Developments in the Forest and Forest Industries Sector Since 1947 in Europe, the USSR and North America. In Commemoration of the 40th Anniversary of the Timber Committee of the UNECE. 1987 ISBN 90-247-3592-0
28. S.V. Kossuth and S.D. Ross (eds.): *Hormonal Control of Tree Growth.* Proceedings of the Physiology Working Group, Technical Session, Society of American Foresters National Convention (Birmingham, Alabama, USA, 1986). 1987 ISBN 90-247-3621-8
29. U. Sundberg and C.R. Silversides: *Operational Efficiency in Forestry.* Vol. 1: Analysis. 1988 ISBN 90-247-3683-8
30. M.R. Ahuja (ed.): *Somatic Cell Genetics of Woody Plants.* Proceedings of the IUFRÓ Working Party S2.04-07 Somatic Cell Genetics (Grosshansdorf, Germany, 1987). 1988 . ISBN 90-247-3728-1
31. P.K.R. Nair (ed.): *Agroforestry Systems in the Tropics.* 1989 ISBN 90-247-3790-7
32. C.R. Silversides and U. Sundberg: *Operational Efficiency in Forestry.* Vol. 2: Practice. 1989 ISBN 0-7923-0063-7
    Set ISBN (Volumes 29 and 32) 90-247-3684-6
33. T.L. White and G.R. Hodge (eds.): *Predicting Breeding Values with Applications in Forest Tree Improvement.* 1989 ISBN 0-7923-0460-8
34. H.J. Welch: *The Conifer Manual.* Volume 1. 1991 ISBN 0-7923-0616-3
35. P.K.R. Nair, H.L. Gholz, M.L. Duryea (eds.): *Agroforestry Education and Training.* Present and Future. 1990 ISBN 0-7923-0864-6
36. M.L. Duryea and P.M. Dougherty (eds.): *Forest Regeneration Manual.* 1991 ISBN 0-7923-0960-X

FORESTRY SCIENCES

37. J.J.A. Janssen: *Mechanical Properties of Bamboo.* 1991 ISBN 0-7923-1260-0
38. J.M. Bonga and P. Von Aderkas: *In Vitro Culture of Trees.* 1992 ISBN 0-7923-1540-5
39. L. Fins, S.T. Friedman and J.V. Brotschol (eds.): *Handbook of Quantitative Forest Genetics.* 1992 ISBN 0-7923-1568-5
40. M.J. Kelty, B.C. Larson and C.D. Oliver (eds.): *The Ecology and Silviculture of Mixed-Species Forests.* A Festschrift for David M. Smith. 1992 ISBN 0-7923-1643-6

KLUWER ACADEMIC PUBLISHERS – DORDRECHT / BOSTON / LONDON

MIX
Papier aus verantwortungsvollen Quellen
Paper from responsible sources
FSC® C105338
www.fsc.org

If you have any concerns about our products,
you can contact us on
ProductSafety@springernature.com

In case Publisher is established outside the EU,
the EU authorized representative is:
Springer Nature Customer Service Center GmbH
Europaplatz 3, 69115 Heidelberg, Germany

Printed by Libri Plureos GmbH
in Hamburg, Germany